International Encyclopaedia of ENERGY RESOURCES

International Encyclopaedia of ENERGY RESOURCES

— *Volume 1* —

— *Edited By* —
Dr. M.P. Singh
Head of the Department of Forest Sciences
Birsa Agricultural University
Ranchi - 834 006 (Jharkhand)

2005
DAYA PUBLISHING HOUSE
Delhi - 110 035

ISBN 81-7035-357-2 (SET)
ISBN 81-7035-358-0 (Vol. 1)
ISBN 81-7035-359-9 (Vol. 2)
ISBN 81-7035-360-2 (Vol. 3)
ISBN 81-7035-361-0 (Vol. 4)

Published by : **Daya Publishing House**
1123/74, Deva Ram Park
Tri Nagar, Delhi - 110 035
Phone: 27383999
Fax: (011) 23244987
e-mail: dayabooks@vsnl.com
website: www.dayabooks.com

Showroom : 4762-63/23, Ansari Road, Darya Ganj
New Delhi - 110 002
Phone: 23245578, 23244987

Laser Typesetting : **Classic Computer Services**
Delhi - 110 035

Printed at : **Chawla Offset Printers**
Delhi - 110 052

PRINTED IN INDIA

Preface

This book is for people who still think that there are energy problems ahead, and who are not satisfied with an energy future dominated by coal and nuclear power.

Fortunately, the time has come when one can examine other alternatives seriously–in particular, energy conservation and renewable energy–without being relegated at once to the outer fringes. Southern California Edison, Pacific Gas and Electric, and numerous other progressive utilities, having embraced these alternatives, are a bit too formidable to be drummed out of the corps. The Office of Technology Assessment, the Gas Research Institute, and the Harvard Business School likewise are not thought of as outposts of a lingering counterculture.

Still, not enough people in the energy supply industries, in government, or in private agencies concerned with energy policy are aware of the full range of possibilities in energy conservation and in the numerous renewable energy technologies. For them, this book is an attempt to lay out, in an ordered fashion, the basis in fact, logic, and judgment, for an energy -efficient, renewable energy future.

One cluster of views about energy may be characterised as "supply oriented." The response to the oil shortage was quite expectedly framed in terms of increasing energy supply, especially in forms other than oil. The basic premises motivating supply-side arguments included the following assumptions:

Energy demand will grow with output. A healthy and growing economy requires a growing supply of energy. In early formulations the rate of energy supply growth was about equal to the growth rate of the economy.

While this volume does provide a great deal of information about energy sources, its more appropriate use is as a stimulus for discussion. Towards this end the book often raise more questions than they answer. The complexity of the ethical dilemmas posed by energy sources is such that we are still trying to identify what questions should be asked. The volume recognises that the first step in making the types of ethical decisions which are increasingly demanded of us is searching for the right questions. The book vary greatly in both length and orientation. This will make it useable by a wide range of groups–from students in the classroom, to professionals such as scientists, medical doctors, business people, lawyers, and ethicists, as well as environmental and other socially active groups. It is hoped that this volume will contribute in a better way to a society.

Dr. M.P. Singh

Contents

Introduction

The source of more than 90 percent of the energy we use is from fossil fuels, oil, coal, and natural gas. It took billions of years for these supplies to be created. While there is yet no shortage, the amount of resources remaining is still finite. And we have little more than a century to develop adequate, practical, and inexpensive technologies to replace some of these sources. It is time to sound a tocsin to urge us to prepare for the inevitable—that an energy crisis due to scarcity will occur, whose consequences are not as easily overcome as adjusting to increased fuel prices.

Fossil fuels would seem to be a gift from the gods to earthlings. The availability of this source of energy has brought unimaginable comfort and prosperity to many of the inhabitants of our planet. At first glance these fuels seem to be cheap, clean, safe, and almost infinitely available. The case for continuing their use as the engines of our economy seems overwhelming. Indeed, it appears to be the energy policy of the United States to rely on fossil fuels for the foreseeable future. *The Clean Air Act of 1993* emphasizes ways of making fossil fuel utilization less burdensome for the environment, and of increasing the efficiency of the devices using fossil fuels. It does not provide for actively supporting projects seeking early replacements for these energy sources.

Upon closer examination, this source is neither as infinite nor as safe, clean, or cheap as one might suppose. The reserves of oil are expected to last only for decades or perhaps as much as a century at the present level of use. The known reserves of natural gas are expected to last equally as long but information about the untapped reserves is not as accurate as is that for oil. Although the reserves of coal are not as short as those of petroleum, still they are a finite resource that, with use, will ultimately disappear. Coal, which is must more plentiful, is a killer. It has taken the lives of about 100,000 miners since the beginning of the century. In addition it is responsible for the many cases of black lung disease among miners. The fine airborne particles of coal dust cause many upper respiratory diseases including lung cancer. Burning coal produces environmental gases that destroy lakes and forests. Obtaining coal by strip mining causes an environmental disaster.

The use of petroleum products in automobiles and trucks is responsible for smog, low-level ozone (an upper respiratory irritant), and excess carbon monoxide in the atmosphere. Oil spills, each of which seems like a preventable accident, are clearly hazards that are part of the oil economy. And they have not been a boon to marine life.

State and federal environmental protection agencies have done a commendable job in dealing with some, but not all, of the environmental challenges presented by fossil fuel use. However, one challenge of fossil fuel use cannot be addressed except by eliminating fossil fuels as an energy source or at least tempering their use. The burning of any fossil fuel results in increasing the carbon dioxide in the atmosphere, an effect that a United Nations scientific advisory committee has stated will lead to global warming, with serious climatic and economic consequences, perils that will be more extensively dealt with in this text. Global warming may have already begun. The future of fossil fuels will be a mixed blessing for the Earth and its inhabitants.

The energy produced by fossil fuels is not as cheap as it seems. The market price does not reflect the political and diplomatic costs inherent in its use. One-fourth of the global reserves of oil are in the Middle East, a warren of satrapies that are difficult to deal with. The flow of oil is so important to the economy of the United States that it is necessary to invest billions to maintain a fighting force should the flow be interrupted. That such an investment is necessary was adequately demonstrated in Operation Desert Storm during the Persian Gulf War against Iraq. And it justifies keeping a reaction force on the ready in that region. Desert Storm was fought because it seemed as if the flow of oil to the United States and worldwide could be seriously interrupted. Imagine the danger to the peace on Earth when the supplies of petroleum are seriously diminished. Although global warming poses a more immediate threat, the finiteness of the supplies of oil may be even more dangerous in the long run. More than half of the oil used in the United States has to be imported. About half of these imports come from the Middle East. They affect the balance of trade of the United States adversely and have serious budgetary consequences. None of these considerations, despite their effect on the pocketbooks of the citizens, is included in calculating the cost of using fossil fuels. In 1998 the reaction force was strengthened in response to Iraq's refusal to allow unfettered inspections of potential sites that might be loci for the manufacture of nuclear or biological weapons of mass destruction.

Although public policy has focused on maintaining fossil fuels as a source of energy, there has been some support for the development of alternatives. The cost of generating energy commercially using these alternatives is rapidly decreasing. Still, no matter how cheap the alternatives, they would have great difficulty penetrating the energy market to any appreciable extent. For one thing, the market price of fossil fuels is unreasonably cheap. Their cost for the most part reflects only the cost of recovery and delivery, not manufacture. For another reason, the capital investment in the fossil fuel economy has been largely amortized and new technologies would require substantial capital investments to compete. For still another, powerful financial interests are invested in fossil fuels, and they will fight to keep these entrenched as long as possible. The tactics for this fight have already been revealed and will be discussed in the text. For there to be alternatives, there has to be some initial subsidization.

The prevailing public opinion is that the best strategy for dealing with the fossil fuel problem is to encourage conservation. Bowing to this popular point of view this book will discuss the variety of ways in which conservation can lengthen the time for fossil fuels to continue as our preeminent source of energy. With conservation in place for fossil fuels and their use discouraged, a window of opportunity exists while they are inexpensive and plentiful, to begin to replace these carbon-based fuels by other technologies as quickly as possible. With the best of intentions it may take a generation or more to introduce extensive use of alternative energy sources.

Since the crisis of the1970s, India energy policy has changed from just another agenda item to a significant policy issue. Energy is not only vital to our economy and national security, but also has an extremely complex and dynamic relationship to other domestic policies such as transportation, housing, environment, social policies, and foreign trade. As a result of the 1973 oil crisis, as well as the tenfold increase in oil prices imposed by the Organization of Petroleum Exporting Countries (OPEC), four administrations have searched for a national energy policy.

Because of laws already on the books at the time of the Arab oil embargo, Congress, rather than the President, had the major role in crafting energy policy. Government controls had been holding down the price of domestically produced oil since 1971, when price controls were imposed as part of an overall anti-inflation effort. Natural gas prices had been regulated at the wellhead since 1954, except when gas was sold within the producing state.

The question of oil and gas pricing became the vehicle for putting the nation's energy problem squarely on the legislative agenda.

While congressional power was not confined to pricing, that was the most important existing authority. Confronted with the new energy era, the legislative branch had to decide how the government should proceed in developing new energy habits among Indians. Gas-guzzling cars were wasteful, but the consumer had traditionally been relatively independent from government rules. Windows could be caulked to save energy, but the bureaucracy had no easy way of dictating that it has to be done.

Also to be considered was how to encourage domestic production and use of both conventional fuels, such as uranium and coal, and new alternative sources, such as liquids from coal, energy from the sun and wind, and oil from shale. Utilities could be ordered to burn coal but there had to be exceptions. In the political arena, the list of exceptions grew and grew. Nuclear power was a possible solution, but the problems of radioactive waste disposal kept coming up. Solar energy sounded plausible, but as a practical matter seemed limited.

The public was confused about energy in the 1970s, skeptical about shortages, and ready to blame the oil companies for the nation's woes. The public's confusion about the energy situation was reflected in Congress where the lack of a consensus about the actual existence of the problem made it even harder to agree on a solution. Congress shied away from making decisions about energy.

The key energy challenge facing India today is preventing bottlenecks in energy supply from constraining economic growth. Due to rising consumption and higher world oil prices, the Indian government is faced with debilitating oil import costs. Between 1992 and 1997, India estimates that it needs 140,000 megawatts (MW) of new electricity generation capacity. This shortfall is anticipated to increase as foreign investors continue to encounter difficulty in moving ahead with their independent power projects.

Natural gas is India's most important potential alternative to coal. India is planning to make widespread use of natural gas in power generation and in the industrial and residential sectors. Projected demand will require large volumes of gas pipeline and liquefied natural gas (LNG) imports.

India's oil demand has risen to 1.6 million barrels per day (b/d), or more than twice domestic output. Almost half of India's trade deficit is due to petroleum imports, the cost of which also limits capital that could be invested in the economy. India is hoping to restrain demand by gradually raising product prices to international levels.

India's heavy reliance on highly polluting coal makes development and installation of Clean Coal Technology (CCT) a high priority. However, CCT is expensive to implement and questions have arisen on how to spread the costs of such technology.

India plans to add to its current nuclear capacity of 1.7 gigawatts (GW), with plans to have 5.7 GW of nuclear generating capacity operating or under construction by 2000.

In 1995, coal accounted for 63.3 percent of India's primary energy production. In the same year, petroleum accounted for 18.6 percent, hydroelectricity 8.9 percent, natural gas 8.2 percent, and nuclear power 1 percent.

India's electricity is generated overwhelmingly by coal (70 percent). Hydroelectricity ranks a distant second (about 25 percent), followed by natural gas, nuclear power, oil, and renewable, which account for the remaining 5 percent.

The current fuel mix is expected to change slightly through the forecast period ending in 2010. Coal (65%) is projected to remain roughly the same as in 1995, while hydro (14%) and natural gas (10%) will have higher shares of total production. Oil production will decline sharply, however, to only a 9 percent share.

Overall, India's energy production was around 8.8 quadrillion Btu (quads) in 1995. By 2010, India's energy output is expected to reach 16.4 quads. In comparison, China's total energy production was 11.7 quads in 1970, 35.6 quads in 1995, and is forecast to rise to 64 quads by 2010.

The book is giving information of all countries of the world in chronological order. The details of different sectors of energy *viz.*, oil, Gas, Power, Natural Gas and other details of energy is covered to make it more comprehensive and complete in all manner.

1

Afghanistan

General Background

Afghanistan currently is governed by a transitional administration headed by Hamid Karzai, who took office after a U.S.-led coalition defeated the previous Taliban government, which had provided sanctuary in Afghanistan for the terrorist group al-Qaeda. After more than two decades of war and chaos, and three years of drought in the late 1990s, Afghanistan's primarily agricultural economy is in very poor condition.

Foreign aid has begun to flow in to the transitional government, and pledges of assistance now total about $4.5 billion. Afghan agriculture, which is believed to make up about 80 per cent of the country's economic output, has benefitted over the last year from increased rainfall.

Still, the hurdles to recovery in Afghanistan are high. The transitional government has limited authority in much of the country, and competing regional power centers remain. The country's infrastructure also is in very poor condition. Commerce is inhibited by roads in need of repair, and in many places, existing electricity and telephone lines are inoperable.

Afghanistan recently replaced its currency. "Old Afghani" notes were exchanged in September 2002 for "New Afghani" notes, at a ratio of 100-to-1. This move was intended to give credibility to a currency which was so devalued that it had become nearly worthless. Use of U.S. dollars or neighboring countries' currencies is still common for many transactions in Afghanistan.

Energy Overview

The Soviets had estimated Afghanistan's proven and probable natural gas reserves at up to 5 trillion cubic feet (Tcf) in the 1970s. Afghan natural gas production reached 275 million cubic feet per day (Mmcf/d) in the mid-1970s. The Djarquduk field was brought online during that period boosted Afghan natural gas output to a peak of 385 Mmcf/d by 1978-79. However, sabotage of infrastructure

by the anti-Soviet mujaheddin fighters limited the country's total production to 290 Mmcf/d, an output level that was held fairly steady until the Soviet withdrawal in 1989. After the Soviet pullout and subsequent Afghan civil war, roughly 31 producing wells at Sheberghan area fields were shut in pending the restart of natural gas sales to the former Soviet Union.

At its peak in the late 1970s, Afghanistan supplied 70 per cent -90 per cent of its natural gas output to the Soviet Union's natural gas grid via a link through Uzbekistan. In 1992, Afghan President Najibullah indicated that a new natural gas sales agreement with Russia was in progress. However, several former Soviet republics raised price and distribution issues and negotiations stalled. In the early 1990s, Afghanistan also discussed possible natural gas supply arrangements with Hungary, Czechoslovakia, and several Western European countries, but these talks never progressed further. Afghan natural gas fields include Djarquduk, Khowaja Gogerdak, and Yatimtaq, all of which are located within 20 miles of the northern town of Sheberghan in Jowzjan province. In 1999, work resumed on the repair of a distribution pipeline to Mazar-i-Sharif. Spur pipelines to a small power plant and fertilizer plant also were repaired and completed. Mazar-i-Sharif is now receiving natural gas from the pipeline. A training center for natural gas workers is being reopened in Mazar-i-Sharif with Russian assistance. The possibility of exporting a small quantity of natural gas through the existing pipeline into Uzbekistan also is reportedly being considered.

Soviet estimates from the late 1970s placed Afghanistan's proven and probable oil and condensate reserves at 95 million barrels. Oil exploration and development work as well as plans to build a 10,000-bbl/d refinery were halted after the 1979 Soviet invasion. A very small amount of crude oil, about 300 barrels per day (bbl/d), is produced at the Angot field in the northern Sar-i-Pol province.

Petroleum products such as diesel, gasoline, and jet fuel are imported, mainly from Pakistan, Turkmenistan, and Uzbekistan. Turkmenistan also has a petroleum product storage and distribution facility at Tagtabazar near the Afghan border, which supplies northwestern Afghanistan.

Besides oil and natural gas, Afghanistan also is estimated to have 73 million tons of coal reserves, most of which is located in the region between Herat and Badashkan in the northern part of the country. Although Afghanistan produced over 100,000 short tons of coal annually as late as the early 1990s, as of 2000, the country was producing only around 1,000 short tons.

Afghanistan's power grid has been severely damaged by years of war, and only about 6 per cent of its population currently has access to electricity. Transmission lines from the Kajaki Dam in Helmand province near Kandahar were hit by an airstike in November 2001, but were repaired in early 2002. On several occasions since then, however, power to Kandahar has been cut off by attacks on the transmission lines. Other operational dams include the Dahla Dam in Kandahar province, the Breshna-Kot Dam in Nangarhar province, which has a generating capacity of 11.5 MW, and the 66-MW Mahipar dam which supplies Kabul.

Neighboring countries also supply electricity to some of Afghanistan's border regions. Turkmenistan supplies electricity to much of northwestern Afghanistan, including Mazar-i-Sharif and Herat. This arrangement was affirmed in an agreement signed in August 2002 between the Karzai government and Turkmenistan, continuing an earlier agreement between the Taliban government and Turkmenistan. Uzbekistan also supplies electricity to the northern area around Mazar-i-Sharif, supplementing a small local gas-fired power plant. Uzbekistan resumed its supply arrangement in August 2002, after having terminated supplies of electricity in 1999 during the period of Taliban rule. Iran also supplies electricity to Afghanistan, in some areas directly adjacent to the Afghan-Iranian border in Herat, Farah, and Nimroz provinces.

Afghanistan as an Energy Transit Route

Due to its location between the oil and natural gas reserves of the Caspian Basin and the Indian Ocean, Afghanistan has long been mentioned as a potential pipeline route, though in the near term, several obstacles will likely prevent Afghanistan from becoming an energy transit corridor. Unocal had pursued a possible natural gas pipeline from Turmenistan to Pakistan in the mid-1990s, but pulled out after the U.S. missile strikes against Afghanistan in August 1998. The new Afghan government under President Karzai has tried to revive the pipeline plan, and talks have been held between the governments of Afghanistan, Pakistan, and Turkmenistan in 2002 on the issue, but a signing ceremony for a framework agreement between the governments has been delayed until at least December 2002.

Given the obstacles to development of a natural gas pipeline across Afghanistan, it seems unlikely that such an idea will make any progress in the near future, and no major Western companies have expressed interest in reviving the project. The security situation in Afghanistan is one obvious major risk, and the tensions between India and Pakistan make it unlikely that such a pipeline could be extended into India, which unlike Pakistan has sufficient immediate demand for imported natural gas to justify a project of such magnitude. Financial problems in the utility sector in India, which would be the major consumer of the natural gas, also could pose a problem.

Energy Infrastructure at a Glance

Oil

Angot Oilfield: Produces a small quantity of crude oil; located in Sar-i-Pol province

Natural Gas

Sheberghan Area Gas Fields: The Djarquduk, Khowaja Gogerak, and Yatimtaq natural gas fields are all located within 20 miles of Sheberghan.

Pipeline to Mazar-i-Sharif: A pipeline connects these natural gas fields to Mazar-i-Sharif. Gas is used for a small power plant, a fertilizer plant, and domestic use.

Local Pipelines: Small local pipelines near the natural gas fields distribute natural gas in small quantities to nearby villages and Sheberghan

Electricity

Kajaki Dam: Located in Helmand province near Kandahar; transmission lines to Kandahar repaired in early 2002, after being damaged by airstrikes in November 2001.

Mahipar Dam: Installed capacity of 66 MW. Near Kabul. Operational but currently lacking adequate water.

Naghlu Dam: Installed capacity of 100 MW. Operational. Provides most of the electricity used in Kabul.

Darunta Dam: Installed capacity of 11 MW. Operational. In Nangarhar province near Jalalabad.

Sarobi Dam: Installed capacity of 22 MW.

Dahla Dam: Kandahar province. Operational.

Mazar-i-Sharif Power Plant: Small natural gas-fired power plant near Mazar-i-Sharif, partially operational at under 30 MW.

MAP OF AFGHANISTAN'S NATURAL RESOURCES

Note: *This listing of Afghanistan's energy infrastructure was compiled from information available in press and media sources, and should not necessarily be considered comprehensive. Only facilities which have been reported to be functional or under repair have been included.*

2

Algeria

Algeria is important to world energy markets because it is a significant oil and gas producer and exporter. Algeria also is a member of OPEC and an important, growing energy source for Europe.

General Introduction

Following years of civil war and continuing political unrest, Algeria now is experiencing a significant economic upturn, in large part aided by strong oil and natural gas export revenues since 1999. Real gross domestic product (GDP) growth is expected to reach 4.3 per cent in 2003, following

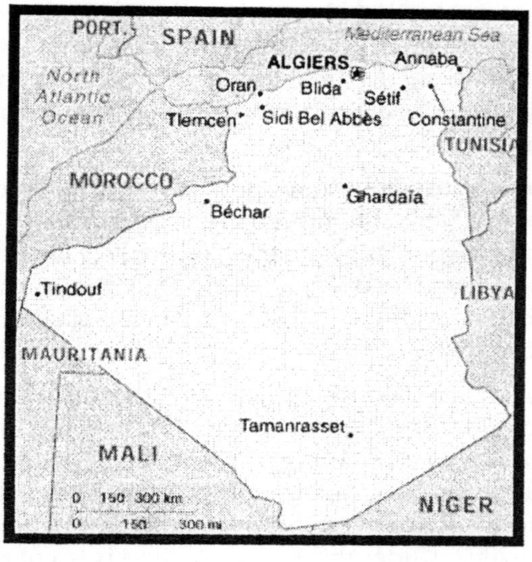

estimated growth of 3.4 per cent in 2002. The sharp increase in oil export revenues which Algeria enjoyed beginning in early 1999 caused the country's foreign reserves to rebound sharply (to nearly $18 billion by late 2001, compared to $12 billion at the end of 2000), external debt to fall (to the lowest level in a decade), hard currency reserves to rise (to $22 billion by end-2003), the current account balance to improve dramatically, and pressures on government finances to decrease. In addition, moves towards liberalization and deregulation have proven helpful.

Despite the recent good news, Algeria continues to face serious economic, social, and political problems, including: high unemployment (officially 27 per cent, but possibly much higher); continued political violence by Islamic fundamentalists and others; labor unrest; a large black market (possibly 20 per cent of the country's GDP); continued weakness in the non-oil economy (a severe drought hurt the agricultural sector in 2000, and heavy flooding struck northern Algeria in November 2001, killing 764 and causing an estimated $300 million in damage); and slow progress on economic reform efforts (largely due to opposition by labor unions and the armed forces). Recently, there has been a flareup of protests by the country's restive Berber minority demanding greater autonomy, increased employment opportunities, and better living conditions. The unrest has centered on the Kabyle region of northeastern Algeria.

In late December 2002, Algeria adopted its budget for 2003, based on an extremely conservative assumption of $19 per barrel for Algerian oil (currently, world oil prices are hovering around $30 per barrel). Algeria remains highly dependent on oil and natural gas exports, which account for more than 90 per cent of Algerian export earnings, and about 30 per cent of GDP. The country's new budget also sets out an ambitious GDP growth target of 6.5 per cent, a small depreciation of the dinar, and curbs on state spending. Algeria's top economic goal is to reduce the country's unemployment rate. The key to achieving this appears to be privatization (including promises to sell off 100 state firms), deregulation, restructuring of public enterprises, and attraction of foreign investment. To date, however, only slow progress has been made in these areas.

In December 2001, the International Monetary Fund (IMF) issued its annual "Article IV" assessment of the Algerian economy, urging that the government proceed with privatization and banking reform, while lowering tariffs aimed at protecting domestic industry and reducing dependence on hydrocarbons. The IMF praised the Algerian government for its strong fiscal discipline (and careful monetary policy as well), and for allowing the dinar to depreciate against the dollar. Finally, the IMF pointed out that high oil prices provide Algeria with an opportunity to make progress on implementing reforms and addressing the country's many problems.

In late 2001, an important new hydrocarbons reform bill was introduced, but progress possible stalled in 2002. The bill would open Algeria's all-important energy sector to private (including foreign) investment, although state oil and gas company Sonatrach most likely would remain in public hands. The law faces opposition from trade unions and others, and already has been watered down somewhat from its original form, while Energy and Mines Minister Chekib Khelil has stated that "it is not necessary to privatize" Sonatrach.

In December 2002, Algeria signed a cooperation pact with the European Free Trade Association (EFTA), providing for expanded and liberalized trade with EFTA members (Iceland, Liechtenstein, Norway, and Switzerland). Algeria also is pursuing membership in the World Trade Organization, with the latest negotiations concluded in Geneva during November 2002. In late 2001, Algeria and the EU reached an Association Agreement after years of negotiations, and the deal was ratified by the European Parliament in October 2002. Under the accord, Algeria is to cut tariffs on EU agricultural and industrial products over the next 10 years. In exchange, the EU will eliminate duties and quotas on many Algerian agricultural products.

President Abdelaziz Bouteflika, elected President on April 15, 1999 for a 5-year term, has attempted to implement plans for national reconciliation and economic reforms (*i.e.*, deregulation, privatization). More than 100,000 rebels, soldiers and civilians have died in Algeria's civil war, which began in 1992 following the military's nullification of a national election won by the Islamic Salvation Party. On July 13, 1999, President Bouteflika offered amnesty to rebel groups, and on September 16 a national referendum was held in which voters approved the offer. Although the government claimed that nearly 80 per cent of rebels (including members of the Islamic Salvation Army) accepted amnesty, the level of violence appeared to rise once again, with the most violent groups apparently stepping up attacks. In August 2000, President Bouteflika replaced Prime Minister Ahmed Benbitour with Ali Benflis. Parliamentary elections were held in May 2002, resulting in a strong showing for the president's party, the FLN.

Oil

Although oil was first discovered in Algeria at the Hassi Messaoud oil field in 1956, Algeria is considered to be underexplored. Algeria's National Council of Energy believes that the country still contains vast hydrocarbon potential. Over the last few years, significant oil and gas discoveries have been made, largely by foreign companies (in partnership with state-owned Sonatrach, as required by current Algerian law). Sonatrach and its foreign partners hope to increase Algeria's crude oil production capacity significantly over the next few years. In order to accomplish this, Algeria will require significant amounts of foreign capital and expertise. Energy Minister Chekib Khelil has stated that his goal is "to double the number of companies operating in Algeria over the next five years" to 40 companies. Khelil also has expressed his view that the industry needs to be restructured in order to survive, and that new regulatory bodies independent of the Energy and Mining Ministry might be needed as well.

Official estimates of Algeria's proven oil reserves remain at 9.2 billion barrels. With recent oil discoveries, plans for more exploration drilling, improved data on existing fields, and use of enhanced oil recovery (EOR) systems, proven oil reserve estimates are expected to be revised upward in coming years. Algeria should also see a sharp increase in crude oil exports over the next few years due to a rapid shift towards domestic natural gas consumption and planned increases in oil production by Sonatrach and its foreign partners. Approximately 90 per cent of Algeria's crude oil exports go to Western Europe, with Italy as the main market followed by Germany and France. The Netherlands, Spain and Britain are other important European markets. Algeria's Saharan Blend oil, 45° API with 0.05 per cent sulfur and negligible metal content, is among the best in the world.

As mentioned above, the Algerian parliament is considering a law which would restructure the state oil company, Sonatrach (and Sonelgaz, the state utility) in order to attract private international investment. One possibility would be for Sonatrach to remain the national oil company but eventually be forced to compete for new projects. Non-core subsidiaries of Sonatrach also could be privatized if the law passes. In January 2001, Algeria's oil and gas industry labor unions announced their opposition to any government plans to open up the country's hydrocarbon sector to foreign investors.

Oil Production

Algeria's average crude oil production during the first 10 months of 2002 was 839,000 bbl/d. Together with 430,000 bbl/d of lease condensate and 190,000 bbl/d of natural gas plant liquids, Algeria averaged 1.46 million bbl/d of total oil production during January-October 2002. Algeria's crude oil production quota was set at 780,000 bbl/d as of February 1, 2003, significantly below the country's crude oil production capacity of 1.1 million bbl/d (note: only crude oil production is subject to the OPEC quota). Algeria had estimated net oil exports (including crude oil, lease condensate, and natural gas liquids) of around 1.25 million bbl/d in 2002, most of which went to Europe and the United States. Domestic oil consumption is around 200,000 bbl/d.

In coming years, it is likely that Algeria's oil production capacity will be increasing rapidly. Already, Algeria plans to raise its crude oil production capacity to 1.5 million bbl/d by 2004 (and 2 million bbl/d within 10 years), and has begun pressing its case for a higher OPEC quota. Much of Algeria's increased production capacity will come from foreign independent oil companies, such as Amerada Hess (the El-Gassi field), Anadarko (Berkine, Ourhoud), Burlington Resources (Block 405), BHP Billiton (ROD), and Cepsa (Ourhoud).

By far, the largest oil field in Algeria is Hassi Messaoud, located in the center of the country, which produces about 400,000 bbl/d of 46° API crude, down from 550,000 bbl/d in the 1970s, but up from 300,000 bbl/d in 1989. The Hassi Messaoud area contains an estimated 6.4 billion barrels, or about 70 per cent of the country's proven oil reserves. Sonatrach operates Algeria's other major oil fields, including Rhourde el-Baguel (Algeria's second largest oil field, located to the northeast of Hassi Messaoud), Tin Fouye Tabankort Ordo, Zarzaitine (30,000 bbl/d), Haoud Berkaoui/Ben Kahla, el-Gassi el-Agreb and Ait Kheir. The Hassi R'Mel gas field (north of Hassi Messaoud, south of Algiers) also produces around 18,000 bbl/d of 46.1° API crude.

In April 2000, Amerada Hess announced that it had acquired (for $55 million) the Gassi el-Agreb Redevelopment Project from Sonatrach. Amerada Hess will form a joint operating company with Sonatrach, to be called Sonahess, and will invest $500 million over 5 years to enhance recovery from the el-Gassi, el-Agreb, and Zotti fields. Currently, the three fields produce around 30,000 bbl/d, and the redevelopment project aims to increase production to 45,000 bbl/d by late 2003.

Algeria's oil sector, unlike that of most OPEC producers, has been open to foreign investors for more than a decade. One of the largest joint ventures in Algeria is the partnership between Anadarko, Lasmo and Denmark's Maersk Oile to develop the Hassi Berkine South oil field. Oil from the field's Block 404 was first produced in May 1998, and the field had production capacity of around 285,000 bbl/d as of mid-2002.

BHP has stated that it will spend $190 million on oil field development at the "ROD" integrated oil development project in the Berkine Basin in eastern Algeria. Production is expected to commence in 2004 at 35,000 bbl/d, and peak at 80,000 bbl/d. In July 2000, several companies (Burlington Resources, Talisman, and Sonatrach) announced that they would develop the MLN field in Block 405a. MLN is expected to produce around 35,000-40,000 bbl/d when completed. Exploration success rates in the Berkine Basin have been high, and several billion barrels of oil may lie within 15 miles or so of the area.

In early January 2003, Sonatrach announced that it had brought the 1-billion-barrel Ourhoud oil field online, ahead of schedule, with initial production of 75,000 bbl/d. The field, which is slated to reach 230,000 bbl/d by the end of February 2003 when all three oil treatment trains come online, is divided into three blocks operated by Anadarko (Block 404), Cepsa of Spain (Block 406a), and Burlington Resources (Block 405). When it reaches full capacity, Ourhoud will raise Algeria's crude oil production capacity from 1.1 million bbl/d to 1.3 million bbl/d. Total development costs at Ourhoud total $1.3 billion.

Although Algeria has experienced a significant influx of foreign investment in recent years, it still has many oil fields in need of additional foreign capital and EOR investment. Halliburton has an eight-year contract to provide EOR services and boost production at Hassi Messaoud, for instance, which saw production fall sharply beginning in the mid-1980s. Algeria's second largest oil field,

Rhourde El Baguel, already has received foreign investment to boost its production capacity. Rhourde El Baguel contains about three billion barrels of 42.6° API oil, of which less than 450 million barrels has been produced since 1963. In February 1996, Arco (now owned by BP) signed a $1.3-billion production sharing agreement (PSA) with Sonatrach to increase production at the field. BP expects to raise the field's output from 27,000 bbl/d to 125,000 bbl/d by 2010.

In July 2002, Sonatrach awarded six new exploration contracts as part of Algeria's third licensing round. Companies receiving blocks included Anadarko (Berkine Basin Block 403c/e); Petrovietnam (Oued Mya blocks 433a-416b); RWE-DEA, Cepsa and Edison (Reggane blocks 351c-352c); Gaz de France (Sbaa Basin blocks 352a-353); TotalFinaElf and Cepsa (Timimoum blocks 325a-329); and Cyprus-based Medex Petroleum (Illizi blocks 226-229b and 242). In November 2001, a tender for development of five oil and natural gas projects in the Illizi Basin received 14 bids from foreign companies such as BP, TotalFinaElf, Amerada Hess, and BHP Billiton. Besides Illizi itself, projects include Zarzaitine, El Adeb Larache, Rhourde Nouss, Tinrhert. Two other companies working in the Illizi basin include Ireland's Tullow Oil and Japan's Teikoku Oil (Block 222b).

Sinopec reportedly was awarded a $525 million contract in October 2002 to help increase the crude oil recovery rate at Zarzataine, near Hassi Messaoud. In November 2002, the Kuwait Foreign Petroleum Exploration Company (KUFPEC) and Anadarko announced a partnership to explore the Berkine Basin. KUFPEC has not been active in Algeria for over 10 years.

Downstream

Algeria has four oil refineries, with combined capacity of 450,000 bbl/d. The 30,000-bbl/d Hassi Messaoud plant supplies products to southern Algeria, as does the smaller 7,000-bbl/d In Amenas plant. The 60,000-bbl/d Algiers refinery processes crude from Hassi Messaoud. Finally, the coastal 60,000-bbl/d Arzew refinery, which uses Algerian Saharan blend as feedstock, produces both heavier and light products for domestic consumption and export. In January 2001, Algeria issued a tender for a new refinery in the central Adrar region near the Sbaa basin. Algeria also is looking at upgrading the In Amenas refinery.

Although Algeria has a substantial petrochemical and fertilizer industry, low capacity utilization rates mean continued reliance on imports. The majority of Algeria's petrochemical plants are located at Annaba (a 550,000-ton- per-year (t/y)–ammonium phosphate fertilizer plant and ammonium nitrate and nitric acid complex), Arzew (365,000 t/y ammonia, 146,000 t/y urea, and 182,500 t/y ammonium nitrate), and Skikda (a 130,000 t/y high-density polyethylene unit, 120,000-t/y ethylene cracker, and a substantial aromatics complex). Sonatrach has undertaken a number of petrochemical and fertilizer expansion projects, including a new methyl tertiary butyl ether (MTBE) complex and a polyester resin complex.

Algeria uses seven coastal terminals for crude oil, refined product, NGL, and liquefied natural gas (LNG) exports. These are located at Arzew (Algeria's largest crude oil export port), Skikda (Algeria's second largest crude oil export port), Algiers, Annaba, Oran, plus the Tunisian facilities of Bejaia and La Skhirra. Arzew handles about 40 per cent of Algeria's total hydrocarbon exports (including all of its NGL exports), and Algeria has ambitious plans for the port area. Among other things, the government would like to build a petrochemicals complex at Arzew, as well as a condensate refinery and desalination plant. Work also needs to be done to maintain and upgrade Arzew's crude oil loading capacity. A refurbishment project on the port began in 1998. Skikda port is limited to 80,000-ton tankers and will require dredging and other maintenance work in order to accommodate larger tankers.

Natural Gas

Commercial production of natural gas began in 1961. Algeria has 160 trillion cubic feet (Tcf) of proven natural gas reserves, primarily associated (with oil), ranking it in the top 10 worldwide. Sonatrach estimates that Algeria's ultimate gas potential is around 204 Tcf, and natural gas (including natural gas liquids) accounted for about 60 per cent of Algeria's total hydrocarbons production in 2000. Algeria also is a major natural gas exporter, accounting for one-fifth of EU natural gas imports in 2000 (Russia accounted for 39 per cent in that year). As of 2002, Algeria's total natural gas export capacity, via pipeline and LNG tanker, was over 2 Tcf per year. This is expected to increase rapidly in coming years as major new gas fields, export pipelines, and LNG facilities come online. Algeria's goal is to export 3 Tcf per year or more by 2010. Algeria is a founding member of the Gas Exporting Countries' Forum, a loose group of 15 gas producing countries formed in Tehran in May 2000.

Algeria's largest gas field is the super-giant Hassi R'Mel, which initially held proven reserves of about 85 Tcf. Hassi R'Mel accounts for around 1.35 billion cubic feet (Bcf) per day, or about a quarter of Algeria's total dry gas production. The remainder of Algeria's gas reserves are located in associated and non-associated fields in the southeast, and in non-associated reservoirs in the In Salah region of southern Algeria. The Rhourde Nouss region holds 13 Tcf of known reserves in the Rhourde Nouss, Rhourde Nouss Sud-Est, Rhourde Adra, Rhourde Chouff, and Rhourde Hamra fields. Smaller gas reserves are located in the In Salah region (5-10 Tcf) as well as at the Tin Fouye Tabankort (TFT)(5.1 Tcf), Alrar (4.7 Tcf), Ouan Dimeta (1.8 Tcf), and Oued Noumer fields. Four plants at Arzew and Skikda, owned by Sonatrach, liquefy gas for export.

Algeria's natural gas pipeline export capacity includes around 900 Bcf/y via the 667-mile Trans-Mediterranean (Transmed, renamed Enrico Mattei) line from Hassi R'Mel via Tunisia and Sicily to mainland Italy, and 350 Bcf/y via the 1,013-mile Maghreb-Europe Gas (MEG, renamed Pedro Duran Farell) line via Morocco to Cordoba, Spain, where it ties into the Spanish and Portugese gas transmission networks. Given that over the past decade, natural gas has been the fastest growing fuel source in the EU, with natural gas–mainly imported–expected to account for 26 per cent of EU energy consumption by 2010, Algeria has plans to increase its natural gas export capacity significantly in coming years. In August 2001, Sonatrach awarded ABB a $93 million contract to build a natural gas compressor station on the Pedro Duran Farrell (MEG) line in order to raise capacity to 460 Bcf/y by late 2004 (and 650 Bcf/y by 2006). There also are plans to expand Transmed capacity to more than 1 Tcf per year.

One complication in Algeria's natural gas export strategy to Europe has been EU liberalization, which has complicated the legality of traditional "destination clauses" for gas deliveries. Such clauses prevent the offtaker of the gas from reselling it to another EU state, and this has complicated Algeria's attempts at signing agreements with EU purchasers, such as ENEL.

In late July 2001, Spain's Cepsa and Algeria's Sonatrach agreed to move ahead with a new natural gas pipeline (Medgaz) linking Algeria directly to Spain. Since then, several other companies–BP, Endesa, Eni, Gaz de France, and TotalFinaElf–have acquired 12 per cent shares in Medgaz. In September 2002, the consortium completed a study of the line's feasibility.The 350-530 Bcf/y Medgaz line most likely would go from Hassi R'Mel through the port of Arzew to Almeira, Spain. Medgaz could be operational by 2006. In November 2002, Cepsa said that it had signed a letter of intent to purchase 35 Bcf/y of natural gas via Medgaz.

In December 2001, Sonatrach signed a deal with Italy's Enel and Germany's Wintershall on a feasibility study of another new natural gas pipeline, this one from Algeria under the Mediterranean Sea to Sicily and onwards to the Italian mainland and also southern France. Initial capacity on this line could be in the 280-350 Bcf per year range. Possible routes include one to Sardinia and Corsica, entering mainland Italy at La Spezia. An alternative route would run through Sardinia to the Italian mainland at Castiglione della Pescaia.

Another possibility that has been mentioned recently is a Trans-Sahara natural gas pipeline from Nigeria, across the Sahara, and north through Algeria to the Mediterranean coast. The pipeline could cost $5-$7 billion and utilize the Medgaz line to Spain.

Aside from exports to Italy, Spain, Portugal, etc., Algeria has a policy of using its natural gas reserves as a source of domestic energy and as a raw material for the petrochemical industry. Algeria consumes around 400 Bcf of natural gas per year. Approximately 95 per cent of the country's electricity is generated by natural gas.

Development of the In Salah region is one of the lynchpins in Algeria's plan to increase its natural gas exports. In February 2000, BP Amoco (now BP) and Sonatrach signed a $2.5 billion deal to develop seven of the twelve existing fields in the In Salah region, including the Garat al-Bafinat, Teguentour, Krechta, Reg, In Salah, Hassi Moumeme, and Gour Mahmoud fields. These fields contain estimated dry natural gas reserves of 5 Tcf, with a potential for 10 Tcf total. In addition, the joint venture, called In Salah Gas, will appraise existing wells and explore for new gas reserves in the In Salah region. In Salah Gas is the first major natural gas joint venture between Sonatrach and a foreign partner. Production from the region originally was expected to come online in late 2003, after the drilling of up to 200 production wells and construction of a $1 billion, 48-inch pipeline link to Hassi R'Mel. However, progress has been slowed due to several factors, including EU rules on natural gas re-exports and slower-than-expected natural gas demand growth in potential importing countries such as Spain, and production now is expected to begin in 2004. Main engineering work on In Salah is to be performed by Japan's JGC; U.S.-based Kellogg, Brown, and Root; Bechtel; and Sonatrach subsidiary Enefor.

In May 1997, In Salah Gas sealed its first natural gas sales deal with Italian electricity generator Enel. The deal enables In Salah Gas to take over an existing contract to supply Enel with 141 Bcf/y of gas. Sonatrach will continue supplying the Italian power giant with natural gas supplies until In Salah is ready. The deal represents In Salah's first step towards achieving its sales goal of 318-388 Bcf/y. Besides Enel, the venture is also marketing gas to other potential clients in Europe, Turkey and North Africa.

Besides In Salah, three other important Algerian natural gas and condensates projects are Ohanet, In Amenas, and Gassi Touil. Ohanet is located in the Illizi province on the northern edge of the Sahara desert about 60 miles west of the Libyan border. Ohanet is being developed–at a cost of around $1 billion–by Australia's BHP (with a 45 per cent share), Woodside Petroleum (15 per cent), Japan Ohanet Oil and Gas, Swiss-Swedish ABB, and U.S.-based Petrofac. Production from Ohanet is expected to begin in late 2003 or 2004, and is to include natural gas, natural gas liquids, and liquefied petroleum gas (LPG). The development project includes construction of a natural gas processing plant, with capacity of 30,400 bbl/d of condensate, 27,700 bbl/d of LPG, and 665 million cubic feet (Mmcf)/day of natural gas–as well as a pipeline. BHP is to operate the fields in partnership with Sonatrach.

In November 2002, Sonatrach and BP signed a deal to develop natural gas production in the In Amenas region. The $1.8 billion project is due to come onstream in 2005 and to produce around 900 million cubic feet per day of "wet" (*i.e.*, associated wtih oil) natural gas, plus 50,000 bbl/d of condensate and LPG. The project also includes construction of three pipelines to carry the hydrocarbons to the Sonatrach distribution system at Ohanet.

In May 2002, Algeria issued and international tender for development of Gassi Touil, which is believed to contain natural gas reserves of 9 Tcf.

Liquefied Natural Gas (LNG) Exports

With the start-up of the Arzew GL4Z plant in 1964, Algeria became the world's first LNG producer. Algeria was the second largest exporter of LNG (behind Indonesia) in 2000, with 19 per cent of the world's total, exported mainly to Western Europe (France, Belgium, Spain, Turkey) and the United States. Algeria reportedly has a major cost advantage in supplying LNG to Europe compared to regional competitors, with the cost of producing and delivering LNG having come down by an estimated 30 per cent -40 per cent since the 1970s. Desiring to capitalize on this advantage, Algeria is looking at the possibility of building a new 4-5 million-ton LNG train at Arzew.

In 1999, Sonatrach completed a total renovation of its LNG facilities, raising the country's LNG production capacity to around 1 Tcf per year. This refurbishment program focused on the 378 Bcf/year Arzew GL1Z, 378 Bcf/year Arzew GL2Z, and 230-Bcf/year Skikda GL1K plants. Also, Algeria's original 44-Mmcf/d Arzew GL4Z, or "Camel," plant, which was slated for decommissioning by 1997, has been refurbished to keep the plant operational for reserve purposes until at least 2003. Sonatrach plans to expand its exports, especially to Europe. In September 2001, Spain's second largest power company, Iberdrola, purchased a spot LNG cargo from Algeria, the first such purchase by a Spanish utility. In addition to regasification terminals at Barcelona, Cartagena, and Huelva, the *Middle East Economic Digest* reports that Sonatrach is constructing a new LNG receiving facility at Ferrol, in northwestern Spain.

Electricity

Algeria's electricity demand is growing at a rapid, 5 per cent annual rate, and will, according to Sonelgaz, require significant additional capacity–possibly 8,000 MW by 2010–in coming years (currently, Algeria has around 6,000 MW of installed power generating capacity). This likely will require billions of dollars worth of investments in new generating capacity, plus transmission and distribution infrastructure (*i.e.*, lines and sub-stations). In order to accomplish this, Algeria's government hopes to attract foreign capital.

In January 2002, legislation passed by Algeria's parliament ended Sonelgaz's monopoly over electric power generation, converted the company into a joint-stock company, and cleared the way for Algeria's first independent power projects (IPPs). However, further legislation which would allow Sonatrach to operate along commercial lines is stalled at the moment. Possible IPPs include: 1) the 2,000-MW, combined-cycle Hadjret En Nouss plant at Skikda, scheduled for completion in 2003-4, with 1,200 MW of the plant's capacity slated to serve export markets in Europe and 800 MW slated for domestic consumption; 2) the 1,200-MW Terga plant near Oran Tipasa, scheduled for completion in 2005-6; and 3) the 1.200-MW, combined-cycle Koudiat Draouch plant near Annaba, scheduled for completion in 2006-7. In the nearer term, the 440-MW Hamma natural gas turbine plant in Algiers is moving ahead toward commissioning in early 2002. The plant is being built by Ansaldo Energia. Private financing is also planned for a three-by-100-MW gas turbine unit at Hassi Messaoud. Future plans include natural-gas-fired power plants at Ain M'Lila, Arzew, M'Sila, Tighemt, and more.

In May 2001, Sonatrach and Sonelgaz established a joint venture–the Algerian Energy Company (AEC)–to export electricity. Among other projects, AEC is examining the feasibility of establishing a trans-Mediterranean power link to Italy, with the study expected to be completed in July 2003. In December 2001, Sonelgaz signed a joint venture agreement with Italian power grid manager GRTN on the possibility of constructing an undersea power cable to export electricity to Europe via Sardinia or Sicily. In November 2001, Sonelgaz signed a similar deal with Spain's power group Red Electrica de Espana to build an underwater power line between Algeria and Spain. Currently, Algeria has two links to the Moroccan electricity grid and supplies over 550 gigawatthours (GWh) of electricity to Morocco.

Sonatrach has a $107 million contract with Anadarko and Italy's GE Nuovo Pignone to build the country's first privately financed natural-gas-fired power plant at Hassi Berkine. GE Nuovo Pignone, a subsidiary of General Electric, will also provide a gas treatment system, liquid fuel gas turbine storage and services.

In July 2002, Sonatrach and Sonelgaz formed a new, renewable energy joint venture company, called New Energy Algeria (NEAL). NEAL will look at development of solar, wind, biomass, and photovoltaic (PV) energy production. One project reportedly under consideration is a 120-MW hybrid natural gas/solar power plant and a wind/diesel/PV facility at Timimoun.

Country Overview

Independence: July 5, 1962 (from France)

Population (7/02E): 32.3 million

Location/Size: North Africa/919,595 sq. miles, more than one-quarter the size of the United States

Major Cities: Algiers (capital), Constantine, Annaba, Arzew, Skikda, Oran, Ghardaia, Bechar, Ouargla, Touggourt

Languages: Arabic (official), French, Berber dialects

Ethnic Groups: Arab (84 per cent), Berber (16 per cent), European (less than 1 per cent).

Religions: Sunni Islam (state religion) 99 per cent, Christianity and Judaism 1 per cent

Defense (1999E): Army (105,000), Navy (7,000), Air Force (10,000), Paramilitary Forces (181,200). Total armed forces: 303,200.

Economic Overview

Currency: Algerian Dinar (AD)

Market Exchange Rate (1/03E): US$1 = AD 76.5

Gross Domestic Product (at market exchange rates) (2002E): $55.1 billion

Gross Domestic Product (at purchasing power parity rates) (2001E): $177 billion

Real GDP Growth Rate (2002E): 3.4 per cent *(2003F):* 4.3 per cent

Inflation Rate (consumer prices) (2002E): 1.1 per cent *(2003F):* 1.2 per cent

Major Export Products (2002): Petroleum and natural gas

Major Import Products (2002): Industrial equipment; intermediate goods; food; consumer goods; capital goods.

Merchandise Exports (2001E): $14.3 billion

Merchandise Imports (2001E): $11.0 billion

Merchandise Trade Balance (2002E): $3.3 billion

Current Account Balance (2002E): $0.3 billion

Oil Export Revenues (2002E): $11.8 billion

Foreign Exchange Reserves (2002E): $21.8 billion

Total External Debt (2002E): $20.5 billion

Energy Overview

Energy Minister: Chekib Khelil

Proven Oil Reserves (1/1/03E): 9.2 billion barrels

Oil Production (January-October 2002E): 1.46 million barrels per day (bbl/d), of which 0.84 million bbl/d was crude oil, 0.43 million bbl/d was lease condensates, and 190,000 bbl/d was natural gas liquids

Crude Oil Production Capacity (1/03E): 1.1 million bbl/d

Oil Consumption (2002E): 208,000 barrels per day (bbl/d)

Net Oil Exports (January-October 2002E): 1.25 million bbl/d

U.S. Oil Imports from Algeria (January–October 2002E): 275,000 bbl/d

Crude Oil Refining Capacity (1/1/03E): 450,000 bbl/d (according to the *Oil and Gas Journal*)

Natural Gas Reserves (1/1/03E): 159.7 trillion cubic feet (Tcf)

Natural Gas Production (2000E): 2.9 trillion cubic feet (Tcf)

Natural Gas Consumption (2000E): 0.8 trillion cubic feet (Tcf)

Electricity Generation Capacity (2000E): 6.0 gigawatts (95 per cent thermal–mainly natural gas; 5 per cent hydroelecric)

Electricity Production (2000E): 23.6 billion kilowatthours

Environmental Overview

Total Energy Consumption (2000E): 1.23 quadrillion Btu* (0.3 per cent of world total energy consumption)

Energy-Related Carbon Emissions (2000E): 22.7 million metric tons of carbon (0.4 per cent of world total carbon emissions)

Per Capita Energy Consumption (2000E): 39.4 million Btu (vs U.S. value of 351.0 million Btu)

Per Capita Carbon Emissions (2000E): 0.7 metric tons of carbon (vs U.S. value of 5.6 metric tons of carbon)

Energy Intensity (2000E): 23,993 Btu/$1995 (vs U.S. value of 10,918 Btu/ $1995)

Carbon Intensity (2000E): 0.44 metric tons of carbon/thousand $1995 (vs U.S. value of 0.17 metric tons/thousand $1995)

Sectoral Share of Energy Consumption (1998E): Industrial (49.2 per cent), Transportation (22.0 per cent), Residential (28.8 per cent)

Sectoral Share of Carbon Emissions (1998E): Industrial (46.0 per cent), Transportation (26.0 per cent), Residential (27.9 per cent)

Fuel Share of Energy Consumption (2000E): Natural Gas (66.5 per cent); Oil (31.9 per cent); Coal (1.7 per cent)

Fuel Share of Carbon Emissions (2000E): Natural Gas (67.9 per cent); Oil (30.1 per cent); Coal (2.0 per cent)

Renewable Energy Consumption (1998E): 11.9 trillion Btu* (47 per cent decrease from 1997)

Number of People per Motor Vehicle (1998): 19.2 (vs. U.S. value of 1.3)

Status in Climate Change Negotiations: Non-Annex I country under the United Nations Framework Convention on Climate Change (ratified June 9th, 1993). Not a signatory to the Kyoto Protocol.

Major Environmental Issues: Soil erosion; desertification; river and coastal water pollution due to the dumping of raw sewage, petroleum refining wastes, and other industrial effluents; inadequate supplies of potable water.

Major International Environmental Agreements: A party to Conventions on Biodiversity, Climate Change, Desertification, Endangered Species, Environmental Modification, Law of the Sea, Ozone Layer Protection, Ship Pollution, Wetlands. Has signed, but not ratified, the Nuclear Test Ban Treaty.

The total energy consumption statistic includes petroleum, dry natural gas, coal, net hydro, nuclear, geothermal, solar, wind, wood and waste electric power. The renewable energy consumption statistic is based on International Energy Agency (IEA) data and includes hydropower, solar, wind, tide, geothermal, solid biomass and animal products, biomass gas and liquids, industrial and municipal wastes.

Oils and Gas Industry

Major State Companies: Enterprise Nationale pour la Recherche, la Production, le Transport, la Transformation et la Commercialisation des Hydrocarbons (Sonatrach)–State-owned company for exploration, transport and marketing of petroleum, natural gas and related products; Enterprise Nationale de Raddinage des Produits Petroliers (Naftec)–Operates and manages all refineries; Enterprise Nationale de Commercialisation et de Distribution des Produits Petroliers (Naftel)–Domestic product distribution. Societe de Conditionnement, Comercialisation & Transport de Gas Industriels (Cogiz)–produces natural gas by-products.

Oil Export Terminals: Arzew (Algeria's largest crude oil export port), Skikda (Algeria's second largest crude oil export port), Algiers, Annaba, Oran, plus the Tunisian facilities of Bejaia and La Skhirra.

Major Oil Fields: Hassi Messaoud (Algeria's largest oil field), Rhourde el-Baguel (Algeria's second largest oil field, located to the northeast of Hassi Messaoud), Tin Fouye Tabankort Ordo, Zarzaitine, Haoud Berkaoui/Ben Kahla, el-Gassi el-Agreb, Ait Kheir.

Major Natural Gas Fields: Hassi R'Mel, Rhourde Nouss, Rhourde Nouss Sud-Est, Rhourde Adra, Rhourde Chouff, Rhourde Hamra fields. Smaller gas reserves are located in the In Salah region (5-10 Tcf) as well as at the Tin Fouye Tabankort (TFT)(5.1 Tcf), Alrar (4.7 Tcf), Ouan Dimeta (1.8 Tcf), and Oued Noumer fields.

Oil Refineries (crude refining capacity bbl/d, 2003E): Skikda (300,000), Algiers (60,000), Arzew (60,000), Hassi Messaoud (30,000)

LNG Facilities: Arzew GL4Z, Arzew GL1Z, Arzew GL2Z, Skikda GL1K

Selected Foreign Energy Company Involvement: Agip, Amerada Hess, Anadarko, Burlington Resources, BHP Billiton, BP, Cepsa. Conoco, Edison, ENI, ExxonMobil, Gaz de France, Halliburton, Lasmo, Louisiana, Maersk, Neste Oy, Oryx, PetroCanada, Petrovietnam, Phillips, Ranger, Repsol, Sasol, Samsung, Sun Oil, Talisman, TotalFinaElf, Tullow, Wintershall, YPF.

3

Angola

Angola is sub-Saharan Africa's second largest oil producer and production may reach 1 million barrels per day by 2003. Major offshore oil finds have made Angola a leading area for hydrocarbon exploration in sub-Saharan Africa.

General Background

Angola is beginning its recovery from a 27 year civil war that began shortly after the nation achieved independence from Portugal in 1975. Hundreds of thousands of people have been killed in the strife, Africa's longest running conflict in the post-colonial era. Angola's civil war has ravaged the non-mineral sectors of the country's economy, destroyed much of its infrastructure, and displaced an estimated 2.5–4 million people. One million people have been displaced since the civil war resumed in 1998 between the government and forces of the National Union for the Total Independence of Angola (UNITA). A cease-fire was signed in April 2002 following the death of UNITA leader Jonas Savimbi; however, the peace process is not yet complete. Foreign governments and non-governmental organizations have provided large amounts of aid, although some western governments cite Angola's oil wealth as a sign that the country should pay for its own recovery. Angola was elected as the African regional representative on the UN Security Council, an appointment that will last two years, from 2003 to 2004.

Angola's President Jose Eduardo dos Santos has announced he will not run in the next election. President dos Santos stated that the next elections could be held in 2002 or 2003, though 2003 now seems the more likely date. Angola's first multi-party elections were held in 1992, though the renewal of hostilities has led to the postponement of subsequently scheduled elections.

The Angolan economy is highly dependent on its oil sector, which accounts for about half of the country's Gross Domestic Product (GDP) and over 90 per cent of export revenues. Angola maintains positive GDP growth rates due to the strength of its oil sector, which has very few linkages, however,

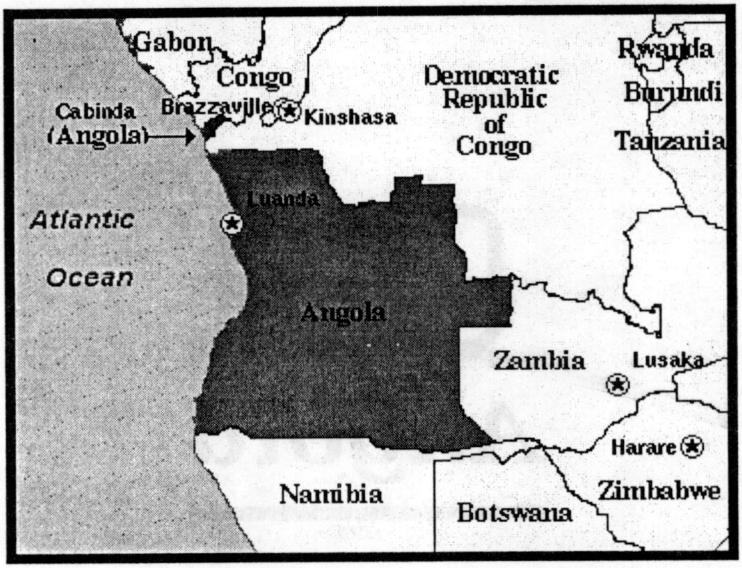

to other sectors of the economy. Angola's real GDP grew by 3.4 per cent in 2001 and has grown at an 11.4 per cent annual rate during the first half of 2002.

In spite of Angola's expanding economy, the majority of Angola's 10 million people live in poverty, and its external debt is estimated to have been $9.6 billion at the end of 2001. That same year, inflation was 152.5 per cent, a substantial improvement over the 325 per cent inflation in 2000. Inflation was expected to decrease to 103.6 per cent in 2002, although the final figure is not yet available. In April 2000, Angola signed a nine-month economic monitoring program agreement with the International Monetary Fund (IMF) that requires Angola to undertake a sustained program of economic reforms before it is considered for a formal loan agreement. Under the agreement with the IMF, Angola has promised to allow outside auditors to examine the way it spends oil money. In July 2001, the IMF stated that, although several economic targets were not met, the Angolan government has made progress in several areas including price liberalization, exchange rate stability and reductions in inflation. The Angolan government expects a visit from the IMF technical mission later this year, to discuss technical aspects of providing assistance to organizations responsible for financing macroeconomic and humanitarian projects.

Over the past year, allegations have arisen that suggest large portions of Angola's oil revenue have been embezzled or channeled to military purposes. According to the IMF, about $1.2 billion of Angola's oil revenue for the year 2001 is "unaccounted for." Though the government strongly denies these allegations, NGOs and others accuse the Angolan government of using the missing money to support its efforts to fight UNITA.

Oil

Angola is sub-Saharan Africa's second largest oil producer behind Nigeria, with the majority of its crude oil production located offshore Cabinda. Crude reserves also are located onshore around the city of Soyo, offshore in the Kwanza Basin north of Luanda, and offshore of the northern coast. Angola's crude oil generally is of high quality, with an API gravity ranging from 32° to 39.5° and sulfur content from 0.12 per cent to 0.14 per cent.

The Angolan province of Cabinda faces a situation similar to the Niger Delta states in Nigeria. Cabinda produces more than half of Angola's oil and accounts for nearly all of its foreign exchange earnings. Political tensions are high in some areas of Cabinda as separatist groups demand a greater share of oil revenue for the province's population. The separatist groups often kidnap foreign nationals in an attempt to draw attention to their independence claims. The province receives about 10 per cent of the taxes paid by ChevronTexaco and its partners operating offshore Cabinda. In September 2002, the Angolan government announced that it was prepared to open talks with Cabindan separatist groups and offer the province some measure of autonomy, but ruled out the prospect of complete independence.

Angola's national oil company, Sonangol, was established in 1976. A hydrocarbon law passed in 1978 made Sonangol sole concessionaire for exploration and production. Associations with foreign companies are in the form of joint ventures (JVs) and production sharing agreements (PSAs). The top foreign oil companies operating in Angola are ChevronTexaco, TotalFinaElf, and ExxonMobil.

Production

Angola's crude oil production has increased by nearly 600 per cent since 1980. Crude oil production averaged 742,000 barrels per day (bbl/d) in 2001. In the first eight months of 2002, crude oil production averaged 897,000 bbl/d, and production is expected to top one million bbl/d in 2003. Block Zero (Area A, Area B, and Area C), located offshore the enclave of Cabinda, accounts for the majority of Angola's crude oil production, although large new reserves have been discovered offshore mainland Angola. The ChevronTexaco subsidiary, Cabinda Gulf Oil Company (CABGOC), is the operator of the Block Zero fields, with a 39.2 per cent share in the JV. Other partners include Sonangol (41 per cent), TotalFinaElf (10 per cent) and ENI-Agip (Agip 9.8 per cent). Angola's largest producing oil fields are Takula (Area A), Numbi (Area A), and Kokongo (Area B). ChevronTexaco was fined $2 million for environmental damage in August 2002, stemming from a pipeline leak in June 2002. The company estimated that it was still losing around 12,000 bbl/d in crude oil, since it refused to use the pipeline until measures could be taken to prevent further leaks.

In August 2001, CABGOC and its partners announced the commencement of production from the Nemba Norte platform. The platform, whose installation was completed in March 2001, initially produced 20,000 bbl/d. Nemba Norte has yet to reach peak production of about 50,000 bbl/d, but will have produced an average of about 42,000 bbl/d for the year 2002. Nemba Norte crude is transported by pipeline to the Malongo Terminal for export. It is anticipated that production levels on Block Zero will be maintained or possibly raised to 600,000 bbl/d, by the implementation of further field development and enhanced oil recovery. CABGOC plans to invest nearly $4 billion in field development activities over the next few years.

Construction of the Xikomba deepwater development in Block 15 was begun by ExxonMobil's subsidiary Esso in August 2002. Discovered in 1999, the Xikomba field is located 230 miles northwest of Luanda, and is estimated to have recoverable reserves of around 100 million barrels of oil; at its peak it is expected to produce 80,000 bbl/d. Production will consist of 9 subsea wells that will be connected to a Floating, Production, Storage, and Offloading (FPSO) vessel, and the first oil is expected to be extracted near the end of 2003. Procurement, fabrication, and installation of flexible flowlines and risers that are to be connected to the FPSO, as well as the installation of various subsea equipment will be handled by Technip-Coflexip. Sonangol is concessionaire of the block, while Esso holds a 40 per cent interest, BP Exploration Angola Ltd has a 26.67 per cent share, Agip Angola Exploration B.V. holds a 20 per cent interest, and Den Norske Stats Oljeselskap has the remaining 13.33 per cent.

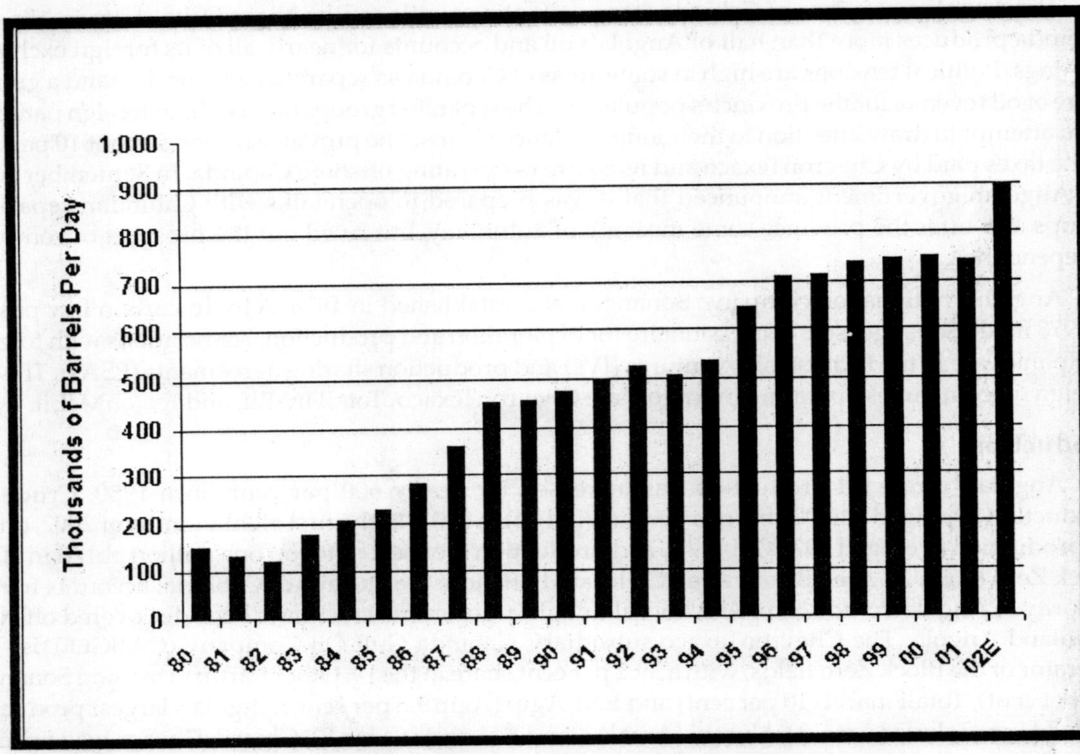

ANGOLA'S CRUDE OIL PRODUCTION, 1980-2002E

CABGOC is the operator of Angola's first producing deep-water block. In January 2000, CABGOC announced that production had begun on the Kuito field in Block 14. First oil from the Kuito field was achieved only 15 months after the award of the contract, making it the fastest cycle time of any project of its kind in sub-Saharan Africa. The Kuito field had initial production of 80,000 bbl/d, and a peak production level of 100,000 bbl/d is targeted. Other discoveries on the block, Belize, Benguela, Tomboco, and most recently Tombua, are being developed jointly, and the first four are expected to come onstream in 2005. Initial output should flow at about 140,000 bbl/d and increase to 200,000 bbl/d when all fields are onstream. The French engineering firm Technip was awarded the contract for the construction of a 42-wellhead drilling/production platform for the Benguela and Belize discoveries. The Tomboco discovery, which is further offshore, will probably be tied into this production facility. CABGOC has a 31 per cent interest on Block 14, and is joined by Sonangol (20 per cent), Agip (20 per cent), TotalFinaElf (20 per cent), and Petrogal (9 per cent).

Exports

Angola's crude oil exports to the United States were 313,000 bbl/d during the first eight months of 2002. Angola was the ninth largest supplier (second largest non-OPEC supplier outside the Western Hemisphere) of imported crude to the United States in 2001. Angolan crude is also exported to markets in Europe, Latin America and Asia.

Angolan exports to Asian countries have grown rapidly in recent years. Recent trade figures suggest that China's oil imports from Angola have grown by more than 400 per cent in 2001. China's demand for Angolan crude is primarily a result of its adoption of stricter environmental standards that will place a premium on lower-sulfur West African crudes. Angolan oil exports to other Asian destinations have also increased. Exports to South Korea went from $97 million in 1998 to $588 million in 1999.

Exploration

The success in offshore discoveries in Angola has sparked interest in Angola's exploration blocks. ChevronTexaco and TotalFinaElf may set up a consortium to conduct oil exploration activities along the border between Angola and the Republic of Congo–Brazzaville. When the announcement was made in October 2002, the Angolan oil minister, Jose Botelho de Vasconcelos said it was still too early to speculate about the size of existing reserves, and that the process was still being studied. However, the consortium is expected to be in place by the end of 2002. The two nations have agreed to explore the area jointly to avoid conflict over which nation has legal rights to the reserves.

ExxonMobil's subsidiary, Esso Exploration Angola Ltd and Sociedade Nacionale de Combustiveis de Angola made a deepwater oil discovery on Angola's Block 15 in September 2002. This is the 13th oil discovery on that block, which is located 240 miles northwest of Luanda. The well, named Reco Reco-I, tested at a rate of 2,640 bbl/d and was drilled in 4,718 feet of water to a total depth of 12,460 feet.

BP and Sonangol made their first oil discovery in the ultra-deep waters of Block 31 in September 2002. The well tested at a maximum rate of 5,357 bbl/d, although the full extent of the Plutao discovery is still being evaluated. This is the second well BP has drilled in Block 31, the first one being a dry hole. BP has a 26.7 per cent stake in the block, Esso holds a 25 per cent stake, 20 per cent is retained by Sonangol, 13.3 per cent is held by Statoil, 10 per cent by Marathon, and 5 per cent by TotalFinalElf.

Also in September 2002, Canadian Natural Resources Ltd. reached a 4 year production sharing deal with Sonangol EP, which will allow the Calgary-based firm to explore for oil in the deep waters of Block 16 off the coast of Angola. The 1.2 million acre block is located 72 miles offshore, and water depths in the area range from 300 meters to 1,500 meters. Canadian Natural will be the operator and will retain a 50 per cent working interest in the block. Odebrecht Oil and Gas Angola will have a 30 per cent working interest, while Sonangol EP will retain the remaining 20 per cent.

It should be noted that not all exploration efforts have been successful. Following the drilling of a number of dry holes in the ultra-deep waters of the Kwanza basin and elsewhere, some firms may review their oil exploration strategies in Angola. Having spent more than $1 billion in signature bonuses and wildcat wells covering four exploration blocks, the dry holes are a major disappointment for firms operating there, including ChevronTexaco (Gabela-1), TotalFinaElf (Mariposa-1), BHP (Iona-1 and Kangandala-1), Agip (Leao-1), and ExxonMobil (Eova-1). Though ExxonMobil's Semba-1 well did find oil, it was not commercially viable, and was subsequently abandoned.

Several significant discoveries have been made on deepwater offshore Block 17, located northwest of Luanda. The Girassol field was discovered in 1996 in 4,500 feet (1,365 meters) of water. Girassol is estimated to contain 725 million barrels of recoverable reserves. TotalFinaElf, operator with 40 per cent interest in the Block 17 PSA, and its partners ExxonMobil (20 per cent), BP (16.67 per cent), Statoil (13.33 per cent), and Norsk Hydro (10 per cent) decided to fast-track the development of the Girassol field. Production began in 2001, and by 2002 was producing 120,000 bbl/d, more than half of its expected peak production of 200,000 bbl/d. Additional discoveries have been made on Block 17, including Dalia (1997), Rosa (1998), Lirio (1998), Tulipa (1999), Orquidea (1999), Cravo (1999), Camelia (1999) and Jasmin 1 (2000). The Dalia field, with an estimated 1 billion barrels of recoverable reserves, will be developed using a FPSO. First oil from Dalia is expected by the end of 2004.

The Mavacola-1 discovery on Block 15 was announced in May 2001. Mavacola-1, which flowed at a test rate of 3,200 bbl/d, is the eleventh discovery made on the block since the Kissanje discovery was made in February 1998. Other recent discoveries on Block 15 include Vicango-1 (2001), Mbulumbumba-1 (2001), Batuque (2000), Saxi-1 (2000) and Mondo (2000). Block 15 contains an estimated recoverable hydrocarbon potential of over 3.5 billion barrels of oil equivalent. Plans for multiple developments on the block are being developed. The Kizomba-A project will develop the Hungo (1998) and Chocalho (1999) discoveries. An FPSO vessel will be utilized for the Kizomba-A project, with production scheduled to start before the end of 2004. Production is targeted to eventually reach 250,000 bbl/d. The Kizomba-B project, also projected to reach 250,000 bbl/d, will develop the Kissanje, Marimba (1998), Dikanza (1998) and Xicomba (1999) discoveries. First oil from Kizomba-B is not expected before the end of 2005. ExxonMobil (40 per cent), the operator, BP (26.67 per cent), Agip (20 per cent), and Statoil (13.33 per cent) are the participants on the Block 15 PSA. The Kizomba-C project will also encompass multiple fields, though it is still in the planning stages. So far, the $3.5 billion project is expected to develop the Batuque, Saxi, and Mondo fields, and may begin producing oil by 2006 or 2007.

Refining and Downstream

The Fina Petroleos de Angola refinery in Luanda has a crude oil processing capacity of 39,000 bbl/d. The refinery is a joint venture between Sonangol (36 per cent), TotalFinaElf (61 per cent) and private investors (3 per cent). TotalFinaElf, operator of the Luanda refinery, plans to raise capacity to 60,000 bbl/d by 2004. Two processing units will be enlarged and a third scrapped to boost efficiency. The refinery needs to adapt by 2005 to meet product specifications and deadlines set by SADC. These include phasing out lead and increasing the octane content in gasoline. The refinery produces almost all of Angola's domestic requirements of gasoline, kerosene and jet fuel, as well as a small amount of products for export.

Angola is developing plans for a new 200,000-bbl/d refinery, to be located in the central coastal city of Lobito. The majority of products refined at the new facility (80 per cent) would be exported regionally. At a conference in October 2002, Sonangol's managing board chairman stated that preliminary work is proceeding smoothly. Following a one year delay, it is anticipated that construction will commence before the end of 2003, with the refinery coming onstream by 2007. Equity in the refinery has not been finalized, but Sonangol will likely retain a 50 per cent share, with 20 per cent going to a foreign technical partner, 20 per cent to a foreign financial partner, and the remaining 10 per cent to entities in the South Africa Development Community. Sonangol is investing between $3 and $5 billion in the refinery and is using Dresdner Kleinwort Wasserstein as its financial adviser. Diesel and gasoline produced in the refinery will meet technical and environmental specifications required in targeted markets such as the United States, Western Europe and South Africa.

Three firms, Sonangol (through its subsidiary Sonangol Distribution), TotalFinaElf, and Sonangalp, a joint venture between Sonangol (51 per cent) and Petrogal (49 per cent), provide product distribution and marketing in Angola. Angola's small market size and lack of infrastructure have hindered Sonangol's plans to attract additional foreign companies to the country's downstream market.

Natural Gas

Angola has estimated natural gas reserves of 1.6 trillion cubic feet (Tcf). New discoveries could push Angola's proven gas reserves to 9.5 Tcf, and possibly as high as 25 Tcf. The majority (approximately 85 per cent) of natural gas produced in Angola is flared, the remainder is reinjected to aid in oil recovery and processed in the production of LPG. World Bank studies estimate that Angola accounts for 30 per cent of the natural gas flared in Africa. The government is developing strategies to reduce natural gas flaring and increase commercial usage of natural gas. CABGOC has initiated two zero-flare fields, Nemba and Lomba, and plans to make Kuito the third.

ChevronTexaco and Sonangol have agreed to develop an LNG (liquefied natural gas) project that would convert natural gas from offshore oil fields to LNG for export. The facility will process natural gas from offshore Blocks 1, 2, 3, 4, 15, 16, 17, and 18. The facility, which will be located in Luanda adjacent to the existing refinery, will initially consist of one LNG train with a capacity to process 4 million tons per year of LNG. The site is sufficiently sized for the plant to expand to additional LNG trains. Changes to the project plans and geographical difficulties have pushed the start-date for construction back to 2004, with the $2-billion LNG project expected to be in service by 2007. Sonangol and ChevronTexaco each initially held a 50 per cent interest in the project, but it was announced in July 2001, that BP, ExxonMobil and TotalFinaElf were joining the consortium. Equity was realigned in 2002 to reflect the partners' responsibilities, with ChevronTexaco retaining 32 per cent, Sonangol 20 per cent, and BP, ExxonMobil, Norsk Hydro, and TotalFinaElf taking 12 per cent apiece. The firms are the respective operators on Blocks 15, 17 and 18. ChevronTexaco will remain the LNG project's operator. The LNG project could spur the development of other gas-related projects including: gas-fired generation facilities in the Luanda area, industrial usage of gas as a fuel (plans call for the Luanda cement plant to convert to natural gas), and expanding the domestic market for LPG.

Electricity

Angola's electricity generating capacity as of January 1, 2000 stood at 586 megawatts (MW). Although capacity is split evenly between thermal (oil & gas-fired) and hydroelectric units, hydroelectric facilities generate more than two-thirds of Angola's electricity. It is estimated that only 15 per cent of Angola's population has access to electric power.

Angola has announced plans for the major rehabilitation of its power sector infrastructure. Significant portions of the country's generation and transmission facilities have been damaged during the civil war. Angola intends to recover the productive capacity of the Empresa Nacional de Electricidade (ENE), the state-owned electric utility, by rehabilitating its hydropower stations. Of the six stations, only three (Cambambe, Biopo, and Matala) are functioning. The other dams (Mabubas, Lumaun, and Gove) have been severely damaged. The government plans to create a national grid by linking the three regional electricity grids and establishing linkages with neighboring countries. This project, coupled with the power plant rehabilitation, could provide the basis for Angola becoming a regional exporter of electricity. Rehabilitation work on the Matala Dam (51 MW) was completed in September 2001, and work on the Cambambe facility (130 MW) is scheduled to be completed in 2002.

Angola's energy minister, Luis Filipe da Silva, announced in September 2002 that the government plans to build a new dam in Dala municipality that will supply the Moxico province with electricity. Construction is expected to begin in 2003, and until the dam is operational, Moxico's capital town, Luena, will receive a 1.5 MW generator to increase its power supply.

The Brazilian construction company Odebrecht, which is currently rehabilitating the Matala facility, completed construction of a new hydroelectric facility at Capanda on the Kwanza River. In

August 2002, the Office of Middle Kwanza Exploitation began filling up the new dam's reservoir in preparation for electricity production scheduled to begin in early 2003. Work on the 520-MW plant began in the mid-1980's, but was suspended due to the civil war. When fully operational, the Capanda project will nearly double Angola's electricity generating capacity.

Russia's largest diamond-producing firm, Alrosa, has plans to build a 16MW dam on the Tchicapa River in the Lunda Sul Province to provide energy for its prospecting operations. The new dam could be finished in as little as 30 months and will be constructed in partnership with ENE, Angola's state-run electricity company. At a cost of $45 million, the dam is expected to provide energy for the second phase of Alrosa's diamond prospecting project at Catoca. The dam will also supply energy to Lunda Sul. Angola's Lunda Sul and Lunda Norte provinces are said to be important diamond-producing regions that have enormous hydropower potential.

Country Overview

Independence: November 11, 1975 (from Portugal)

Population (July 2002E): 10.6 million

Location/Size: Southern Africa, bordering the Atlantic Ocean (on the west), the Democratic Republic of the Congo (on the north and east); Zambia (on the east); and Namibia (on the south)/1,246,700 square kilometers (481,350 square miles), slightly less than twice the size of Texas

Major Cities: Luanda (capital), Huambo, Lobito, Soyo

Languages: Portuguese (official), over 60 local languages including Umbundu, Kimbundu, Kikongo, Tchokwe and Ovambo

Ethnic Groups: Ovimbundu 37 per cent, Kimbundu 25 per cent, Bakongo 13 per cent, other 22 per cent

Religion: Traditional beliefs 47 per cent, Roman Catholic 38 per cent, Protestant 15 per cent

Defense (1999E): Army (106,000), Navy (2,000), Air Force (6,000), Paramilitary Forces (15,000)

Economic Overview

Currency: New Kwanza (NKz)

Exchange Rate (11/25/02): US$1 = 50.73 NKz

Gross Domestic Product (GDP) (2002E): $8.5 billion

Real GDP Growth Rate (2001E): 3.4 per cent *(2002E):* 11.2 per cent

Inflation Rate (2001E): 152.6 per cent *(2002E):* 103.6 per cent

Current Account Balance (2001E): -$2.0 billion

Major Trading Partners: United States, South Korea, Belgium, Portugal, China, France, Taiwan, South Africa

Major Export Products: Crude oil, diamonds, refined petroleum products

Major Import Products: Consumer goods, military and capital goods, machinery, vehicles and spare parts

Total External Debt (2001E): $10.5 billion

Energy Overview

Proven Oil Reserves (1/1/02E): 5.4 billion barrels

Oil Production (2001E): 742,000 barrels per day (bbl/d), all of which is crude oil

(January-August 2002E): 897,000 bbl/d, all of which is crude oil

Oil Consumption (2001E): 31,000 bbl/d *(2002E):* 31,000 bbl/d

Net Oil Exports (2001E): 711,000 bbl/d *(2002E):* 866,000 bbl/d

Refining Capacity (1/1/02E): 39,000 bbl/d

Natural Gas Reserves (1/1/02E): 1.6 trillion cubic feet

Natural Gas Production (2000E): 20.5 billion cubic feet (bcf)

Natural Gas Consumption (2000E): 19.8 bcf

Electric Generation Capacity (1/1/00E): 586 megawatts

Electricity Generation (2000E): 1.2 billion kilowatthours (59.7 per cent hydroelectric, 40.3 per cent thermal)

Environmental Overview

Total Energy Consumption (2000E): 0.09 quadrillion Btu* (0.03 per cent of world total energy consumption)

Energy-Related Carbon Emissions (2000E): 3.6 million metric tons of carbon (0.1 per cent of world carbon emissions)

Per Capita Energy Consumption (2000E): 6.9 million Btu (vs. U.S. value of 351.0 million Btu)

Per Capita Carbon Emissions (2000E): 0.3 metric tons of carbon (vs. U.S. value of 5.6 metric tons of carbon)

Energy Intensity (1999E): 9,900 Btu/$1990 (vs. U.S. value of 12,638 Btu/$1990)**

Carbon Intensity (1999E): 0.35 metric tons of carbon/thousand $1990 (vs. U.S. value of 0.19 metric tons of carbon/$1990)**

Fuel Share of Energy Consumption (2000E): Oil (67.2 per cent), Natural Gas (23.0 per cent), Coal (0.0 per cent)

Fuel Share of Carbon Emissions (2000E): Natural Gas (64.4 per cent), Oil (32.8 per cent), Coal (0.0 per cent)

Status in Climate Change Negotiations: Non-Annex I country under the United Nations Framework Convention on Climate Change (ratified May 17, 2000). Not a signatory to the Kyoto Protocol.

Major Environmental Issues: Overuse of pastures and subsequent soil erosion attributable to population pressures; desertification; deforestation of tropical rain forest (in response to both international demand for tropical timber and to domestic use as fuel), loss of biodiversity; soil erosion contributing to water pollution and siltation of rivers and dams; inadequate supplies of potable water.

Major International Environmental Agreements: A party to Conventions on Biodiversity, Desertification and Law of the Sea.

The total energy consumption statistic includes petroleum, dry natural gas, coal, net hydro, nuclear, geothermal, solar and wind electric power. The renewable energy consumption statistic is based on International Energy Agency (IEA) data and includes hydropower, solar, wind, tide, geothermal, solid biomass and animal products, biomass gas and liquids, industrial and municipal wastes.

Oil and Gas Industries

Organization: State-owned Sociedade Nacional de Combustiveis de Angola (*Sonangol*) oversees offshore and onshore oil operations in Angola.

Major Oil Fields: Takula (Block Zero), Numbi (Block Zero), Kokongo (Block Zero), Pacassa (Block 3), Cobo/Pambi (Block 3)

Major Refineries (1/1/02 Capacity): Fina Petroleos De Angola–Luanda (39,000 bbl/d); a new 200,000 bbl/d refinery at Lobito is scheduled to begin construction in 2003

Major Oil Terminals: Luanda, Malango (Cabinda), Palanca, Quinfuquena

Foreign Oil Company Involvement: Ajoco, BHP, BP, Canadian Natural Resources, ChevronTexaco, Daewoo, Engen, ENI-Agip, ExxonMobil, Falcon Oil, Fortum, Gulf Energy Resources, INA-Naftaplin, Lacula Oil, Marathon Oil, Mitsubishi, Naftgas, Naphta-Israel, Norsk Hydro/Saga, Occidental, Ocean Energy, Pedco, Petrobras, Petrofina, Petrogal, Petro-Inett, Petronas, Phillips, Prodev, Shell, Statoil, Teikoku, TotalFinaElf.

4

Antarctica

Introduction

The continent of Antarctica is home to a uniquely beautiful and harsh environment that has changed little in the last 30 million years. The continent, approximately twice the size of Australia, lies mainly within the Antarctic circle and is surrounded by ocean. It is covered almost entirely by a sheet of ice and snow which has an average thickness of approximately 6,500 feet, comprising slightly more than 90 per cent of the world's ice and 70 per cent of the world's freshwater resources. This thick sheet of ice–also known as the ice cap–reflects most of the heat generated by the sun back into the atmosphere, leaving the continent with an annual mean temperature of -57° Fahrenheit. Antarctica is the world's coldest and most pristine environment. The continent is also the driest and windiest landmass on Earth; with winds reaching speeds of up to 200 miles per hour and areas where rain has not fallen in millions of years. Antarctica also is the highest continent on earth, with an average elevation of 7,380 feet.

Oil Exploration

Antarctica's serenely primitive wilderness faces an uncertain future as debate continues over the question of tapping into the continent's wealth of mineral resources. Beginning in the early 1950s, scientists began to notice the value inherent in such an unusual and largely undiscovered continent and began establishing research stations there. By 1958, the International Geophysical Year (IGY) saw twelve countries establish over 60 bases, both on and around the Antarctic continent, with some countries claiming parts of Antarctica as sovereign territory. The countries initially involved in the scientific body known as the Scientific Committee on Antarctic Research (SCAR) included the United States, Chile, Argentina, Norway, France, Australia, New Zealand, the United Kingdom, South Africa, Belgium, Russia and Japan. SCAR now includes Germany and Poland as members of the organization which aims to coordinate a number of scientific and working group programs and to further facilitate

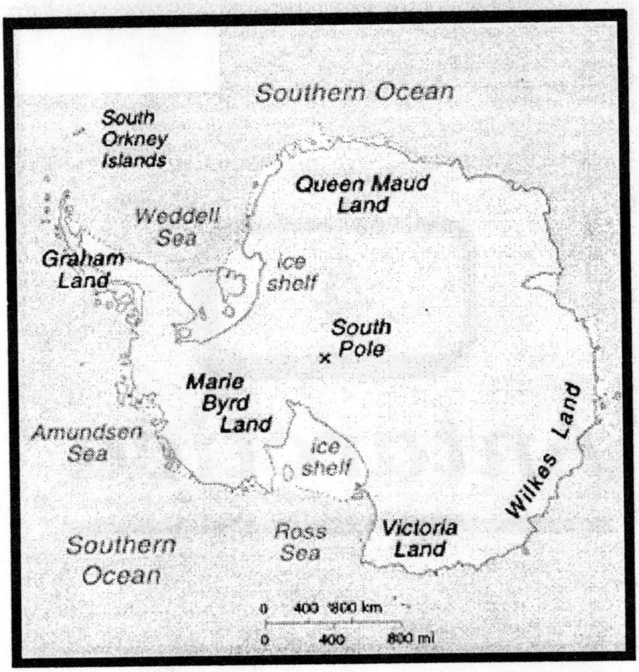

the sharing and planning of scientific information relating to Antarctic research, such as Climate Change.

The successful establishment of SCAR and the IGY in Antarctica was due in large part to cooperation between the countries involved, and led directly to the signing of the Antarctic Treaty in 1959, which has administered Antarctic affairs since 1961 when it officially entered into force. The Antarctic Treaty, signed during mounting Cold War tensions, successfully banned all military activity, nuclear testing, and the dumping of radioactive materials on the continent. The 1991 Protocol on Environmental Protection to the Antarctic Treaty, also known as the Madrid Protocol, entered into force in 1998 and serves as an additional mechanism for ensuring the protection of the Antarctic environment. The Madrid Protocol goes further than the original treaty as it designates Antarctica as a natural reserve devoted to peace and science and places a moratorium on mining and drilling for oil for a minimum of 50 years. The Protocol sets forth basic principles and detailed, mandatory rules which apply to all human activities in Antarctica.

The call for an environmental protocol to the Antarctic Treaty came after scientists discovered large deposits of natural resources such as coal, natural gas and offshore oil reserves in the early 1980s. Antarctica is considered to be part of the theoretical super-continent known as Gondwanaland, which separated near the end of the Paleozoic era and consisted of South America, Africa and Australia. And, because it once was completely covered in vegetation, many scientists believe it may hold one of the last supergiant oil fields yet to be discovered. The continental shelf of Antarctica is considered to hold the region's greatest potential for oil exploration projects, and although estimates vary as to the abundance of oil in Antarctica, the Weddell and Ross Sea areas alone are expected to possess 50 billion barrels of oil–an amount roughly equivalent to that of Alaska's estimated reserves. However, Antarctica's extreme conditions make oil field accessibility in many areas economically problematic.

Nevertheless, following the energy crisis of the 1970s, several oil companies looked to Antarctica as a possible solution to future world oil shortages by announcing plans to exploit the continent's resources. The necessary conditions for economically-sound oil production projects were beginning to ripen along with high oil prices and demand, and improved drilling technology. The prospect that Antarctica's fragile wildnerness could be tainted as a result of oil exploration and drilling activities resulted in the mobilization of several conservation groups who were intent on preserving the continent's status as the most pristine in the world.

On January 28, 1989 the *Bahia Paraiso*, an Argentine navy transport ship hauling supplies and tourists, ran aground approximately two miles off the coast of Antarctica in the vicinity of Palmer Station. Although no one aboard the ship was injured, the wreck proved to be a setback for the nearby coastal ecosystem, as a 30-foot gash in the ship's double-walled hull released some 250,000 gallons of diesel fuel and other petroleum products into the surrounding area. The effects from the fuel spill on the local flora and fauna were mostly limited to various sea bird, krill and moss populations, with few populations seeing mortality rates greater than 20 per cent. Because the *Bahia Paraiso* spill was reportedly the first known accident of its kind in the Antarctic region, the accident alarmed environmental groups, which viewed the incident as a foreshadowing of future accidents if trends in tourism and ship transport were to continue at their current pace along the continental fringe.

The devastating March 24, 1989 *Exxon Valdez* oil spill in Alaska's Prince William Sound two months later sent an even stronger alarm around the world to dozens of international environmental organizations for the need to protect Antarctica's unique environment from similar accidents. Although avoiding altogether the issue of sovereignty claims, the development of the 1991 Protocol on Environmental Protection to the Antarctic Treaty is viewed as a model for future environmental treaties which employ the precautionary principle towards natural resource exploration.

Climate Change

Because of its influence on world weather and climate patterns, Antarctica lies at the heart of the debate on climate change and has become the premier location in which to study the effects of global warming. Over the course of the past 50 years, Antarctica's average year-round temperature has warmed by about 3-4 degrees Fahrenheit–more than 10 times the average worldwide increase during that period. As a result of these increases in temperature (which are not necessarily the result of global warming), the Antarctic continent has experienced changes in its landscape and ecology. Scientists are also concerned that with such dramatic increases in the icy continent's temperature, significant rises in global sea levels could occur if Antarctica's thick polar ice sheet melts. The West Antarctic ice sheet alone, if melted, could raise average sea levels around the world by about 20 feet, resulting in the flooding of low-lying coastal zones.

Despite the numerous scientific research bases located on the continent, Antarctica produces negligible amounts of greenhouse gases and is not considered to be a significant contributer to the problem of global warming. In 1998, Antarctica produced 0.06 million metric tons of carbon, 100 per cent of which came from the burning of petroleum fuel products. Carbon dioxide is the chief culprit among the greenhouse gases, which are blamed for contributing to a process called global warming. The theory of global warming states that as carbon dioxide and other 'greenhouse gases' continue to accumulate in the earth's atmosphere, the earth's natural warming process known as the 'greenhouse effect' is intensified, thus resulting in higher global temperatures.

The effects of climate change on Antarctica vary from migrations of seal and penguin populations to other parts of the continent to abrupt changes in the glacial landscape. In March, 2000 the Larson B ice shelf in northern Antarctica–an area the size of Delaware–broke away from the continent and retreated into the sea. Similarly, in January, 1995, the Larson A ice shelf calved away from the continent, disappearing into the sea and bringing with it part of an Argentine base camp. Ice shelf instability is created as a result of higher sea and air temperatures, and a number of northern ice shelves have displayed similar trends such as the Wordie, Muller and Prince Gustav Channel shelves.

Perhaps most vulnerable to the effects of Antarctic climate change are the continent's wildlife populations. Increasing temperatures, less ice, and more snowfall have altered the patterns and habitats of several Antarctic species, such as the southern fur and elephant seals and Adelie and Chinstrap penguins.

Although the past 20 years have seen an increase in the numbers of Chinstrap penguins on the continent, the opposite is true for the Adelie penguin. As temperatures have increased, removing significant portions of sea ice, krill (a tiny shrimp-like crustacean) populations in certain areas have decreased significantly due to a reduction in the amount of their food source–algae–which grows on the bottom of sea ice. As krill is one of the mainstays of the Adelie penguin diet, the Adelie penguin population has decreased by almost 50 per cent in some areas. Rising Antarctic temperatures also have resulted in more snowfall on the continent; making breeding difficult for the ice-loving Adelie penguins. Seal populations, on the other hand, have benefited from the reduction in sea ice as they prefer breeding and feeding in open waters. Several seal populations in Antarctica have increased their numbers by over 300 per cent in the last 20 years.

Energy Use in Antarctica

Approximately 4,000 scientists and researchers live in Antarctica during the austral summer, with about half of these being American. During the winter months, however, this number is greatly reduced to a total of about 500 people.

Although the number of people living on the Antarctic continent remains quite small, scientists have been blamed for pollution problems ranging from sewage leaks to oil spills. In July, 1995, approximately 21,000 gallons of oil were spilled in Argentina's Marimbo base as the result of a poorly constructed fuel storage system, which consisted of large rubber fuel bladders placed directly on the snow. Similar incidents have occured on other international scientific bases, including Russia's Bellingshausen base (known to be one of the continent's more contaminated sites) and the United States' McMurdo Station. However, as more attention has been placed on the continent of Antarctica such spills are rare, and scientists have done much to clean up and improve waste management procedures on the continent.

Antarctica in the 21st Century

Despite the 1991 Madrid Protocol's 50-year moratorium placed on the exploitation of Antarctica's natural resources, future economic and population pressures could pose significant environmental threats to the continent of Antarctica. An expected and dramatic increase in world demand for energy over the next 30 years may expose the continent to countries and markets looking for alternative petroleum supplies. If Antarctica is indeed eventually opened for oil exploration activities, which many believe it will be, the potential for oil pollution occurrences such as oil tanker spills, the dumping of waste oil, natural oil seeps and well blowouts will rise substantially.

5

Arab Maghreb Union

The Arab Maghreb Union (AMU), which includes Mauritania, Morocco, and Tunisia (Note: due to their size and importance, Algeria and Libya are covered in separate reports): is an important oil and gas producer, exporter, and transit center to southern Europe. Natural gas production from the region is increasing rapidly, and new pipelines to Spain and Italy are being planned.

Table 5.1: Economic and Demographic Indicators for the Arab Maghreb Union

Country	*Gross Domestic Product (Nominal GDP), 2002E (Billions of U.S. $)*	*Real GDP Growth Rate, 2002 Estimate*	*Real GDP Growth Rate, 2003 Projection*	*Per Capita GDP, 2002E*	*Population 2002E (Millions)*
Algeria	$55.10	3.40%	4.30%	$1,762	32.3
Libya	$16.60	1.80%	2.90%	$3,001	5.4
Mauritania	$1.00	3.70%	4.50%	$343	2.8
Morocco	$36.60	3.80%	4.30%	$1,235	31.2
Tunisia	$21.10	2.00%	4.30%	$2,155	9.8
Regional Total/Average	**$130.40**	**3.10%**	**4.10%**	**$1,600**	**81.5**

MAURITANIA

Real GDP growth in Mauritania is expected to be a strong 5.0 per cent in 2003, following estimated growth of 5.0 per cent in 2002. Inflation is forecast at only 3.0 per cent, up slightly from 2.5 per cent in 2002, and the country appears to moving forward with a privatization and economic reform program. In March 2002, for instance, four companies (ONE of Morocco, Union Fenosa, Vivendi, and Anglo-American) submitted bids on the country's state power company.

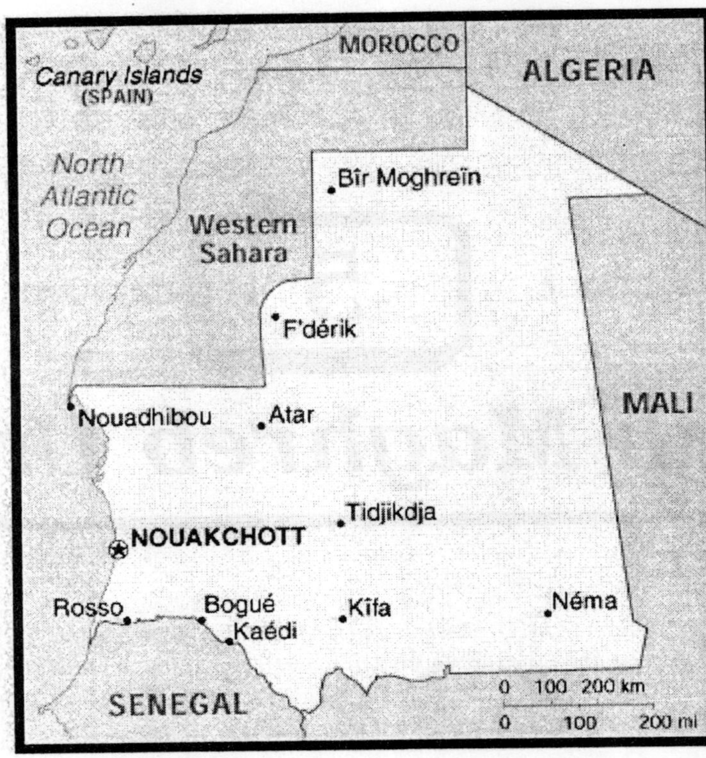

Despite these positive signs, Mauritania remains a poor country (more than half the population lives on less than $1 per day) with a rapidly growing population and many social and economic problems. Mauritania also remains heavily reliant on foreign aid and debt relief (in March 2000, the Paris Club eased Mauritania's debt service burden as a reward for the country's moves towards economic reform). In January 2002, Mauritania was hit by torrential rains which caused significant damage to crops, livestock, and grazing land in southern regions of the country.

President Maaouya Ould Sid'Ahmed Taya (who came to power in 1984) continues to push ahead with economic reforms efforts. As a consequence, Mauritania has attracted significant amounts of donor funds and also managed (in Juner 2002) to reach the completion point under the World Bank/ International Monetary Fund (IMF) heavily indebted poor countries (HIPC) initiative. This qualifies Mauritania for full debt relief, and should reduce its external debt-service payments by around 50 per cent. In general, Mauritania has achieved a solid economic performance record according to the IMF.

The World Trade Organization's (WTO's) evaluation of Mauritania points to several challenges for the country's economic progress. These include: supply-side constraints (*i.e.*, a limited number of exportable products); infrastructure problems; and limited human resources. The WTO points to the country's mining sector as offering great potential, while pointing out that the manufacturing and tourism sectors are underdeveloped. Currently, mining and fishing account for the vast majority of Mauritania's exports. Foreign aid also continues to be very important. In April 2002, the EU and

Mauritania signed a strategic cooperation agreement setting forth guidelines for disbursement of $144 million in European Development Funds over the next five years.

Energy

Mauritania has not traditionally been thought of as a promising hydrocarbons area, but this may now be changing as exploration has yielded some positive signs. In May 2001, for instance, a consortium of energy companies led by Australia's Woodside Petroleum made a significant oil discovery with its Chinguetti-1 well, located in deep waters off the country's southwest coast, and in October 2002, the Chinguetti 4-4 appraisal well confirmed the presence of oil-bearing geological structures. In addition, the Banda prospect in Chinguetti is believed to contain as much as 100 million barrels of oil plus several trillion cubic feet of natural gas. In general, it is believed that Mauritania may contain significant oil deposits off its coast. Foreign oil companies operating in Mauritania include: British Borneo, Dana Petroleum, Elixir, Fusion, Hardman, and Woodside Petroleum.

The overwhelming majority (99 per cent) of Mauritania's commercial energy consumption in 1999 was oil, all of which was imported. Mauritania also consumes a significant percentage of "non-commercial" (*i.e.*, wood, biomass) energy.

MOROCCO

King Hassan II, ruler of Morocco since 1961, died on July 23, 1999, and was succeeded by his son, Mohammed VI. The King is believed to be an advocate of economic reform and political liberalization, but, generally speaking, economic reform has proceeded slowly in recent years. One example is liberalization of the country's domestic oil market, promised in 1997 but not yet carried out. State-owned sugar and tobacco companies reportedly are slated for sale in 2003.

Morocco's real gross domestic product is expected to grow by around 4 per cent in 2003, up from 2.2 per cent growth in 2002 but down sharply from the 7.6 per cent growth achieved in 2001. In November 2002, Morocco was afflicted by heavy rains and severe flooding which caused major infrastructure damage near Casablanca. This flooding has now occurred two years in a row, following three years of drought, during which the country's large agricultural sector (accounting for around a fifth of GDP and more than two-fifths of the workforce) was adversely affected. Morocco's vulnerability to erratic rainfall patterns, and therefore agricultural output, has encouraged the government in its attempts at economic diversification, particularly towards manufacturing and services (including tourism). Prime Minister Driss Jettou's government has initiated a $100 million emergency program to repair damaged infrastructure and protect vulnerable areas from future floods.

Prior to the terrorist attacks of September 11, 2001, Morocco had a strong and growing tourism sector, although this appears to have been set back somewhat in the aftermath of that event. Tourism also is being hurt by the general perception of Middle East political instability. On the positive side, Morocco's hard currency reserves were up as of July 2002, while the trade balance also has improved as exports have risen.

In 2000, an oil and gas discovery (of unknown magnitude) in the Talsint region near the border with Algeria raised hopes that Morocco could add another important asset to its economy, help cut the country's energy import bill (now around $1-$1.5 billion per year) and also attract new investment to the country. Morocco also has an expanding manufacturing base.

Morocco maintains relatively tight fiscal and monetary policies, with the 2003 budget calling for only a small increase in spending, and this has helped sharply reduce the country's fiscal deficit. Morocco's inflation rate (consumer prices) is expected to average around 3.0 per cent in 2003, up

slightly from 2.5 per cent in 2002. On January 9, 2002, King Mohammed VI announced plans for a series of economic and administrative measures aimed at promoting foreign investment in Morocco.

Morocco's current five-year economic plan, which runs from 1999 through 2004, calls for promoting job creation (unemployment is a serious problem in Morocco, at 13 per cent or higher), exports and tourism, accelerating the country's privatization process (as of late 2001, the program–begun in 1993–had largely stalled, although a 35 per cent share in Maroc Telecom was sold off to France's Vivendi), upgrading of infrastructure, and reducing social inequalities (especially between urban rich and rural poor). The government also has considered imposing a value-added tax (VAT), as well as direct taxes on business and individual income.

Morocco has pursued an economic reform program supported by significant lending from the World Bank and International Monetary Fund (IMF) since the early 1980s. This reform program has, at the IMF's urging, liberalized the foreign exchange regime, lowered tariffs and other trade barriers, reformed the banking system, restrained government spending, reduced the foreign debt burden (in part through "debt-for-equity" swaps, in part through refinancing), and encouraged foreign investment (now permitted in most sectors of the economy). Morocco also has signed several agreements with the European Union on economic cooperation, including one establishing a free trade zone for industrial goods over a 12-year transition period. The European Free Trade Association (EFTA), as it is known, came into effect on December 4, 1999, following ratification by both sides.

Morocco traditionally has had almost no energy reserves, but does contain the world's largest phosphate reserves, and produces significant amounts of fertilizers and phosphoric acid. Morocco imports around 90 per cent of its energy needs, including significant amounts of oil (about $900 million in 1999) and coal. The government is encouraging increased coal and hydroelectricity production. Coal imports and phosphate exports travel largely through the deepwater port of Jorf Lasfar, which is also the site of a $1-billion phosphoric acid plant and a coal-fired power plant.

As of early 2003, the decades-old dispute between Morocco and the Polisario Liberation Front over the Western Sahara continues. A referendum on the future of the territory, a former Spanish colony which has rich phosphate deposits, was scheduled for January 1992 under U.N. auspices, but has yet to be held. In December 1999, the U.N. Security Council voted to keep the U.N. mission in Western Sahara through February 2000. On February 18, 2000, U.N. Secretary General Kofi Annan stated that the core problem of determining who is eligible to vote on the question of independence for the Western Sahara region "could.prevent the holding of the referendum" indefinitely. In January 2001, the Polisario said that it remained in a state of war with Morocco over the Western Sahara. In November 2001, the king made the first visit to the Western Sahara by a Moroccan monarch in 16 years.

Oil and Natural Gas

Morocco contains insignificant proven oil reserves of 1.6 million barrels, although most sedimentary basins (especially offshore on the Atlantic continental shelf or in deep waters off the shelf) in the country have not been explored, and Morocco is actively pursuing expansion of its upstream oil and natural gas sector. At present, the country relies on imports for nearly all of its oil consumption needs (around 150,000 bbl/d), except for about 200 barrels per day (bbl/d) from its Sidi Rhalem field, located in the Essaouira Basin. This could change in coming years with a possibly significant discovery in August 2000 by Lone Star, a local subsidiary of US-based Skidmore, of oil at Talsint, near the border with Algeria. Initial reports indicated that Talsint contained as much as 20 billion barrels equivalent of oil and natural gas. This was soon lowered to 1-2 billion barrels, then to 100 million barrels. Many oil analysts are skeptical that Talsint contains even this much recoverable oil and gas, and so far, Lone Star's well (Sidi Belkacem #1) at Talsint only has proven reserves of 10 million barrels or less of total hydrocarbons (the exact mix of oil and natural gas is unknown).

Besides Talsint, other areas of Morocco which are being explored include: 1) the Loukos South Offshore block in the Atlantic Ocean north of Rabat, in which Lone Star holds acreage; 2) the Al Hoceima-Nador Offshore area in the Mediterranean, in which Conoco holds a reconnaissance license; 3) the Cap Draa Haute Mer Offshore area south of Agadir, in which Kerr-McGee, Enterprise Oil, and Energy Africa hold shares; 4) offshore blocks (Safi Haute Mer and Ras Tafeilney) south of Rabat, in which U.S. Vanco Energy Co. and the U.K.'s Lasmo have oil exploration agreements; 5) a block at Labrouj in the center of the country, for which Lone Star has a contract; 6) an area around Ounoura, on the Atlantic coast near Essaouira, and for which Lone Star plans to invest $50 million; 7) five deepwater blocks (Rimella A-E) on the Atlantic coast south of Agadir, for which Shell Maroc has signed an 8-year exploration agreement; 8) the Tiznit offshore area, in which Sweden's Taurus Oil and Energy Africa have exploration permits; and 9) the western Prerif, for which Anschutz and Enterprise Oil have exploration licenses. In late 2000/early 2001, Morocco's state oil company ONAREP (Office National de Recherches et d'Exploitation Petrolieres) opened bidding on Morocco's first offshore licensing round. The bidding--for exploration permits on eight blocks in the Atlantic Ocean offshore the Rabat-Safi area--has been extended several times to allow more oil companies to study the seismic data.

In October 2001, Morocco signed two controversial oil exploration deals, both offshore the disputed Western Sahara region. The two contracts were with: 1) TotalFinaElf for the Dakhla Offshore zone; and 2) Kerr-McGee for a 44,000-square-mile area off the Western Sahara coast. These deals would mark the first such authorizations by Morocco in the disputed territory, which is believed to be potentially rich in oil reserves. The Polisario Front, which has fought Morocco for years over the Western Sahara region, protested the contracts to the United Nations, which stated that the contracts were not illegal in and of themselves, as long as the resources were developed for the benefit of the people living in the region. The Polisario also called on the European Union to cancel what it calls the "illegal and illicit" contract between Morocco and TotalFinaElf.

Currently, Morocco produces small volumes of natural gas from the Gharb Basin in the north, including the Meskala field just north of Essaouira. In January 2000, an energy ministry spokesman said that Morocco would soon issue new tax exemptions and other incentives to help attract foreign investors to explore for oil in Morocco. In March 2000, Morocco modified its hydrocarbons law in order to, among other things, offer a 10-year tax break to offshore oil production firms and to reduce the government's stake in future oil concessions (to a maximum of 25 per cent).

In February and May 2002, Morocco protested that an offshore exploration area granted by Spain to Repsol actually was in Moroccan territorial waters. The area, over which Spain's cabinet approved nine, six-year licenses, is located close to Spain's Canary Islands, and about 56 miles northwest of Morocco's coastal Tarfaya region.

In October 1998, International Petroleum Investment Company (IPIC) of Abu Dhabi and Spanish Energy Company Cepsa formed a 50-50 joint venture called Cepsa Maghreb to market and distribute petroleum products and LPG in Morocco. Products are to be supplied from Cepsa's refineries in Spain, and French company Vitogaz is building an LPG import terminal and tank farm near Casablanca.

Morocco has two refineries (Samir Oil Refinery at Mohammedia and Sidi Kacem) which have combined capacity of 154,901 bbl/d. In late November, severe flooding led to a massive fire at the Samir Refinery, which produces around 80 per cent -90 per cent of the country's refined petroleum products. In early December 2002, Samir officials said that full repairs could cost $150 million and take 9-13 months, although restoration to 60 per cent capacity reportedly occurred very quickly.

Refining company Samir reportedly is planning to invest around $600 million over the next 5 years to modernize its refineries and to expand capacity. Samir has a 125,230-bbl/d refinery at Mohammedia (near Casablanca), which it plans to expand to 160,000 bbl/d (at a cost of $20 million). On September 1, 2000, Morocco decided to increase the price of petroleum products in the country by 4.8 per cent -10.8 per cent, largely in response to increased world oil prices, as well as a weaker currency against the dollar.

Morocco contains only small natural gas reserves–around 100 billion cubic feet (bcf), with about 85 per cent located in the Essaouira Basin, and the rest in the Rharb and Pre-Rif basins. This does not count any possible additional reserves at Talsint. Besides its own domestic gas reserves, Morocco is a major transit center for Algerian gas exports across the Strait of Gibraltar to Spain, via the 300-bcf/year, Maghreb-Europe Gas (MEG) pipeline (also known as the Pedro Duran Farell pipeline). In November 1999, Spanish company Natural Gaz SDG announced its intention to invest up to $400 million in a natural gas distribution network for Morocco.

Coal and Electricity

Morocco produces coal from its one mine at Jerada, and imports coal from the United States, Colombia, and South Africa. Coal, most of which is imported from South Africa, is consumed largely by power plants at Mohammedia and Jorf Lasfar.

With electricity demand increasing at around 7 per cent annually, Morocco is planning several new power plants. Currently, Morocco has installed power generating capacity of around 4,000 MW, slated to expand to 5,700 MW by 2004. Morocco recently completed a major expansion of its $1.5-billion, coal-fired Jorf Lasfar plant (Africa's largest independent power plant) with the addition of two, 348-megawatt (MW), units by U.S.-based CMS Energy Corp. In early February 2001, CMS announced that Unit 4 had been completed, raising Jorf Lasfar's total power generating capacity to 1,356 MW, enough to supply more than half of Morocco's total power requirements. Units 1, 2, and 3 of Jorf Lasfar were completed in 1994, 1995, and June 2000, respectively. The plant is located on the Atlantic coast about 80 miles southwest of Casablanca. Currently, Morocco has 25 hydroelectric power plants, 5 steam thermal facilities, nine turbine stations, and numerous diesel plants.

Morocco's National Office of Electricity (ONE) has an ambitious plan to supply the countryside with electricity by 2010 (currently, only around 15 per cent of the rural population has access to electricity). Morocco plans to spend around $3.7 billion on energy projects through 2003, and part of this money will go towards building power plants and towards rural electrification. As part of this plan, a new, $500-million, 350-400-MW combined-cycle power plant, to be fired by natural gas from the MEG pipeline, is planned. The plant, Morocco's third independent power project (IPP), is to be built at Tahaddart, near Tangier in northern Morocco. In November 2001, Spain's Endesa signed an agreement to build and operate the 334-MW plant. Other future power plants are planned for Aon Beni Mathar (180 MW), Dhar Sadane (75 MW), Sendouk (65 MW), and Tarfaya (60 MW). In October 2000, the United States and Morocco signed an agreement on cooperation in energy efficiency and renewable energy.

In September 2000, Spain and Morocco, which put in place a 350-MW interconnection in 1998, agreed to study the potential for further linkage of the two countries' electricity grids. In December 2001, Endesa became the first Spanish company to import electricity from Morocco, while a deal was signed with Spain's Union Fenosa as well. Morocco reportedly supplies as much as 25 per cent of Spain's power consumption needs.

ONE also is looking at to increase renewable energy's share of Morocco's total power production considerably in coming years. In June 2002, Morocco and the United States signed a convention on development of "clean" energy sources. ONE is looking at hydroelectric pumped storage (*i.e.*, the 230-MW Afourir project in the Azilal region south of Rabat), solar power (widely used in many remote, rural areas of the country), as well as at building wind power parks throughout the country (including a 140-MW windfarm at Tangier, a 50-MW farm near Tetouan, and a 60-MW facility at Tarfaya). In May 2002, a consortium led by Total Energie won a contract to supply 16,000 rural homes with solar power. Morocco's Energy and Mines Minister, Mustapha Mansouri, said that this project will "reduce the pressure on woodlands, whose over-exploitation currently threatens the environmental balance." The project also should help Morocco achieve its goal of electrifying 80 per cent of the country's villages (up from around 50 per cent now) by 2008. Morocco also has expressed interest in nuclear power for desalination and other purposes, and in September 2001 signed an agreement with the United States establishing the legal basis for construction of a 2-MW research reactor just east of Rabat.

TUNISIA

Tunisia's real gross domestic product (GDP) is expected to grow by a solid 4.3 per cent in 2003, up from 2 per cent growth in 2002. Inflation is expected to remain low in 2003, most likely below 2 per cent. Privatization is moving ahead slowly, despite President Ben Ali's call in 2001 for an acceleration in the process. The country's tourism sector has suffered since the September 2001 terrorist attacks in the United States, as well as the April 2002 bombing of an ancient synagogue at El Ghriba on the island of Djerba. Meanwhile, agriculture has been adversely affected by drought, and Tunisian exports have been hit by slow economic growth in Europe. Tunisia's unemployment rate remains at high levels (officially 15 per cent, but unofficially much higher).

The Tunisian government, headed by President Zine al-Abedine Ben Ali (now in the midst of his third, five-year term, which runs through October 2004) and his Constitutional Democratic Rally (RCD) party, appears determined to continue liberalizing and reforming the country's economy, although at a pace which does not provoke popular unrest. In April 2002, Tunisia's parliament passed a new constitution that will allow Ben Ali to stand for a fourth, 5-year term in 2004.

The IMF is pushing Tunisia to accelerate its structural reform efforts (*i.e.*, reducing the fiscal deficit by cutting spending), increase privatization, liberalize trade, and diversify its economy. The government, however, is concerned about moving too quickly and stirring up social pressures. For 2002-2006, the government is looking to increase domestic and foreign investment, with a target of $34 billion for the period. The IMF's June 2002 report on Tunisia generally praised the country's "well-coordinated macroeconomic, structural and social policies while urging further reforms and aiming for economic growth of 6 per cent annually over the next five years.

A 1995 European Union (EU)-Tunisia Association Agreement went into effect in 1998, under which Tunisia is to dismantle its customs barriers to European goods over a 12-year period. On January 1, 2000, tariff barriers were lifted for European manufactured goods on "List 4" (consumer goods which have an equivalent in Tunisia), exposing Tunisian consumer goods to EU competition. Around 75 per cent -80 per cent of Tunisia's exports go to the EU. To replace lost customs revenues, Tunisia is attempting to improve its collection of taxes domestically. In May 2001, Egypt, Jordan, Morocco and Tunisia agreed to set up a free trade zone ahead of the 2010 target for trade barriers to end in the Euro-Mediterranean area.

Oil

During the first 10 months of 2002, Tunisia produced around 78,000 barrels per day (bbl/d) of oil, nearly all of which was crude oil (3,000 bbl/d was natural gas liquids). This represents a 32 per cent decline from the country's peak oil output, of 114,000 bbl/d, in 1992. Meanwhile, domestic petroleum demand is increasing rapidly, and Tunisia became a net oil importer in 2000 for the first time in over 20 years. Because Tunisia's refining capacity is low, the country exports crude oil and imports refined products. Currently, Tunisia has over 300 million barrels in proven oil reserves, and few new discoveries have been made in recent years.

Tunisia's state-owned oil company is Enterprises Tunisienne d'Activites Petrolieres (ETAP). ETAP is seeking to encourage exploration, especially in the northern part of the country, the northern coast, and the Gulf of Hammamet. It believes the development of new, smaller fields in cooperation with independent companies is the way to accomplish this. Tunisia reformed its hydrocarbons laws in August 2000, in hopes of attracting just such upstream investment. One of the most important provisions is a reduction in the tax rate from 75 per cent to 50 per cent for foreign firms if ETAP takes a 40 per cent share of the concession. Royalties are fixed at 10 per cent for oil and 8 per cent for gas.

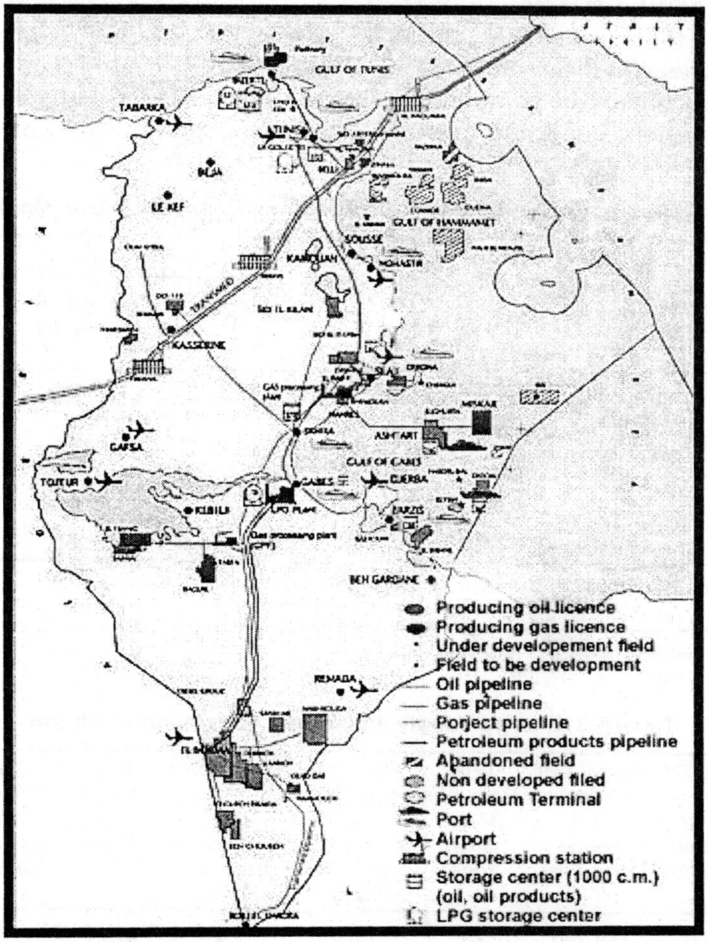

In addition to obtaining self-sufficiency in production for Tunisia, ETAP is pursuing overseas exploration and production. The company is working in Syria with Preussag of Germany to develop small oilfields and has signed an oil cooperation agreement with Iraq. ETAP has a joint venture with Sonatrach of Algeria to explore a border area and also with Libya's National Oil Company exploring an offshore block.

Tunisia's largest oilfield is El Borma, discovered in 1964 near the Algerian border, and operated by Agip since then. Ashtart (production around 15,000 bbl/d) is the only other oilfield with proven and estimated reserves over 100 million barrels. About 75 per cent of Tunisian oil production comes from these two fields and the Sidi El Kilani field. However, production at all three fields has been declining steadily. The Al Manzah field started production at 4,000 bbl/d in October 2000, with Centurion of Canada the operator.

As of December 2002, ETAP reportedly was awaiting tenders on 20 blocks, including onshore blocks C1-C4 (central Tunisia) and S1-S5 (south Tunisia), plus offshore blocks N1-N4 (northern Tunisia) and E1-E7 (eastern Tunisia). Fields currently under development include Isis, al Manzah, Belli, and Ashtart. In addition, there is ongoing work at the following blocks and permit areas: Grombalia, Kerkouane, Jenein North, Chorbane, Borj el Khadra, Baraka South, Kebili, El Hamra, Bazma, Jorf, Fejaj, Anaguid, Mellita, Robbana, Alyane, Tatouine, and North Kairouan. Companies involved include Agip, Anadarko, Anschutz, Centurion, Coparex Netherlands, Eni, Eurogas, Kufpec, Lasmo, Nuevo, Petro-Canada, and Preussag.

Table 5.2: Energy Consumption and Carbon Dioxide Emissions in the Arab Maghreb Union, 2000

Country	*Total Energy Consumption (Quadrillion btu)*	*Petro-leum*	*Natural Gas*	*Coal*	*Nuclear*	*Hydro-Electric*	*Other Rene-wable Electric*	*Net Electricity Imports*	*Carbon Dioxide Emissions (Million metric tons of carbon)*
Algeria	1.23	31.9%	66.5	1.7	0.0	0.0	0.0	0.0	22.7
Libya	0.58	66.1%	33.1	0.0	0.0	0.0	0.0	0.0	10.9
Mauritania	0.05	100%	0.0	0.0	0.0	0.5	0.0	0.0	0.9
Morocco	0.42	72.4%	0.4	21.3	0.0	3.1	0.0	0.0	7.8
Tunisia	0.29	57.0%	42.9	0.0	0.0	0.4	0.0	0.0	5.2
Regional Total/Average	**2.57**	**50.3%**	**44.2**	**4.3**	**0.0**	**0.6**	**0.0**	**0.0**	**4.5**

Note: Percentages may not add up to 100 per cent due to rounding.

Table 5.3: Energy Supply Indicators, Arab Maghreb Union

Country	*Crude Oil Reserves, 1/1/03 (Million Barrels)*	*Natural Gas Reserves, 1/1/03 (Billion Cubic Feet)*	*Coal Reserves (Million Short Tons)*	*Petroleum Production, 2002E (Thousand Barrels Per Day)*	*Natural Gas Production, 2000 (Million Short Tons)*	*Coal Production, 2000 (Million Short Tons)*	*Electric Generating Capacity, 2000 (Gigawatts)*	*Crude Oil Refining Capacity, 1/1/03 (Thousand Barrels Per Day)*
Algeria	9,200	159,700	44	1,460	2.9	0.02	6.0	450
Libya	29,500	46,400	0	1,377	0.2	0.0	4.6	343
Mauritania	0	0	0	0	0	0.5	0.1	0
Morocco	1.6	43	6	1.5	0	3.1	4.1	155
Tunisia	308	2,750	42.9	78	0.1	0.4	2.0	34
Regional Total/Average	**39,010**	**208,893**	**50**	**2,916.5**	**3.2**	**0.6**	**16.8**	**982**

Tunisia has five oil terminals on the Mediterranean coast. The largest of these is La Skhirra, on the Gulf of Gabes. A 22,000-bbl/d, 78-mile pipeline between the Sidi El Kilani oilfield and La Skhirra was inaugurated in March, 2001. In addition, a line from Algeria carries oil from that country's southern, In Amenas region, for export. Other Tunisian oil terminals include Ashtart, Gabes, Sfax, and Bizerte. In addition, anew terminal at Bejaia (140 miles east of Algiers), with the capacity to accommodate crude tankers as large as 300,000 tons, is scheduled to be constructed by late 2003 or early 2004.

Natural Gas

Tunisia has 2.8 trillion cubic feet (Tcf) of proven natural gas reserves, with around two-thirds located offshore. In 2000, Tunisia produced 66 billion cubic feet (Bcf) of natural gas. The majority of the country's gas output comes from the 1-Tcf Miskar non-associated gas field, discovered in 1975 by Elf and now operated by BG, in the offshore Amilcar permit. Tunisia has four other small producing natural gas fields (El Franning, El Borma, Baguel, and Zinnia) that together produced almost all of the remainder of domestic production. In general, the Tunisian natural gas sector is considered more promising than the country's oil sector.

BG, the largest investor in Tunisia's energy sector and the operator of the Miskar field, has a tentative agreement to supply a large proportion of Tunisia's domestic gas requirements through 2020. Miskar, located about 80 miles offshore in the Gulf of Gabes, currently produces 205 million cubic feet per day (Mmcf/d) of gas, which supplies around two-thirds of Tunisia's total gas demand. Miskar production is being expanded in several phases over five years at a cost of $120 million, with production expected to peak at 230 Mmcf/d before a decline sets in around 2009.

Besides Miskar, BG is planning on developing the Hasdrubal natural gas and gas condensate field, also located in the Gulf of Gabes, at a cost of $330 million over a 12-year period. In August 2002, BG announced that it had found oil at the Hasdrubal South West-1 field. The Tunisian government has decided to delay production from the Hasdrubal field until 2007, in advance of the decline of the Miskar field. BG is conducting exploration and is considering development of the Jugutha gas field. Finally, BG is looking at developing compressed natural gas as a vehicle fuel, especially for public transport, in Tunisia.

The 850-Bcf-per-year-capacity Trans-Mediterranean (TransMed) pipeline transports Algerian natural gas to Sicily, crossing the Mediterranean seabed from Cap Bon. Tunisia receives royalties (5.25 per cent -6.75 per cent of the gas' value, in cash or in kind) from this pipeline as payment for access through its territory. A natural gas pipeline also is set to link Libya and Tunisia, with an agreement signed in 2001.

Electricity

Tunisian power demand is growing rapidly, at an estimated 7 per cent annually. Recent census figures indicate that around 94.6 per cent of Tunisian homes had access to electricity in 1999, up sharply from 85.9 per cent in 1994. Around 95 per cent of Tunisian power generating capacity is natural-gas fired, with 4 per cent oil-fired and 1 per cent hydroelectric. Tunisian generating capacity is expected to reach 3,540 MW by 2006, up from 2,480 MW in early 2002.

The $261-million, 471-megawatt (MW), Rades combined cycle (natural gas and diesel-fired) power project is scheduled to come online shortly. A consortium comprised of U.S.-based PSEG (60 per cent), and Japan's Marubeni (40 per cent) is developing the project on a 20-year build-own-operate-transfer (BOOT) basis. Rades–Tunisia's first independent power plant–is to supply 20 per cent -25 per cent of the country's power needs. Tunisia's state-owned gas and power company, STEG (Societe Tunisienne de l'Electricite et du Gaz) has signed a 20-year power-purchase agreement with the consortium building Rades. Under this deal, STEG will by 100 per cent of Rades power output while PSEG manages the facility.

Besides Rades, BG is planning to build a $200-million, 500-MW power plant near Sfax that will utilize natural gas from the Miskar, and eventually, Hasdrubal fields. BG will likely partner with STEG or ETAP, and the facility should be operational in 2006. Also, a 27-MW, gas-fired power plant is being built at Zarsis, next to the El Biban gas field. Previously, El Biban natural gas was flared, but once complete, the power plant will take around 6 million cubic feet per day of El Biban gas as feedstock.

In July 2002, the European Investment Bank (EIB) approved a 150 million Euro loan aimed at upgrading Tunisia's power transmission network. Main elements include high-voltage power lines, underground cables, power transformers, and telecommunications for network control and transmission. Tunisia's power grid is in need of upgrading both to meet domestic demand as well as to increase reliability as part of Tunisia's ongoing integration into Europe's power grid. Currently, Tunisia is connected to Algeria's power grid, and is in the process of being linked to Libya as well.

6

Argentina

Argentina is a significant Latin American energy producer and consumer. It is a net energy exporter, primarily to neighboring Brazil and Chile.

General Background

After almost four years of recession, the Argentine economy ground to a halt in December 2001 as Buenos Aires defaulted on its approximately $140 billion debt. Civil unrest swept the nation as citizens were locked out of the country's banks and unemployment reached a record high of 18 per cent. On December 20, 2001, President Fernando de la Rua resigned and an interim administration took the helm. Over the next ten days, two interim Presidents were sworn in and shortly thereafter resigned in the face of gathering economic and political problems. On December 31, 2001, Senator Eduardo Duhalde was appointed by the Argentine Congress to serve as President, making Duhalde the country's fourth President in a month.

The most pronounced of President Duhalde's crisis management measures was to abandon the country's 1991 Convertibility Law, which had pegged the Argentine Peso 1:1 to the U.S. dollar for eleven years. The peso was initially devalued by approximately 30 per cent, but later depreciated more sharply after being allowed to float freely, losing approximately 70 per cent of its value. Seeing the potential boon to exporters as a result of devaluation, the Duhalde administration also implemented a 20 per cent tax on many exports, including oil and many oil derivatives, In addition, the Duhalde administration restricted bank withdrawals across the country thereby staving off a potentially massive drain of deposits as well as mitigating the country's growing inflation.

Argentina's economy continues to suffer in 2002. The Argentine Peso has been trading around 3.6:1 with the U.S. dollar since March 2002, causing significant losses in the private sector, as well adding some distorition to dollar-denominated macro-economic indicators. (See Devaluation and Utilities and Economic Overview) The Argentine real gross domestic product (GDP) shrank by

approximately 16 per cent during the first half of 2002, the largest economic contraction in the country's history. Concomitant with the shrinking economy, inflation for the year is soaring and is forecast to reach an unprecedented 30.7 per cent for 2002. While some observers speculate that the worst of the financial collapse is behind the country, others project that the export tax will restrict those industries that could propel economic growth, namely agriculture and energy, and economic malaise will continue into the medium-term.

Argentina's next elections are scheduled for March 2003. The ballot will include candidates for the country's Presidency, as well as half of the lower house and one third of the Senate. The Duhalde administration has been in negotiations with the International Monetary Fund (I.M.F.) for months, and some observers expect an agreement on short-term assistance to be reached in November 2002.

Oil

Argentina, with around 2.9 billion barrels of proven oil reserves, is a significant player in Latin American oil markets. It was the fourth-largest oil producer in Latin America in 2001, behind Mexico, Venezuela, and Brazil. The country was the region's third-largest oil consumer, after Brazil and Mexico, and also the third-largest exporter, behind Venezuela and Mexico. Exports go primarily to Chile, Brazil, Uruguay and Paraguay, with small amounts also bound for the U.S. Gulf coast.

Sector Organization

Argentina's oil sector is completely privatized. Repsol-YPF, Argentina's largest oil company, was formed in 1999 when Spanish oil company Repsol bought Argentina's formerly state-held YPF. Although exploration and production activity in Argentina are completely open to the private sector, Repsol-YPF retains a position of dominance. Pressure to reduce its presence has resulted in several asset swaps with other major Latin American oil companies, including Brazil's Petrobras, Chile's ENAP, and Argentina's (historically privately owned) Pérez Companc. In 2001, the three companies accounted for approximately 80 per cent of Argentina's oil production. In October 2002, Perez Companc announced the sale of a controlling interest of its shares to Petrobras.

Exploration and Production

Argentina's production increased rapidly throughout the 1990s, allowing exports to grow from negligible levels in the 1980s to around 900,000 barrels per day (bbl/d) in the late 1990s. Although the country is widely considered to be too mature an oil producer to offer significant exploration potential, there are opportunities for further development of existing reserves. Most of Argentina's oil is produced in two onshore basins: Neuquén (with over 40 per cent of proven oil reserves) in western Argentina,

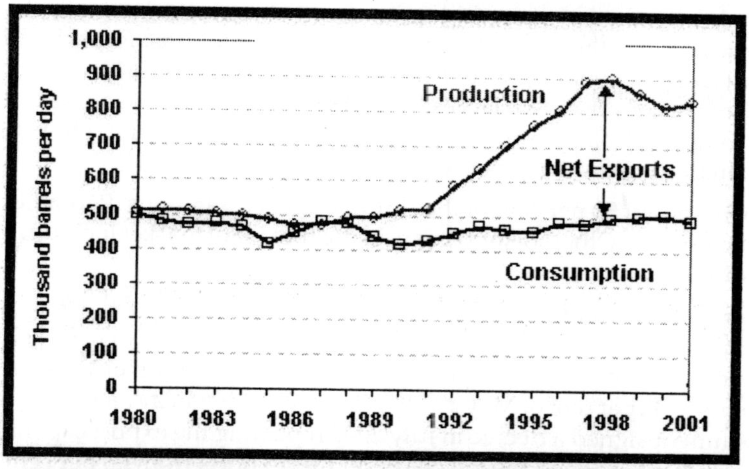

ARGENTINE OIL PRODUCTION AND CONSUMPTION, 1980-2001

and Golfo San Jorge, in the southeast. Other oil producing areas include the Noroeste, Cuyana, and Austral basins.

After reaching peak production in 1998, the Argentine upstream oil sector suffered in the late 1990s when the Asian financial crisis precipitated a plunge in world oil prices and a worldwide slowing of exploration and production development. Despite higher oil prices in 2001 and 2002, Argentine oil's rebound has been relatively slow, primarily due to uncertain economic and political conditions. Some observers have speculated that 2002's financial crisis and the government's economic policies, (particularly a 20 per cent tax imposed on oil exports, as well as a briefly imposed oil export cap) have significantly impacted oil production this year. Observers estimate that Argentine oil production may have dropped by as much as 21,000 bbl/d in January 2002 when compared to the previous year.

Some exploration continues in Argentina despite political risks. After failing to attract any takers at a December 2001 government auction of exploration licenses, private firms Pecom Energia (Argentina) and Devon Energy Corporation (U.S.) sold their assets to new foreign investors including Petrobras (Brazil), TotalFinaElf (France) and Wintershall (Germany) during the summer of 2002. In July 2002, Pioneer Natural Resources (U.S.) announced that it will resume drilling for oil in Argentina after suspending operations for 16 months. In August 2002, Petrobras announced a new hydrocarbon discovery in the Rio Negro province. Exploratory drilling continues to determine the size of the discovery and the characteristics of the oil.

Transportation

Argentina's three major crude oil pipelines all start at Puerto Hernández in the Neuquén basin. Two pipelines are domestic, transporting crude oil north to the Luján de Cuyo refinery near Mendoza and east to Puerto Rosales on the Atlantic. Argentina's one international crude oil pipeline, the Estenssoro-Pedrals (formerly called the Transandino), runs over the Andes, to a refinery in Chile. Repsol-YPF, Alberta Energy, Unocal, and ENAP hold stakes in the pipelines. Argentina also supplies Paraguay and Uruguay with crude oil via tanker.

Downstream

Three companies control the downstream oil sector in Argentina: Repsol-YPF, Esso, and Shell. Repsol-YPF accounts for just over 50 per cent of the country's total crude oil refining capacity, mainly from its ownership of the country's two largest refineries, La Plata and Lujan de Cuyo. The remaining capacity is split among Shell (13 per cent) and Esso (16 per cent), with one refinery each, and the five refineries of three small companies (17 per cent).

Repsol-YPF owns about 50 per cent of the country's service stations. Pérez Companc also has an extensive fuel retail network. Other major players in the fuel retail market include multinationals such as Shell and ExxonMobil.

Among the consequences of the 2002 financial crisis was a fuel oil shortage as the currency devaluation encouraged exports and severely restricted the importation of petroleum products. At the peak of the domestic petroleum products crisis in June 2002, the government announced an export cap that limited oil exports to 36 per cent of May production levels. After heavy lobbying from the petroleum industry, the government signed a decree in July 2002 repealing the export cap. In late October 2002, the Argentine government reduced the tax on diesel fuel exports from 20 per cent to 5 per cent after reaching an agreement with oil companies to guarantee the supply of oil to the domestic market.

Natural Gas

Argentina has the third largest proven reserves of natural gas in Latin America (after Venezuela and Mexico), at over 27 trillion cubic feet (Tcf), and probable reserves are much higher. Argentina overtook Venezuela to become the region's second-largest gas producer in 1999 (Mexico is the largest producer). The largest producing basins are Neuquén, Austral, and Noroeste, with the Neuquén basin alone accounting for about 60 per cent of Argentine gas production. Production in the Austral basin is constrained by the capacity of its pipeline to Buenos Aires. Supply of Argentine gas currently exceeds combined domestic and export demand.

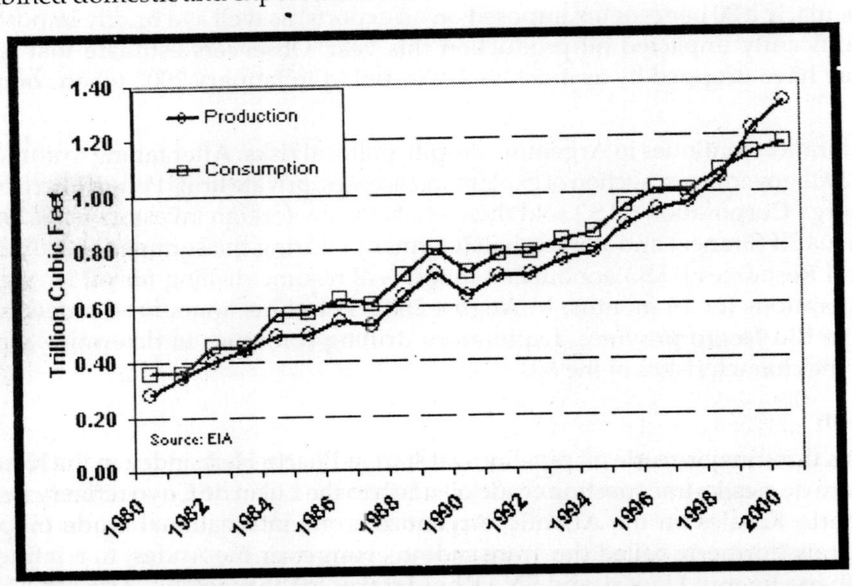

ARGENTINE NATRUAL GAS PRODUCTION AND CONSUMPTION, 1980-2000

Industry Organization

Argentina's natural gas industry–production, transportation, distribution, and marketing–is in the hands of the private sector and is operated within a competitive market structure. There are no restrictions on imports or exports. In the early 1990s, then-state monopoly *Gas del Estado* was split into two pipeline companies, *Transportadora de Gas del Sur* (TGS), and *Transportadora de Gas del Norte* (TGN), along with eight distributors operating within a regulated monopoly. The company was then privatized in accordance with the 1992 Gas Law, which restricted cross-ownership in production, transmission, and distribution activities. The law also created a regulatory body, ENARGAS, for the downstream sector. ENARGAS sets rates for natural gas transmission and distribution carriers operating under a non-discriminatory "open access" system. Although the downstream sector of the natural gas industry is now highly competitive, Repsol-YPF retains a dominant position in the upstream sector.

TGS is South America's largest pipeline company. The company delivers two-thirds of the nation's total gas consumption (to southern Argentina and greater Buenos Aires) and is 70 per cent -owned by U.S.-based Enron and Pérez Companc. Transmission activities are carried out by TGS and TGN, which cannot sell gas, but must guarantee free access to their pipelines. Large users can purchase gas from transmission and distribution companies or negotiate prices directly with producers. Metrogas is the country's largest distributor.

Natural gas companies have been particularly challenged by Argentina's financial crisis. Many companies have posted significant losses in 2002 as a result of currency devaluation and a government freeze on tariffs.

Pipelines

Chile is the largest export customer for Argentine gas, with significant exports starting in 1997. The Neuquén basin is the source for all four Argentina-Chile pipelines. The GasAndes pipeline extends 288 miles from Neuquén to central Chile, near Santiago. The GasAndes pipeline has been in operation since 1997 and is majority owned and operated by TotalFinaElf. There are plans to expand the pipelines throughput capacity significantly to reach 494 million cubic feet per day (Mmcf/d) and to extend it to the city of Rancagua.

The Gasoducto del Pacifico opened in November 1999, transporting 140 Mmcf/d over 330 miles to the Bio Bio region in southern Chile. Gasoducto del Pacifico is owned by a consortium of TotalFinaElf (operator), El Paso International, Chile's Gasco, ENAP, and Repsol-YPF. The pipeline services various industrial consumers. Presently, there is not enough demand to satisfy the pipeline capacity. The consortium reportedly is searching for new clients for the pipeline's gas.

The other two Argentine-Chilean pipelines run parallel to each other and supply a market that does not fully utilize the pipelines' capacities. The GasAtacama pipeline, with a 300-Mmcf/d capacity, came onstream in July 1999 and is co-owned by U.S.-based CMS and Endesa. GasAtacama's main customer is the Nopel power plant of the same owners. The NorAndino pipeline has a capacity of just over 280 Mmcf/d and came onstream in November 1999. It is owned by Belgium's Tractabel and U.S.-based Southern Company and supplies two power plants. A landslide in January 2002 ruptured the NorAndino, causing temporary power outages in Argentina and Chile. Emergency natural gas supplies were transported to Chile via the parallel GasAtacama line, prompting the operators to propose a spur line be built to connect the two pipelines to be used in emergencies.

One Argentina-Brazil pipeline has come onstream. The 273-mile, \$250 million, 423-Mmcf/d pipeline connects Paraná, Argentina, to Uruguaiana, Brazil. It provides gas to a \$350-million, 500-

megawatt (MW) AES power plant in Uruguaiana. Service began in early July 2000. Plans to extend the pipeline to Porto Alegre, Brazil, have been severely delayed and expected completion of the project has been pushed back from late 2002, to the second half of 2004.

Additional Argentina-Brazil pipelines are in various stages of the planning process, although recent natural gas discoveries in Bolivia and potential Brazilian discoveries could prevent development of the pipeline projects. These potential Argentina-Brazil pipelines include the Cruz del Sur, Trans-Iguacu, and Mercosur pipelines. The Cruz del Sur would extend to Brazil an Argentine-Uruguayan pipeline that currently is under construction (construction began in March 2001). The pipeline is reportedly almost complete, but has encountered delays as Uruguay has been unable to convert its power generators to natural gas, thus making the initial deliveries infeasible. The Trans-Iguacu pipeline would cross from northern Argentina's Noroeste basin into southern Brazil. The Mercosur pipeline would tap northwestern Argentina's Neuquén basin to Curitiba, Brazil, and could extend to Sao Paulo.

There is one gas import pipeline into Argentina. Bolivia sold gas to Argentina through the Yacimientos-Bolivian Gulf (Yabog) pipeline until 1999, when the contract expired. It remains unclear if or how the pipeline's spare capacity might be used.

Coal

Argentina has very limited coal resources, and coal is not a major component of the country's fuel mix. With only 474 million short tons of coal reserves, the country produced 0.33 million short tons and consumed 1.47 million short tons in 2000. There is a small coal field at Río Turbio in southern Patagonia. Imports come from Australia, the United States, and South Africa.

Electricity

Argentina relies mostly on hydropower and natural gas to fuel its electricity sector. In 2000, the country had 24 gigawatts (GW) of installed generation capacity, of which about 54 per cent was fossil fuel-based (primarily natural gas), 42 per cent hydroelectric, and about 4 per cent nuclear. Electric generation in 2000, which totaled 82.8 billion kilowatthours (bkwh), was 52 per cent thermal, 41 per cent hydropower, and 7 per cent nuclear. Argentina has the third-largest power market in Latin America, behind Brazil and Mexico.

Industry Organization

Argentina has one of the most competitive deregulated power sectors in South America. Since 1991, the government has pursued an aggressive privatization program in tandem with the creation of an open electricity market, which has attracted foreign investors and project partners attracted by the economic growth in the country. Today, power generation capacity generally satisfies domestic demand, but there remain some remote areas that have yet to be connected to the national grid.

The main power regulatory body in Argentina is Ente Regulador de la Energía Eléctrica (ENRE), which oversees all aspects of the industry and mediates disputes among the participants. The functions of generation, transmission, and distribution are open to the private sector, but there are restrictions on ownership of more than one function within the industry. The law guarantees access to the grid in order to create a competitive environment and to allow generators to serve customers anywhere in the country. Argentina's power market, Mercado Eléctrico Mayorista (MEM), is under the administration of the Compañía Administradora del Mercado Mayorista Eléctrico S.A. (CAMMESA), a non-profit organization owned by the participants in the power grid and headed by the Ministry of Energy.

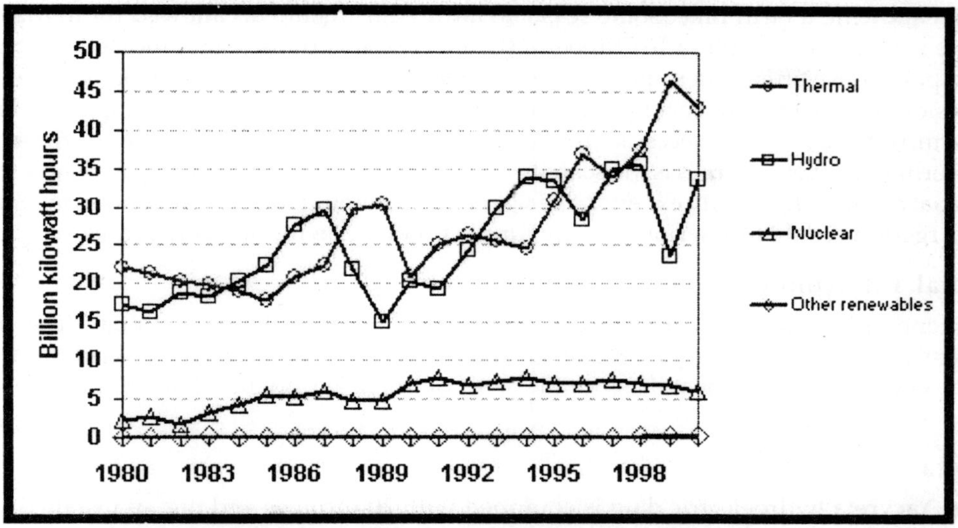

ARGENTINE ELECTRICITY GENERATION, 1980-1998

Generation is carried out by independent companies, including some state-owned companies. Generators earn income from the sale of electricity and from providing reserve capacity to the system. The transmission function is carried out by six primary distribution companies and one extra high voltage transmission company, Compañía Nacional de Transporte Energética en Alta Tensión (Transener). In July 2000, the government sold its 25 per cent stake in Transener.

Smaller distribution companies that share part of their distribution lines operate under a separate organization–Technical Additional Service of the Transportation Function (PFTT), which reports to CAMMESA. Expansions and modifications are established by CAMMESA ensuring maximum stability and security within the system. Transmission companies operate as a regulated monopoly with regulated prices, and are not allowed to sell power. Distribution companies also operate as regulated monopolies within concession areas that are awarded on a competitive basis for 95 years. Tariffs are regulated by ENRE and include connection and usage charges. While most distribution companies have been privatized, some remain in the hands of the provincial governments.

Devaluation and Utilities

Prior to January 2002, utility companies operating in Argentina made sales on the domestic market in U.S. dollars, to which the Argentine Peso was pegged under the country's 1991 Convertibility Law. With the abandonment of the Convertibility Law in January, Argentine utility companies (many of them foreign-owned) suffered sizable losses as the value of the peso fell dramatically. Simultaneous with devaluation, the government imposed a freeze on utility tariffs, holding them at their peso-denominated rate, thus passing the cost of devaluation on to the utility companies. The sinking value of the peso, coupled with a rash of nonpayment as unemployment soars, has severely impacted the natural gas and electricity industries, causing many companies to announce heavy losses or debt default in 2002.

In March 2002, the government set up a special commission to begin negotiations with the private sector on allowing a tariff hike. Some seasonal tariff adjustment was allowed for in May 2002, but intensive negotiations continue. In October 2002, after the government gave indications that it agreed in principal to raise utility rates, thousands of Argentines marched in protest. Some analysts believe that President Duhalde will conclude a conciliatory rate hike, designed to mitigate the crisis in the short-term, but will leave the necessary fundamental restructuring to the next administration. Another, longer-term impact of the crisis is the scarcity of international project financing available in the wake of financial collapse. Electricity experts have warned that without new investments in the transmission sector, Argentina could face serious energy supply problems in the next few years.

Regional Interconnections

Argentina has electricity interconnections with Chile, Brazil, and Uruguay. The InterAndes Transmission Line links an Argentine power station to Chile's northern electric grid. A connection to Brazil has been used more heavily since the onset of Brazil's drought related electricity shortage in 2001.

Yacyreta

The Yacyreta hydroelectric dam is equipped with 20 turbines and has an installed capacity of 3,200 MW. Argentina and co-owner Paraguay share the electricity generated at Yacyreta evenly, with most of Paraguay's electricity being exported to Argentina. Paraguay and Argentina signed the Treaty of Yacyreta in 1973, but the project suffered numerous delays, and the initial phase of construction was not completed until September of 1994. The twentieth turbine was finally set in to motion at Yacyreta in July 1989.

Argentina and Paraguay are planning to build another hydroelectric dam on the Paraná, the 3,000-MW Corpus Critsi dam. The dam would be built about 124 miles upstream from Yacyreta. In May 2000, Argentina and Paraguay signed a working agreement concerning Corpus Cristi. The agreement calls for constructing the dam through international public bidding, and assures that the project's construction will be environmentally sound. There has been considerable opposition to the project among the local community and among environmentalists.

Nuclear Power

Argentina has the most advanced nuclear energy program in Latin America. The 350-MW Atucha I and 650-MW Embalse facilities are fully operational. For the last decade, Argentina has been in the process of building an Atucha II plant. The project is nearly complete, but will require additional funding to finish the job. Construction currently has been halted. The government had intended to privatize all three plants, but privatization had been put on hold indefinitely.

Environment

Notwithstanding its current difficulties, Argentina saw strong economic growth during much of the 1990s. In 2000, Argentina was South America's third largest energy consumer as well as the continent's second largest emitter of carbon dioxide resulting from the consumption of fossil fuels. While developing the economy, the Menem administration (1991-1999) also made significant commitments to environmental protection (see Treaties). However, more needs to be accomplished, leaving President Eduardo Duhalde and his successor to mitigate the consequences of Argentina's rapid economic expansion.

In 2000, Argentina had South America's second highest per capita energy consumption at 73.2 million Btu per person behind only Venezuela, as well as South America's second highest per capita carbon emissions (1.0 metric ton). Conversely, energy consumption per dollar of GDP (energy intensity) in Argentina, and carbon emissions per dollar of GDP (carbon intensity) are both relativiley low compared to other South American countries.

Argentina also faces urban air pollution and industrial pollution problems. While past commitments to the reduction of greenhouse gas emissions and towards increasing the use of renewable fuels augured well for environmental protection, recent political turmoil has left the country's environmental future uncertain. Finding an appropriate balance between economic growth and environmental protection will represent a continuing challenge for Argentina in coming years.

Country Overview

Independence: July 9, 1816 (from Spain)

Population (2002E): 37.8 million

Location/Size: Southern South America/2.8 million square kilometers (1.1 million square miles), about four times the size of Texas

Major Cities: Buenos Aires (capital), Córdoba, La Plata, Mendoza, Rosario, Santa Fe

Languages: Spanish (official), English, Italian, German, French

Ethnic Groups: white (mostly Spanish and Italian) 97 per cent, mestizo, Amerindian, or other nonwhite groups 3 per cent

Religion: nominally Roman Catholic 92 per cent (less than 20 per cent practicing), Protestant 2 per cent, Jewish 2 per cent, other 4 per cent

Defense (8/98): Army (41,000), Navy (20,000), Air Force (12,000), Reservists (375,000), Paramilitary Forces (31,240)

Economic Overview

Currency: Peso

Financial Exchange Rate: US$1 = 3.6 Argentine Pesos (10/29/02)

Nominal Gross Domestic Product (2001E): $267.6 billion *(2002E):* $111.3 billion

Real GDP Growth Rate: (2001E): -4.5 per cent *(2002E):* -13.7 per cent

Inflation Rate: (2001E): -1.1 per cent *(2002E):* 30.7 per cent

Unemployment Rate: (2002E): 22 per cent

Current Account Balance as a per cent of GDP: (2001E): -1.7 per cent *(2002E):* 7.3 per cent

Major Trading Partners: Brazil, United States, Japan, Uruguay, Chile, Germany, France

Major Export Products (2000): Agricultural products (including manufacturing of agricultural products) (55 per cent), industrial products (30 per cent), energy (15 per cent)

Major Import Products (2000): Consumer goods (23 per cent), industrial inputs (including raw materials) (34 per cent), capital goods (43 per cent)

Energy Overview

Proven Oil Reserves (1/1/02E): 2.9 billion barrels

Oil Production (2002E): 818,000 barrels per day (bbl/d), of which about 763,000 bbl/d was crude oil

Oil Consumption (2002E): 483,000 bbl/d

Net Oil Exports (2001E): 335,000 bbl/d

Natural Gas Reserves (1/1/02E): 27.5 trillion cubic feet (Tcf)

Natural Gas Production (2000E): 1.32 Tcf

Natural Gas Consumption (2000E): 1.17 Tcf

Net Natural Gas Exports (2000E): 0.15 Tcf

Coal Reserves (2000E): 474 million short tons (Mmst)

Coal Production (2000E): 0.33 Mmst

Coal Consumption (2000E): 1.47 Mmst

Electric Generation Capacity (1/1/00E): 24 gigawatts (GW)

Electricity Generation (2000E): 82.8 billion kilowattthours (bkwh); conventional thermal 52 per cent, hydroelectricity 41 per cent, nuclear 7 per cent

Environmental Overview

Total Energy Consumption (2000E): 2.7 quadrillion Btu (0.7 per cent of world total energy consumption)

Energy-Related Carbon Emissions (2000E): 36.4 million metric tons of carbon (0.6 per cent of world carbon emissions)

Per Capita Energy Consumption (2000E): 73.2 million Btu (vs. U.S. value of 351.0 million Btu)

Per Capita Carbon Emissions (2000E): 1.0 metric tons of carbon (vs U.S. value of 5.6 metric tons of carbon)

Energy Intensity (2000E): 9,226 Btu/ $1995 (vs U.S. value of 10,918 Btu/ $1995)

Carbon Intensity (2000E): 0.12 metric tons of carbon/thousand $1995 (vs U.S. value of 0.17 metric tons/thousand $1995)

Sectoral Share of Energy Consumption (1998E): Industrial (48.6 per cent), Transportation (23.7 per cent), Residential (18.8 per cent), Commercial (8.8 per cent)

Sectoral Share of Carbon Emissions (1998E): Industrial (44.8 per cent), Transportation (32.7 per cent), Residential (16.2 per cent), Commercial (6.2 per cent)

Fuel Share of Energy Consumption (2000E): Natural Gas (45.2 per cent), Oil (36.3 per cent), Coal (1.5 per cent)

Fuel Share of Carbon Emissions (2000E): Oil (48.1 per cent), Natural Gas (49.3 per cent), Coal (2.5 per cent)

Renewable Energy Consumption (1998E): 393 trillion Btu (0.5 per cent decrease from 1997)

Number of People per Motor Vehicle (1998): 5.6 (vs U.S. value of 1.3)

Status in Climate Change Negotiations: Non-Annex I country under the United Nations Framework Convention on Climate Change (signed June 12, 1992 and ratified on March 11, 1994). Signatory to the Kyoto Protocol (signed March 16, 1998).

Major Environmental Issues: Erosion resulting from inadequate flood controls and improper land use practices; irrigated soil degradation; desertification; air pollution in Buenos Aires and other major cities; water pollution in urban areas; rivers becoming polluted due to increased pesticide and fertilizer use.

Major International Environmental Agreements: A party to the Antarctic-Environmental Protocol, Antarctic Treaty, Biodiversity, Climate Change, Desertification, Endangered Species, Environmental Modification, Hazardous Wastes, Law of the Sea, Marine Dumping, Nuclear Test Ban, Ozone Layer Protection, Ship Pollution and Wetlands. Has signed but not ratified Marine Life Conservation.

The total energy consumption statistic includes petroleum, dry natural gas, coal, net hydro, nuclear, geothermal, solar, wind, wood and waste electric power. The renewable energy consumption statistic is based on International Energy Agency (IEA) data and includes hydropower, solar, wind, tide, geothermal, solid biomass and animal products, biomass gas and liquids, industrial and municipal wastes. Sectoral shares of energy consumption and carbon emissions are also based on IEA data.

Energy Industry

Organization: The energy sector has been privatized and deregulated, with many private sector participants. Oil and gas producers Repsol-YPF, Pérez Companc, Petrolera San Jorge, BP, TotalFinaElf, Austral, Astra, Bridas, Tecpetrol, Pluspetrol. Natural gas pipeline companies have also been privatized–Transportadora de Gas del Norte, Transportadora de Gas del Sur. Electric power companies–numerous producers in a highly competitive generation market, three transmission companies in a regulated monopoly, three regional distributors, municipal utilities.

Major Ports: Bahía Blanca, Buenos Aires, La Plata

Major Oil and Gas Producing Basins: Neuquén, Austral, Golfo San Jorge, Cuyana, Northwest

Major Refineries (Capacity): YPF La Plata (189,000 bbl/d), YPF Lujan de Cuyo (120,000 bbl/d), Shell Buenos Aires (110,000 bbl/d), dSan Lorenzo (37,600 bbl/d), Refinor–Campo Duran (32,000 bbl/d); Total Capacity–639,075 bbl/d

7

Australia

Australia is the world's leading coal exporter and has very large natural gas reserves as well. Australia's proven oil and natural gas reserves have nearly doubled in recent years, though there is much exploration yet to be done. Infrastructure is being developed to bring much more of Australia's natural gas reserves to market. Australia is already the third-largest LNG exporter in the Asia-Pacific region.

General Background

The Australian economy closed 2001 well ahead of expectations. Real gross domestic product (GDP) grew by 4.1 per cent over 2001, well exceeding both international expectations and the global growth rate of 2.8 per cent. The Australian economy, which is heavily dependent on trade, was bolstered in 2001 by the declining value of the Australian dollar as well as by growing demand for raw materials from developing Southeast Asian countries. Internally, expansionary governmental policy over the past few years has fostered significant increases in domestic consumption. As a result, Australia weathered both 1997's Asian financial crisis and the 2001 global economic slowdown much better than many had expected. For the past 10 years, Australia has been the world's fastest growing developed economy.

Australia's Liberal Party, under Prime Minister John Howard, retained power after elections held in November 2001. They remain in coalition with their junior partner, the National Party. The Liberals have stimulated domestic consumption by overhauling the tax system in 2000 and cutting interest rates six times in 2001. Australia's next election is scheduled for late 2003.

Energy

Australia is one of the few OECD countries that is a significant net energy exporter.

Coal is Australia's largest export commodity, and accounts for 44 per cent of the country's total energy needs. Australia is also a net exporter of natural gas. However, the country is becoming increasingly dependent upon foreign oil. A rapidly expanding economy and declining domestic oil production have led some observers to forecast an energy supply crisis in the next 10 years. In addition, stagnating foreign investment, an undersized natural gas market, and difficulty transmitting electricity across the country have led the government, industry leaders, and international observers to call for a new long-term energy strategy.

On June 8, 2001, the Council of Australian Governments convened to lay the groundwork for Australia's new energy strategy. The body acknowledged the country's energy problems and agreed to an independent review of the energy market. Debate over the makeup the review board, however, has paralyzed the reform process, with no immediate solution expected. Once a review body is decided upon, the assessment is expected to take another 12 months.

Among the suggestions put forth by industry and international observers are: 1) new incentives for investment and exploration in Australia's petroleum industry; 2) encouragement for new players into the national natural gas market 3) tax reform that will restore the profitability of pipeline operators; and 4) investment in transmission and infrastructure to provide a reliable supply of electricity.

Oil

Australia produced an average of 633,000 barrels per day (bbl/d) in 2001 and consumed an average of 872,000 bbl/d, resulting in net imports of 239,000 bbl/d. By comparison, net oil imports in 2000 averaged only 54,000 bbl/d. This growing oil deficit has garnered significant interest in recent months and propelled public and private actors to address the country's growing fuel insufficiency.

Australia's expanding oil deficit is primarily a result of demand quickly beginning to outpace supply. Petroleum consumption has grown by only 3 per cent since 1995, but is expected to rise in tandem with Australia's growing economy over the next 20 years. Despite reserves estimated at 3.5 billion barrels in 2002 (a 20 per cent increase over 2001) Australia's Ministry of Industry, Tourism, and Resources, estimates that Australia has only ten years of oil left. According to the Ministry, Australia is using oil three times faster than it is finding it. By 2010, the country is expected to slide from 80 per cent self-sufficiency to 40 per cent.

Oil production had increased gradually since 1980, peaking in the year 2000 at 805,000 bbl/d. In 2001, as was expected, production fell dramatically. Declines were due primarily to decreasing

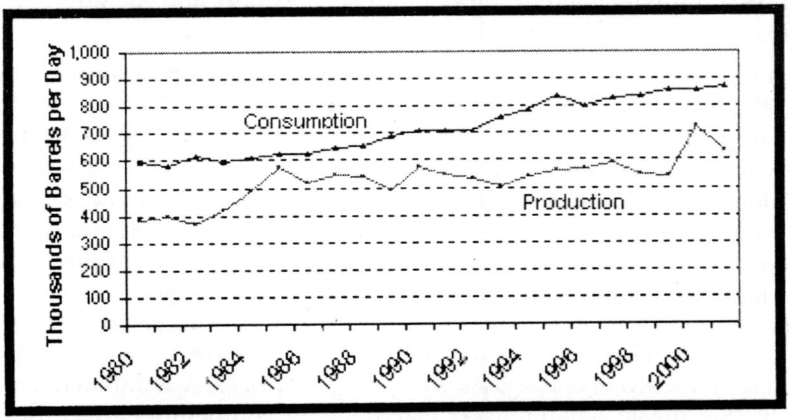

OIL PRODUCTION AND CONSUMPTION IN AUSTRALIA, 1980-2000

production from the Cooper-Eromanga and Gippsland basins. The country's other two oil basins, the Carnarvon and Bonaparte, have both yielded increasing amounts of oil in recent years, but have been unable to keep up with the country's rapidly growing demand.

Observers and players have called for an overhaul of the industry. Critics suggest that a new tax regime, with a defined legal structure and incentives for frontier, deep-water, exploration are crucial to spurring more investment in Australia. Some speculate that solving these issues might unlock the country's large hydrocarbon potential. Unocal, Anadarko, and several Japanese firms have all expressed interest in developing projects in Australia. On March 26, 2002, Apache announced three new hydrocarbon structure discoveries in the Carnarvon Basin. Australian firms Bligh Oil and Minerals Corp. and Roc Oil have also recently announced finds in the nearby Beibu Gulf, located off China's Southwest shore. The Perth Basin, Carnarvon Basin, and Cooper Basin are all potential sites for further exploration.

Australia also has shale oil reserves in the northeastern state of Queensland. Some analysts estimate Queensland could hold up to 30 billion barrels of shale oil. The company developing Queensland's shale oil, Southern Pacific Petroleum/Central Pacific Minerals (SPP/CPM), however, has until recently been unable to capitalize upon this resource due to a struggle with environmental activists from Greenpeace. Since 1998, Greenpeace has been staging public demonstrations in Queensland and pressuring Australian refiners to refuse shale oil. Greenpeace considers shale oil, which is produced by subjecting shale rock to intense heat, to be highly polluting and energy intensive. In 2001, all four major Australian refining firms refused to purchase Queensland's shale oil, despite government excise rebates, forcing the industry to look to the government for support in order to stay afloat. On May 14, 2002, the government granted temporary support by extending existing excise rebates, originally designed only for the domestic sale of shale oil products, to international markets for a period of 12 months. During this time, SPP/CPM will look for international as well as domestic buyers.

Refining

Australian refiners lost $160 million across refining and marketing in 2000. Rising crude oil prices, lower product prices, and increasing competition from neighboring Southeast Asian states all have put tremendous pressure on Australia's downstream petroleum sector. The significant losses suffered by Australian refiners have prompted industry insiders and the government to call for a restructuring of the Australian refining industry.

Australia has ten refineries, dominated by four companies, with a total crude oil distillation capacity of 846,250 bbl/d. The country's three biggest refineries are: British Petroleum Australia's Kwinana refinery with a capacity of 158,500 bbl/d of crude oil; ExxonMobil's Altona refinery with a capacity of 130,000 bbl/d of crude oil; and Caltex's Kurnell refinery with a capacity of 114,000 bbl/d. These refineries, along with the country's other seven, have all faced declining refining gross margins in recent years. According to a 1998 government study of the downstream petroleum industry, Australian refiners are at a significant disadvantage in the global market place. The small size of Australian refineries when compared to larger neighboring facilities concedes a significant economy of scale to their competitors. Also, the country's refineries were built according to a 1970's government mandate that requires new refineries be built to handle Australia's light, sweet crude oil. However, as the premium paid for light sweet crude rises with rising oil prices, Australian refiners are again at a disadvantage in the Southeast Asian market place as their competitors are equipped to purchase heavier, sour crude oils, which are cheaper. These disadvantages, combined with a general oversupply of refining capacity in Asia and the extra cost of transporting crude oil to and from Australia put Australian refiners at a significant disadvantage.

Natural Gas

Australia's natural gas reserves are among the largest in the Asia Pacific region. Proved reserves were estimated at 90 trillion cubic feet (Tcf) as of January 1, 2002, slightly more than double 2001's estimated reserves. Natural gas, however, plays a relatively small role in the country's fuel mix. It is estimated that natural gas will account for 24 per cent of the country's total energy consumption by 2019-2020 and will grow almost twice as fast as other energy sources. In 2000, production of natural gas totaled 1.12 Tcf and consumption totaled 755 billion cubic feet (Bcf). Considering Australia's sizeable natural gas reserves and the relative decline of oil reserves, the Australian government has put an emphasis on natural gas for both domestic consumption and export. In December 2001, the government created the "Gas-to-Liquids Taskforce" charged with increasing the role of natural gas in Australia's economy.

Australia's largest natural gas reserves are in the "Northwest shelf," a collection of fields located in Western Australia 143 miles offshore, and estimated to contain over 30 Tcf of natural gas. Commercialization of the Northwest shelf has long been a dream of the Australian government, but its remote location and deep waters have both impeded production. The Gorgon field, located in the Northwest shelf, is the country's largest single natural gas field, with certified reserves of 9.6 Tcf and estimates as high as 15 Tcf. The field is owned by partially by Chevron (2/7), Texaco (2/7) Shell (2/7) and Exxon (1/7). A consortium of companies with interests in the region have assembled the "Northwest Shelf Consortium," and are currently competing to supply China's soon to be constructed Guangdong LNG import terminal. The Consortium has been short-listed for the project, which entails providing 3 million metric tons of LNG annually beginning in 2005. The South African firm Sasol is also negotiating a large purchase from Gorgon. These new sizeable markets have made possible the development of the high-risk isolated field.

The region's downstream component, the Northwest shelf liquefied natural gas (LNG) plant, may soon be expanded from a 3-train facility to potentially a 4- or 5-train facility as demand for natural gas grows both at home and abroad. In 2001, the plant produced 7.5 million metric tons of LNG. Most of the plant's LNG sales go to Japanese utilities, which have shown increasing demand for Australian LNG products. In May, 2002, Kyushu Electric Power (Japan) agreed to purchase 500,000 metric tons of LNG per year beginning in 2006.

There also have been new natural gas discoveries of the Southern coast of Victoria in the Otway basin. It has been speculated that the Otway basin could hold up to 1.5 Tcf of natural gas. The basin's proximity to major cities Melbourne and Adelaide have sparked the interest of domestic oil companies Woodside and Santos. Two discoveries, Thylacine and Geographe, are estimated to hold up to 2 Tcf, which could be marketed in Australia's Southern states by 2005. Tokyo Gas Company and Mitsubishi have purchased a 40 per cent stake in the Otway basin's Patricia-Baleen gas field from Australian OMV. The group plans to invest $52 million in transportation infrastructure in order to market the gas in Victoria beginning in September 2002.

Australia's natural gas industry also saw its largest emissions trade in December 2001, as the Yellowbank facility in Western Queensland used advanced flaring technology to abate 700,000 metric tons of greenhouse gases. The new technology eliminates the venting of waste methane, a highly polluting greenhouse gas. The plant's owners, Oil Company of Australia and Santos (Australia), were granted emissions reduction credits through the Commonwealth Bank of Australia equal to the amount of foregone waste methane. The Yellowbank facility went on to sell the emissions credits to BP Australia.

Pipelines

Many Australian and international investors have called for a serious overhaul of the country's natural gas pipeline network and its regulatory regime. The existing network was built to carry gas from the country's centrally located fields to urban hubs like Sydney and Melbourne. But with centrally located fields such as the Bass Strait in decline, and offshore projects like the Northwest Shelf, Otway Basin,and Timor Gap on the horizon, Australia will need massive investment in the country's pipeline infrastructure in order to bring new hydrocarbon energy into the grid.

Some new pipeline projects are under way, but a contentious regulatory environment slows development. New projects include the Sea Gas Pipeline, a $300-million, 423-mile line that will take gas from the Otway Basin off Victoria to the Quarintine power station in Adelaide. The Sea Gas Pipeline is operated by Origin and Australian National Power and is scheduled for completion by 2003. Also, a line that crosses the Gulf of Papua, between Papua New Guinea and Australia is in its early stages of construction. In March 2002, a base load for the pipeline was established as Australian Gas Light signed a conditional agreement with ExxonMobil (operators in the Papua New Guinea fields) to buy up to 43 billion cubic feet annually for up to 20 years.

Controversy between pipeline companies and national regulators continues to impede development. While some companies have taken to building pipelines without excess capacity in order to avoid regulation, other have simply cancelled their plans and backed out of the market. Australian Pipeline Trust, Australia's largest pipeline owner, has halted all construction, including its planned $80-million Dubbo-Tamworth pipeline for fear of the regulators. American firm El Paso has also deferred its planned investments until the regulatory environment settles. There are currently numerous cases pending before the Australian Supreme Court concerning pipeline regulation

East Timor

Australia claims oil and natural gas reserves in the Timor Gap, located between Australia and the newly independent state of East Timor. In 1999, the people of East Timor voted in favor of independence from Indonesia. Since this vote, the treaties and contracts that previously governed the Timor Gap's resources have been under re-negotiation. The Timor Gap holds sizeable natural gas reserves and some oil condensate.

There are two main fields in the Timor Gap, the Bayu Undan and the Greater Sunrise Field. The Bayu Undan is situated in 263-feet deep water, 311 miles from the coastal city of Darwin, and could possess as much as 400 million barrels of condensate and 2.4 Tcf of natural gas. In July 2001, negotiators for East Timor, Australia, and Phillips Petroleum (operators in Bayu Undan) agreed that 90 per cent of the royalties earned from Bayu Undan will go to East Timor. This agreement is expected to be codified by an international treaty between East Timor and Australia to be ratified shortly after official East Timorese independence on May 20, 2002. In March of 2002, Phillips made arrangements to sell 3 million tons of LNG annually to Tokyo Electric Power Company and Tokyo Gas company for 17 years beginning in 2006.

The Greater Sunrise field, while possessing significantly larger reserves, has been stalled by infighting amongst the littoral states and investors. The field, which contains up to 9.3 Tcf of natural gas is owned by Woodside (Australia) (33.4 per cent), Phillips (30 per cent), Shell (26.6 per cent), and Osaka gas (10 per cent). About 80 per cent of the field is located on Australian territory, while 20 per cent of it is on Australia-East Timor jointly administered territory. Accordingly, Canberra is demanding 82 per cent of the royalties from the Greater Sunrise field. The final arrangements are yet to be announced. Another debate impeding development over the fields is amongst the investors. Shell wants to employ its newly developed Floating Natural Gas Platform (FLNG) in the Sunrise field, making it the first ever employment of such technology. According to Shell, the $5 billion FLNG terminal could cut costs by 40 per cent. Woodside and Osaka have shown support for Shell's initiative, but Phillips and the government of Australia have expressed strong reservations. The FLNG is designed to export LNG to Japan, China, and North America. The government of Northern Territory has stated that the Greater Sunrise development will leave "no chance for industrial development in Darwin, and no gas for the rest of Australia."

The Timor Sea also contains natural gas in the Evans Shoal, Petrel, and Tern gas fields. Combined, these fields are estimated to contain 4 Tcf of natural gas and are currently operated by the Australian oil major Santos.

LNG

Australia is a growing LNG producer and exporter. It is now the third-largest LNG exporter in the Asia-Pacific region, behind Indonesia and Malaysia. Australia's major LNG project, the NWSP, has created an estimated US$10 billion in business for Australian suppliers since 1980, and produced an extra 80,000 jobs each year of its existence. In October 2000, the government and the Australian Petroleum Production and Exploration Association launched a new policy framework to ensure that Australia remains a competitive and reliable LNG supplier. The agenda commits the government to environmental policies that will not in principle harm the LNG industry and announces a new bylaw for LNG projects with an initial capital expenditure of at least US$50 million to obtain duty-free importation of capital equipment unavailable in Australia and integral to the project.

Coal

Australia is the world's fourth largest producer of coal. In 2000, Australia produced a total of 337.2 million short tons (Mmst) of coal and consumed 144.2 Mmst, resulting in 193 Mmst of coal exported, or 57 per cent of the country's total production. Since 1986, Australia has been the world's largest coal exporter.

Australia's coal deposits are concentrated along the country's western seaboard in the states of Queensland, New South Wales, and Victoria. The Southern state of Victoria and produces mainly thermal coal, or brown coal, which is of lower quality and used primarily for the domestic generation of electricity. New South Wales production comes from Permian black coal, a bituminous type. Some of this coal is used as steaming coal, some is coking coal. Queensland, to the North, contains large deposits of high quality black coal, or coking coal, used in steel production. Most of the country's black coal is exported to neighboring Asian countries. Black coal exports in 2001 totaled approximately 85 Mmst, 77 Mmst of which came from Queensland.

The country's coal industry is dominated by four companies: BHP Billiton, Anglo American (UK), Rio Tinto (Australian-UK), and Xstrata (Switzerland). Xstrata is a new operator in Australia, after purchasing its Australian in assets in 2001 from Swiss mining company Glencore. Australian coal exporters saw record earnings in 2001, benefited by both strong global black coal demand and a weak Australian dollar. In November, 2001, however, Chinese coal suppliers began flooding the market, taking advantage of the record high coking coal prices. Chinese production has since tapered off, but still stands as potential formidable competition for the Australian export industry.

The coal majors have used their strong market position and high world coal prices in recent negotiations with buyers. Japan-Australia price negotiations remain deadlocked in May 2002, and well past their April deadline. Australian suppliers are pushing for higher prices from their Japanese customers, mainly steel manufacturers and utilities. The Japanese, conversely, are suffering from a recession and are seeking lower prices. Last year's price contracts have been rolled over in the interim, but will be nullified come September. Japan accounted for 46 per cent of Australia's black coal exports in 2000.

Other markets for Australia's coal are non-Japan Asia (30 per cent) and Europe (16 per cent). Liberalization of European energy markets has presented attractive opportunities to Australian suppliers. Coal exports to Germany have grown from 292,000 short tons in 1995 to 2.9 million short tons in 2001. Coal is currently Australia's biggest export to the EU.

The Australian Bureau of Agricultural and Resource Economics (ABARE) expects Australia's export coal industry to grow quickly in the short term. ABARE predicts that Australian coal exports will reach 254 Mmst in 2010, 44 Mmst above present levels. Major projects on the horizon include the Hail Creek Coal Project, located in central Queensland, which is a joint venture between Pacific Coal, Marubeni, and Sumitomo. The mine, which will be built and managed by Pacific, will be open cut, and is expected to produce 5.5 Mmst annually by 2003. In 2002, the Chinese coastal city of Zhangjiagang, became a first-time importer of Australian coal, purchasing approximately 44,000 short tons. Also, the April 2002 merger of Anglo-American's subsidiary Anglo-coal Australia with Mitsui coal holdings could also have a significant impact on the industry. The firms have indicated their intention to spend $124 million to join their assets surrounding the Moura mine, located in Queensland's Bowen Basin, to produce a super pit with the potential to produce13 Mmst per year. Finally, AuIron Energy's South Australian Steel and Energy (SASE) project entails the construction of 2.7 Mmst ton per year pig iron plant that will use its own Australian coal and iron ore deposits with the hopes of becoming the world's lowest cost producer of pig iron.

Electricity

As of January 2000, Australia had an electrical generation capacity of 43 million kilowatts (or gigawatts). Approximately 84 per cent of this capacity was thermal (mostly coal) and 14 per cent of it was from renewables (mostly hydro). In 2000, Australia generated 202.7 billion kilowatthours (BkWh) of electricity and consumed 188.5 BkWh. ABARE expects electricity usage to grow by 2.8 per cent per year over the course of this decade.

Sector Organization

In 1996, major reforms were instituted for Australia's electricity industry. Prior to 1996, electric utilities were owned by states, but under reforms many state-owned utilities are being split up and privatized. Victoria and South Australia have already sold and long-term leased (respectively) all of their state-owned electricity utilities. Key to this reform was the creation of the National Electricity Market (NEM). The NEM is a wholesale "pool" operated by the National Electricity Market Management Company (NEMMCO) to which all generators above a certain size are obliged to sell their output at prices determined by the last and highest bid for distribution through regulated transmission networks. NEMMCO is a self-funding company owned by the participant states and the federal government.

Australia has five trading regions, all of which are interconnected, but the NEM does not include the non-connected states of Tasmania, Western Australia, or the Northern Territories. Tasmania is expected to join when its electricity link to the mainland is completed in 2003. New South Wales and Victoria have been combined into a two-state regional market, and it is here that reforms have strongly reduced electricity prices due to overcapacity and strong competition, though prices have begun to rise recently as increasing demand uses previously spare capacity. Since January 2002, residents of Victoria have been able to choose their electricity provider amongst a list of private firms. Reforms did not lower prices much in Queensland or South Australia, although prices overall fell about 11 per cent in the period 1996-2000. However, much of the savings went to large industrial/commercial customers that had the option to choose between retailers.

The NEM seems to successful in encouraging new investment, as about 2,300 MW in additional capacity is being constructed or planned. Also important in sparsely populated Australia is the transmission network development. Reforms have created two types of interconnectors to enhance transmission network development. Regulated interconnectors must pass strict tests in terms of contributing to market development to receive guaranteed rates of return. Unregulated interconnectors derive their income from the price difference between two sides of the interconnector. By January 2003, all electricity consumers will be able to choose between electricity retailers, and prices are expected to fall when this is implemented.

Environment

Energy commodities are a major source of export earnings in Australia and development of these resources in a sustainable manner is a primary policy goal of the country. Although coal is a major component of Australia's primary energy mix, increasing urban air pollution levels are more a consequence of automobile usage than coal consumption. In March 2000, Australia released the Review of Fuel Quality Requirements for Australian Transport, a report attempting to address environmental problems associated with automobile usage in areas such as Queensland, Perth and Western Sidney.

In 2000, Australia contributed 1.5 per cent of the world's total energy-related carbon emissions. Partially because of the greenhouse gas emissions associated with agriculture, the Australian Institute, an independent public policy research center, indicated that if statistics included total greenhouse gas emissions, as opposed to only energy-related emissions, then Australia would have the highest per capita carbon emissions in the developed world.

The Australian government realizes the cost-effectiveness of reducing the environmental impacts of the energy sector. Improving end-use efficiency in the various economic sectors remains a key element of Australia's sustainable energy policy, as does the utilization of renewable energy resources.

Country Overview

Independence: January 1, 1901 (from the United Kingdom)

Population: 19,357,594 (July 2001 est.)

Location/Size: Oceania, continent between the Indian Ocean and the South Pacific Ocean/7,686,850 sq. km (2,971,081 sq. mi), about the size of the contiguous United States

Major Cities: Sydney, Melbourne, Canberra (capital), Brisbane, Perth, Adelaide

Languages: English, native languages

Ethnic Groups: Caucasian (92 per cent), Asian (7 per cent), aboriginal and other (1 per cent)

Religions: Anglican (26 per cent), Catholic (26 per cent), other Christian (24 per cent), non-Christian (11 per cent)

Defense (8/98): Army (25,400), Navy (14,300), Air Force (17,700)

Economic Overview

Currency: Australian Dollar ($A)

Market Exchange Rate (5/24/02): US $1=$A1.79

Nominal Gross Domestic (GDP, 2001E): U.S.$365.8 billion

Real GDP Growth Rate (2001E): 4.1 per cent *(2002F):* 3.8 per cent

Inflation Rate (2001E): 4.3 per cent *(2002F):* 3.0 per cent

Unemployment Rate (2001E): 6.9 per cent *(2002F):* 7.0 per cent

Current Account Balance (2001E): -$15.3 billion *(2002F):* -$16.9 billion

Major Trading Partners: Japan, other Far East, European Union, United States

Major Export Products: crude materials, food and live animals, mineral fuels and lubricants

Major Import Products: machinery and transport equipment, manufactured goods, chemicals

Energy Overview

Proven Oil Reserves (1/1/02E): 3.5 billion barrels

Oil Production (2001E): 632,918 barrels per day (bbl/d)

Oil Consumption (2001E): 872,000 bbl/d

Net Oil Imports (2001E): 239,082 bbl/d

Crude Refining Capacity (1/1/02E): 846,250 bbl/d

Natural Gas Reserves (1/1/02E): 90.6 trillion cubic feet (Tcf)

Natural Gas Production (2000E): 1.12 Tcf

Natural Gas Consumption (2000E): 755 billion cubic feet (Bcf)

Recoverable Coal Reserves (2000E): 90,489 million short tons

Coal Production (2000E): 337.15 million short tons (Mmst)

Coal Consumption (2000E): 144.17 Mmst

Electric Generation Capacity (1/1/00E): 43 million kilowatts (84 per cent Thermal, 14 per cent Hydroelectric)

Electricity Generation (2000E): 202.7 billion kilowatthours

Electricity Consumption (2000E): 188.5 billion kilowatthours

Environmental Overview

Total Energy Consumption (2000E): 4.89 quadrillion Btu* (1.2 per cent of world total energy consumption)

Energy-Related Carbon Emissions (2000E): 96.87 million metric tons of carbon (1.5 per cent of world carbon emissions)

Per Capita Energy Consumption (2000E): 255 million Btu (vs U.S. value of 351 million Btu)

Per Capita Carbon Emissions (2000E): 5.1 metric tons of carbon (vs U.S. value of 5.6 metric tons of carbon)

Energy Intensity (2000E): 10,804 Btu/ U.S.$1995 (vs U.S. value of 10,918 Btu/ $1995)**

Carbon Intensity (2000E): 0.21 metric tons of carbon/thousand U.S.$1995 (vs U.S. value of 0.17 metric tons/thousand $1995)**

Sectoral Share of Energy Consumption (1999E): Transportation (42 per cent) Industrial (37 per cent), Residential (13.5 per cent), Commercial (7.5 per cent)

Sectoral Share of Carbon Emissions (1998E): Industrial (46.4 per cent), Transportation (26.5 per cent), Residential (15.2 per cent), Commercial (11.9 per cent)

Fuel Share of Energy Consumption (2000E): Coal (44.2 per cent), Oil (34.8 per cent), Natural Gas (16.6 per cent)

Fuel Share of Carbon Emissions (1999E): Coal (55.4 per cent), Oil (32.6 per cent), Natural Gas (12.0 per cent)

Renewable Energy Consumption (1998E): 396 trillion Btu* (0.9 per cent increase from 1997)

Number of People per Motor Vehicle (1998): 1.7 (vs U.S. value of 1.3)

Status in Climate Change Negotiations: Annex I country under the United Nations Framework Convention on Climate Change (ratified December 30th, 1992). Signatory to the Kyoto Protocol (April 29th, 1998). Under the Protocol, Australia has agreed to an 8 per cent increase from 1990 emissions levels of a basket of greenhouse gases.

Major Environmental Issues: Soil erosion from overgrazing, industrial development, urbanization, and poor farming practices; soil salinity rising due to the use of poor quality water; desertification; natural habitat of many unique animal and plant species is threatened by clearing for agricultural purposes; the Great Barrier Reef off the northeast coast, the largest coral reef in the world, is threatened by increased shipping and its popularity as a tourist site; limited natural fresh water resources.

Major International Environmental Agreements: A party to the Antarctic-Environmental Protocol, Antarctic Treaty, Biodiversity, Climate Change, Endangered Species, Environmental Modification, Hazardous Wastes, Law of the Sea, Marine Dumping, Marine Life Conservation, Nuclear Test Ban, Ozone Layer Protection, Ship Pollution, Tropical Timber 83, Tropical Timber 94 and Wetlands. Has signed but not ratified, Desertification.

The total energy consumption statistic includes petroleum, dry natural gas, coal, net hydro, nuclear, geothermal, solar and wind electric power. The renewable energy consumption statistic is based on International Energy Agency (IEA) data and includes hydropower, solar, wind, tide, geothermal, solid biomass and animal products, biomass gas and liquids, industrial and municipal wastes. Sectoral shares of energy consumption and carbon emissions are also based on IEA data.

Oil and Gas Industries

Major Oil and Gas Producing Regions: Western Australia; Victoria; South Australia; Queensland; Northern Territory

Major Ports: Sydney; Melbourne; Geelong; Fremantle; Adelaide; Brisbane

Major Oil Fields: Roller, Skate, Bass Strait, Wanea-Cossack, Laminaria, Corallina

Major Gas Fields: Bass Strait, Cooper Basin, North Rankin, Goodwyn, Gorgon

Major Oil Refineries (crude oil capacity): BP Amoco–Bulwer Island (69,825 bbl/d), BP Amoco–Kwinana (158,500 bbl/d), Caltex–Kurnell (114,000 bbl/d), Caltex–Lytton (105,500 bbl/d), Inland Oil

Refiners–Eromanga (1,425 bbl/d), ExxonMobil–Adelaide (74,000 bbl/d), ExxonMobil–Altona (130,000 bbl/d), Shell–Clyde (85,000 bbl/d), Shell–Geelong (110,000 bbl/d)

Coal Industries

Major Coal Producing Regions: New South Wales; Queensland; Victoria; South Australia

Major Export Ports: Newcastle; Hay Point; Gladstone; Port Kembla

8

Azerbaijan

Since becoming independent in 1991, Azerbaijan has attracted significant international interest in its substantial oil and natural gas reserves. Foreign investors are helping the country to develop its rich oil and natural gas reserves in the Caspian Sea basin, and construction of new pipelines may allow Azerbaijan to become a significant energy exporter over the next decade.

General Background

Azerbaijan is one of five Caspian littoral states already capitalizing upon the sea's sizeable, but as of yet mostly untapped, hydrocarbon resources. Azerbaijan's real gross domestic product (GDP) grew by 10.6 per cent to $6.1 billion, in 2002 as foreign investors pushed ahead with major projects in Azerbaijan despite the worldwide economic slowdown. Even with the $3.7 billion which has been invested into the country's oil sector since 1994, Azerbaijan is currently a relatively insignificant world oil and gas producer, although major increases are expected beginning in 2005.

In the meantime, Azerbaijan is struggling to overcome the economic collapse that followed independence from the former Soviet Union, as the country's GDP contracted by almost 60 per cent from 1990 to 1995. As recently as October 2002 it was estimated that almost one half of Azerbaijani citizens earned less than $25 per month. In May 2003, the International Monetary Fund approved renewed funding for its Poverty Reduction and Growth Facility program in Azerbaijan, designed to re-structure the country's economy and alleviate economic hardships. To date, Azerbaijan reportedly has drawn $41 million from the fund.

Azerbaijan's hope for future economic growth rests in large part with successful development of its vast oil and natural gas resources in the Caspian Sea region. Crude oil and oil product exports make up over 70 per cent of Azerbaijan's exports, and oil-related revenue makes up nearly 50 per cent of budget revenues. On December 29, 1999, Azeri President Heydar Aliyev issued a decree creating a State Oil Fund designed to use money obtained from oil-related foreign investment on education,

reducing poverty, and raising the living standards of the rural population in Azerbaijan. As of January 2003, the State Oil Fund reportedly holds $692 million, a $200 million dollar increase over last year.

Heydar Aliyev, who first ruled Azerbaijan between 1969 and 1982 as First General Secretary of the Central Committee of the Azerbaijan Communist Party, became President of the Republic of Azerbaijan in June 1993, making him the country's second president since independence. In his first year as President, a peace was negotiated with neighboring Armenia, which Azerbaijan had been fighting since 1988. As a result the 1994 cease fire, Azerbaijan lost approximately 16 per cent of its territory and many Azeris and Armenians were displaced. Lingering tensions between the two countries have led some analysts and investors to consider parts of the region unstable.

Oil

Azerbaijan's oil production (total liquids) averaged 318,000 barrels per day (bbl/d) in 2002, of which approximately 310,000 bbl/d was crude oil. This represents a 4 per cent increase over 2001 production, and builds upon five consecutive years of growth. At the same time, domestic petroleum consumption in Azerbaijan has fallen since independence, resulting in a growing margin for net petroleum exports. Azerbaijan exported approximately 178,000 bbl/d in 2002, most of which was routed to Russia, Turkey, and Italy.

Estimates of Azerbaijan's proven crude oil reserves range between 7 and 13 billion barrels according to industry journals and government sources. The country's largest hydrocarbon structures are located offshore in the Caspian Sea and account for most of the country's current petroleum production. The majority of Azerbaijan's oil output (64 per cent in 2001) comes from the State Oil Company of Azerbaijan (SOCAR).

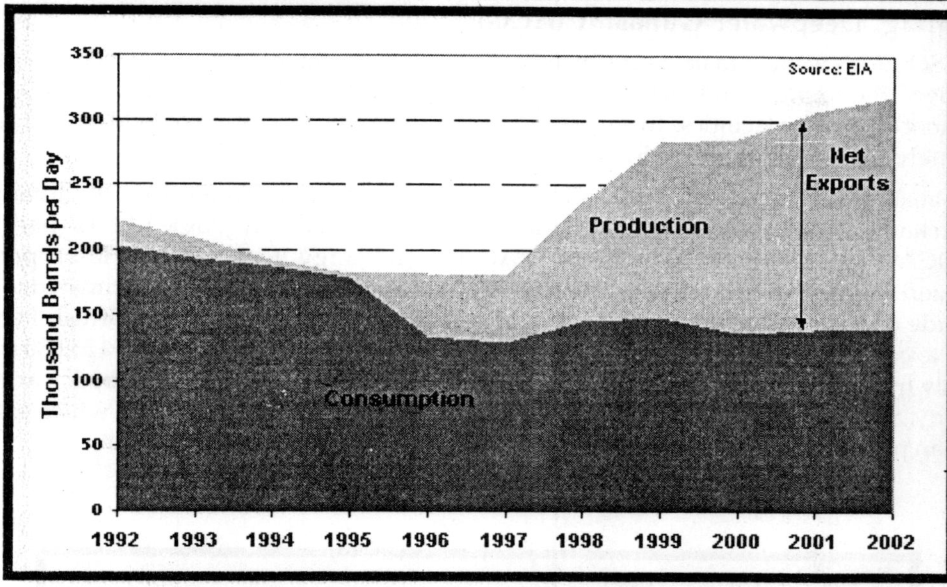

OIL PRODUCTION AND CONSUMPTION IN AZERBAIJAN, 1992-2002

SOCAR

SOCAR was established in September 1992 with the merger of Azerbaijan's two state oil companies, Azerineft and Azneftkimiya. SOCAR and its many subsidiaries are responsible for the production of oil and natural gas in Azerbaijan, for operation of the country's two refineries, for running the country's pipeline system, and for managing the country's oil and gas imports and exports. While government ministries handle exploration and production agreements with foreign companies, SOCAR is party to all of the international consortia developing new oil and gas projects in Azerbaijan. The company (not including subsidiaries) employs 75,000 people and is headed by Natik Aliyev (no relation to President Heydar Aliyev). In January 2003, Azeri President Aliyev issued a decree calling for the re-organization of SOCAR. According to SOCAR president Natik Aliyev, the re-organization should be complete by the end of 2003.

Most of SOCAR's production (65 per cent in 2001) is derived from the offshore field "shallow-water Gunashli," known in the Soviet era as the "28th of April Field," and located 60 miles off Azerbaijan's Absheron Peninsula. Shallow water Gunashli first came online in 1981, but was developed only to a maximum water depth of 400 feet owing to technological constraints. Although foreign firms have shown interest in developing shallow water Gunashli, no deals between SOCAR and its suitors have yet been struck, and press reports indicate that production levels are falling as equipment is in disrepair and the structure is losing reservoir pressure. In August 2002, SOCAR began efforts to rehabilitate shallow water Gunashli independently by adding a new production platform designed to enable 12 new wells. SOCAR also operates 40 other older fields (both on- and off-shore), many of which are considered to be in similar disrepair and have been artificially stimulated for years using water injection. Press reports indicate that inefficiencies owing to aging equipment and largely depleted reservoirs have caused the cost of production of SOCAR's onshore crude oil to reach $18 per barrel.

Azeri, Chirag, Deepwater Gunashli (ACG)

While SOCAR's Soviet-era fields are in decline, an influx of foreign investment since independence has revitalized the country's oil sector through the development of large scale new projects and the refurbishment of existing facilities. To date, Azerbaijan has signed over 20 major field agreements with approximately 30 companies from 15 countries.

Azerbaijan's increase in oil production since 1997 has mainly come from the international consortium known as the Azerbaijan International Operating Company (AIOC) AIOC (partners: BP, Unocal, SOCAR, Inpex, Statoil, ExxonMobil, TPAO, Devon Energy, Itochu, Delta/Hess) operates the offshore Azeri Chirag and deep water Gunashli (ACG) mega-structure, which is estimated to contain proven crude oil reserves of 5.4 billion barrels according to the field's operator, British Petroleum. According to consortium member Statoil, AIOC reached an average daily output of 125,000 barrels in 2002, mostly from the Chirag-1 stationary platform. This production has been dubbed "early oil" by the company, and is currently exported through the "Western Early Oil Pipeline", which extends to the Georgian port of Supsa on the Black Sea.

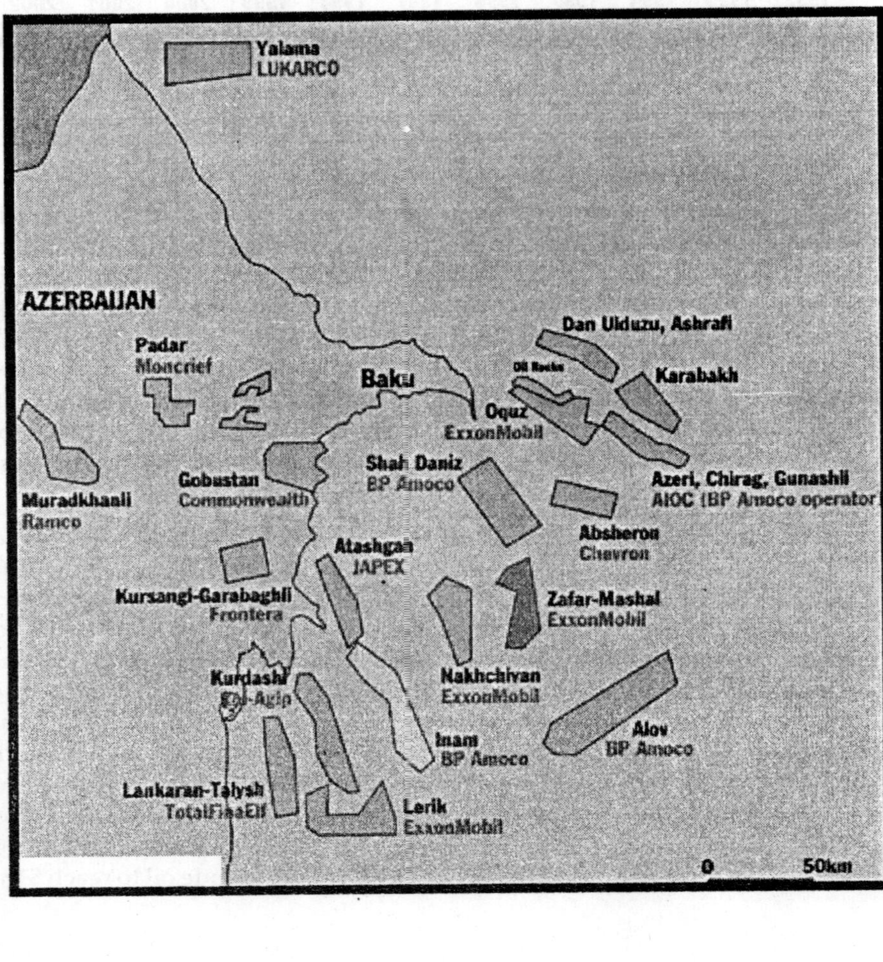

Azerbaijan's main production surge in the next decade is expected to come from the three-phase development of the ACG mega-structure. Production is slated to reach approximately 400,000 bbl/d by 2005 with the full implementation of the first phase. Later implementation of phase 2 is expected by AIOC to boost production to roughly 1 million bbl/d by 2008-09. By some industry estimates, Azerbaijan could be exporting 1 million bbl/d by 2010 (roughly equivalent to 2002 exports from Algeria). Plans for phase 3, which will complete "full field development", are currently being investigated.

Not all of Azerbaijan's foreign investment projects have been successful, with several projects announcing disappointing drilling results and several joint ventures (JVs) and production sharing agreements (PSAs) shutting down. In 2000, Azerbaijan decided to abolish JVs and convert them to the more investor friendly PSAs in an effort to spur increased development.

Oil Exports

Currently, Azerbaijan's only export pipelines are the Baku-Novorossiisk pipeline ("northern route"), which sends Azeri oil to the Russian Black Sea, and the Baku-Supsa pipeline ("western route"), which mainly carries AIOC's "early oil" from ACG to Georgia's Black Sea coast. Oil products such as lubricants also are exported by rail in tank wagons to Georgia's Black Sea ports.

Azerbaijan's options for increasing its oil exports depend to a large extent on the construction of new pipelines. Several oil export pipelines from the Caspian Sea region have been under consideration,

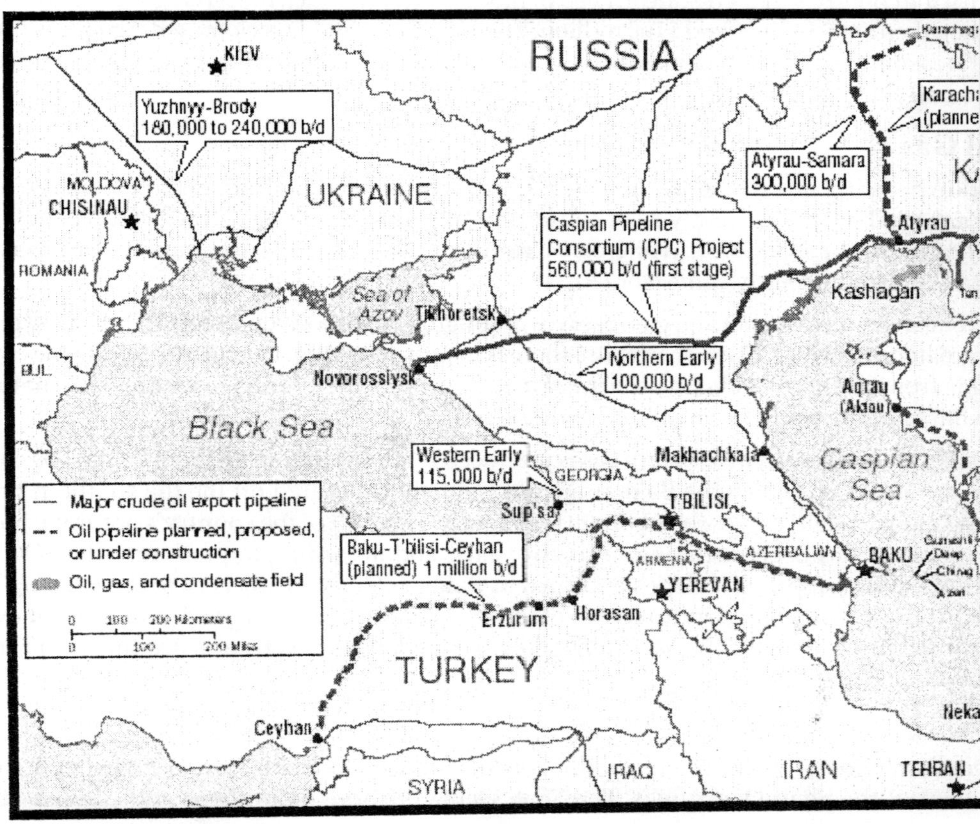

but Azerbaijan has not wavered in its support for the proposed Baku-Tibilisi-Ceyhan pipeline. This so-called "Main Export Pipeline" would export Azeri (and possibly Kazakh) oil along a 1,040-mile route from Baku via Georgia to the Turkish Mediterranean port of Ceyhan, allowing oil to bypass the increasingly crowded Bosporus Straits. Construction of the 1-million-bbl/d pipeline, which is estimated to cost $2.9 billion, is scheduled to be completed in 2004, with oil to begin flowing in 2005.

In addition, Iran, Russia, and Ukraine also have proposed alternative and supplemental oil export routes from Azerbaijan. The European Union's Transport Corridor Europe Caucasus Asia (TRACECA) program has been active in the region since 1993 upgrading and integrating rail and road transportation networks in an effort to promote trade ties with Europe along a West-East corridor. In Azerbaijan, the TRACECA program is providing logistical support to help Azerbaijan's aging rail system meet the country's growing oil transit needs.

Downstream/Refining

Azeri crude oil is refined domestically at two refineries: the Azerineftyag (Baku) refinery, with a capacity of 238,978 bbl/d, and the Azerneftyanajag (New Baku) refinery, which has a capacity of 202,830 bbl/d. With domestic production averaging 318,000 bbl/d in 2002 (and half of that exported as crude oil), Azerbaijan's refineries have been running well below capacity, with overall refinery utilization rates as low as 40 per cent. Heating oil accounts for roughly half of output at Azeri refineries, followed by diesel fuel, gasoline, motor oil, kerosene, and other products.

Both of the country's refineries are in need of modernization, which the Azeri government estimates will cost between $600 million and $700 million. The U.S. Trade and Development Agency is financing a $600,000 feasibility study to be conducted by ABB Lummus for upgrading the two refineries and the specialized oil port of Dubendi.

Natural Gas

Despite boasting one of the world's largest natural gas field discoveries of the last 20 years in Shah Deniz, Azerbaijan is currently a net natural gas importer. Azerbaijan produced 200 billion cubic feet (Bcf) of natural gas in 2001, while consuming roughly 240 Bcf. The majority of Azerbaijan's natural gas imports currently come from Russia.

Azerbaijan has proven natural gas reserves of roughly 30 trillion cubic feet (Tcf), and the potential for even greater reserves. However, because there is insufficient infrastructure to deliver the country's natural gas from offshore fields (the source of the majority of the country's production), natural gas has been flared off instead of being piped to markets.

Virtually all of Azerbaijan's natural gas is produced by SOCAR from offshore fields. The country's leading natural gas producer, the Bakhar oil and gas field, is located off the the southern tip of the Absheron Peninsula and currently accounts for slightly over half of the country's natural gas output. Output at Bakhar has been declining in recent years, and according to press reports, SOCAR has begun efforts to develop a new deposit, known as Bakhar-2 located adjacent to the existing Bakhar field. SOCAR reportedly has plans to route some of the Bakhar-2 natural gas production for export in the near future. In addition to SOCAR, AIOC and other international consortia produce small amounts of associated natural gas as well.

Shah Deniz

Azerbaijan's major natural gas production increases into the future are expected from the development of the Shah Deniz field. The Shah Deniz offshore natural gas and condensate field,

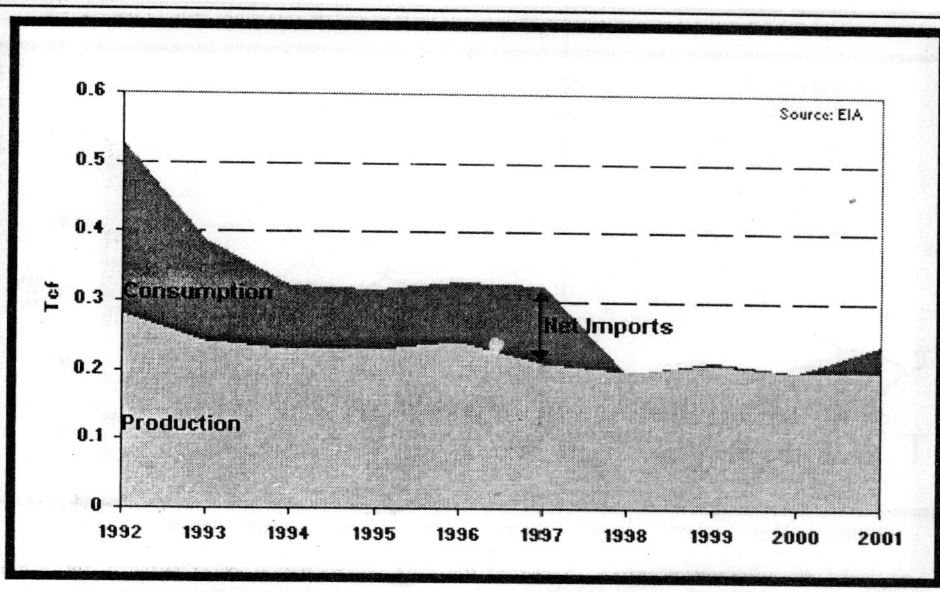

NATURAL GAS PRODUCTION AND CONSUMPTION IN AZERBAIJAN, 1992-2001

which is thought to be one of the world's largest natural gas field discoveries of the last 20 years, contains "potential recoverable resources" of roughly 14 Tcf (400 billion cubic meters) of natural gas according to the project's operator, British Petroleum. However, other industry and trade sources, employing widely different definitions of "reserves", estimate the field's size to be as high as 35 Tcf. Shah Deniz is located in the Caspian Sea, approximately 60 miles southeast of Baku, and is being developed by the Shah Deniz consortium (members: BP, Statoil, SOCAR, LukAgip, NICO, TotalFinaElf, and TPAO). The Shah Deniz production sharing agreement was agreed to in 1996.

The first phase of the Shah Deniz field's development was officially sanctioned on February 27, 2003 and has been estimated to cost over $3 billion. Phase one entails the installation of a new fixed offshore platform, two subsea pipelines to bring the hydrocarbons ashore, and a new onshore gas-processing terminal to be erected adjacent to the existing oil terminal at Sangachal, near Baku. Shah Deniz consortium members expect to begin producing natural gas for export in 2006. According to the partners, once the new infrastructure is in place, Shah Deniz will be capable of producing approximately 296 Bcf of natural gas (8.4 billion cubic meters) per year (roughly on par with 2001 natural gas production from Bahrain) as well as 40,000 bbl/d of condensate. Between 2006 and 2020, Shah Deniz is expected to add 110 million tons of oil equivalent to Azerbaijan's production. Natural gas from Shah Deniz, as well as associated gas from ACG and the Bakhar-2 project, is expected to make Azerbaijan self-sufficient in natural gas as well as to generate significant export revenue.

Natural Gas Exports

Azerbaijan currently lacks any infrastructure for the export of natural gas. However, efforts are currently underway to secure export routes and customers for gas deliveries beginning in 2006.

The main conduit for Azerbaijan's natural gas exports will be the "South Caucasus Pipeline," also known as "Baku-Tiblisi-Erzurum," which will run parallel to the Baku-Tiblisi-Ceyhan oil pipeline for most of its route before connecting to the Turkish infrastructure near the town of Horasan. The

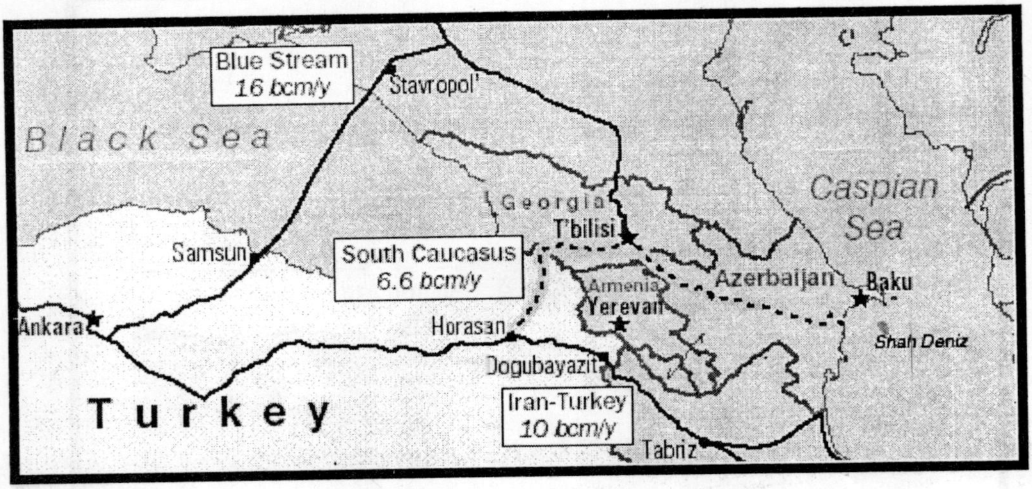

South Caucasus pipeline is designed to have a capacity of 233 Bcf (6.6 Bcm) per year, and is to be completed in time for the Shah Deniz project's first contracted exports in 2006.

On March 12, 2001, Azerbaijan signed its first major natural gas export deal when it concluded an agreement to supply Turkey with natural gas beginning in 2004. In February 2003 the deal was renegotiated owing to project delays, and now calls for exports to begin in 2006 at a rate of 71 Bcf (2 BCM) per year, increasing to reach an average rate of 222 Bcf (6.3 Bcm) per year in 2009. There have been some doubts about Turkey's ability to consume and transit the quantities of gas that the country has agreed to purchase, (for more on natural gas in Turkey see EIA's Turkey Country Analysis Brief). Besides Turkey, other countries which are expected to purchase Azerbaijan's natural gas exports are Georgia and Greece.

Coal

Azerbaijan has no significant coal deposits, nor any measurable domestic coal production or consumption.

Electricity

Azerbaijan's power sector has an installed generating capacity of approximately 5.1 gigawatts (GW), consisting of eight thermal plants (accounting for roughly 80 per cent of generating capacity) and six hydroelectric plants, all of which are owned by the state. Both electric generation and consumption have been relatively flat since independence, with generation totaling 18.2 billion kilowatthours (Bkwh) in 2001, and consumption of 16.6 Bkwh. Because of the country's inefficient distribution network, much of the country's generation is lost in transmission, making Azerbaijan a net electricity importer. In order to supply electricity to all parts of the country (including the Naxcivan exclave), Azerbaijan imports power from Russia, Turkey, Iran, and Georgia.

Built during the Soviet era, Azerbaijan's power infrastructure is in generally poor condition, with minimal public investment and maintenance since independence. The country's economic contraction during the mid-1990s, along with systemic problems–such as prices capped below market rates and frequent non-payment by customers–have left Azerbaijan's power sector without sufficient capital to upgrade aging power-generation facilities. The international donor community has undertaken several projects to restore and add new capacity to Azerbaijan's power sector, including a $53 million loan by the World Bank to complete the construction of the 4,000-MW Yenikand hydroelectric plant (completed in May 2000), and the European Bank for Reconstruction and Development's roughly $21 million loan (in conjunction with the Islamic Development Bank and the European Union) for the reconstruction of the 360-MW Mingechaur hydroelectric station on the Kura River (completed 2001).

In early 2003, Azerbaijan's newly completed Severnaya power plant began operation with the help of Japanese companies Mitsui and Mitsubishi. The 400-MW gas-fired unit will provide power mainly to the Absheron peninsula, and is expected to help raise Azeri electricity generation in 2003 by 6 per cent over 2002 levels. Press reports indicate that Azerenergy is considering numerous plans to develop the country's distribution network and increase its generation capacity.

While state electric company Azerenergy has a monopoly on power generation, the country's national electricity network is divided into five regional grids–Baku; Nakhchivan; North (Sumqayit); South (Ali Bayramli); and West (Ganja–each of which has been opened to foreign investors via open joint stock companies.

Country Overview

Independence: August 30, 1991 (from Soviet Union); National holiday: Independence Day, May 28

Population (7/02E): 7.8 million

Location: Southwestern Asia, bordering the Caspian Sea, between Iran and Russia

Size: 33,436 square miles (slightly smaller than Maine)

Major Cities: Baku (capital), Ganja, Mingechaur, Nakhchivan, Stepanakert, Sumqayit, Yevlakh

Languages (1995E): Azerbaijani (Azeri) 89 per cent, Russian 3 per cent, Armenian 2 per cent, other 6 per cent

Ethnic Groups (1998E): Azeri 90 per cent, Dagestani 3.2 per cent, Russian 2.5 per cent, Armenian 2 per cent (almost all Armenians live in the separatist Nagorno-Karabakh region), other 2.3 per cent

Religions (1995E): Muslim 93.4 per cent, Russian Orthodox 2.5 per cent, Armenian Orthodox 2.3 per cent, other 1.8 per cent. Note: Religious affiliation is still nominal in Azerbaijan; percentages for actual practicing adherents are much lower.

Economic Overview

Currency: Manat

Exchange Rate (5/21/03): U.S. $1 = 4,915 manats

Nominal Gross Domestic Product (GDP) (2002E): $6.1 billion; *(2003F):* $6.9 billion

Real GDP Growth Rate (2002E): 10.6 per cent ; *(2003E):* 9.2 per cent

Inflation Rate (Change in Consumer Prices, 2002E): 2.8 per cent ; *(2003F):* 3.7 per cent

Official Unemployment Rate (2002E): 1.1 per cent

Current Account Balance (2002E): -$642 million

Major Trading Partners: Turkey, Russia, Georgia, Italy, Iran, Ukraine, United Arab Emirates

Merchandise Exports (2002E): $1.9 billion; *(2003E)*: $2.1 billion

Merchandise Imports (2002E): $1.7 billion; *(2003F)*: $1.9 billion

Merchandise Trade Balance (2002E): $243 million; *(2003F)*: $147 million

Major Exports: Oil and natural gas (90 per cent), machinery, cotton, foodstuffs

Major Imports: Machinery and equipment, foodstuffs, metals, chemicals

Foreign Exchange Reserves (excluding gold) (2002E): $650 million

External Debt (2002E): $1.4 billion

Energy Overview

Proven Oil Reserves (1/1/03E): 7-13 billion barrels

Oil Production (2002E): 318,000 bbl/d (of which 310,000 bbl/d was crude oil)

Oil Consumption (2002E): 140,000 bbl/d

Net Oil Exports (2002E): 178,000 bbl/d

Crude Oil Refining Capacity (1/1/03E): 441,808 bbl/d

Natural Gas Reserves (1/1/03E): 30 trillion cubic feet (Tcf)

Natural Gas Production (2001E): 200 billion cubic feet (Bcf)

Natural Gas Consumption (2001E): 237 Bcf

Coal Production (2001E): none

Coal Consumption (2001E): minimal

Electricity Generation Capacity (2001E): 5.1 gigawatts

Electricity Generation (2001E): 18.2 billion kilowatt-hours (Bkwh)

Electricity Consumption (2001E): 16.6 Bkwh

Environmental Overview

Total Energy Consumption (2001E): 0.57 quadrillion Btu* (0.1 per cent of world total energy consumption)

Energy-Related Carbon Emissions (2001E): 9.14 million metric tons of carbon (0.1 per cent of world carbon emissions)

Per Capita Energy Consumption (2001E): 69.7 million Btu (vs U.S. value of 341.8 million Btu)

Per Capita Carbon Emissions (2001E): 1.13 metric tons of carbon (vs U.S. value of 5.51 metric tons of carbon)

Energy Intensity (2001E): 150,572 Btu/$1995 (vs U.S. value of 10,736 Btu/ $1995)**

Carbon Intensity (2001E): 2.43 metric tons of carbon/thousand $1995 (vs U.S. value of 0.17 metric tons/thousand $1995)

Fuel Share of Energy Consumption (2001E): Oil (53 per cent), Natural Gas (44 per cent), Hydroelectric (3 per cent)

Fuel Share of Carbon Emissions (2001E): Oil (61 per cent), Natural Gas (39 per cent)

Status in Climate Change Negotiations: Non-Annex I country under the United Nations Framework Convention on Climate Change (ratified May 16th, 1995). Ratified the Kyoto Protocol on September 28, 2000.

Major Environmental Issues: Local scientists consider the Abseron Yasaqligi (Absheron Peninsula) (including Baku and Sumqayit) and the Caspian Sea to be the most ecologically devastated area in the world because of severe air, water, and soil pollution; soil pollution results from the use of DDT as a pesticide and also from toxic defoliants used in the production of cotton.

Major International Environmental Agreements: A party to the Conventions on Biodiversity, Climate Change, Climate Change-Kyoto Protocol, Desertification, Endangered Species, Marine Dumping, Ozone Layer Protection. Has signed, but not ratified: none.

The total energy consumption statistic includes petroleum, dry natural gas, coal, net hydro, nuclear, geothermal, solar and wind electric power. The renewable energy consumption statistic is based on International Energy Agency (IEA) data and includes hydropower, solar, wind, tide, geothermal, solid biomass and animal products, biomass gas and liquids, industrial and municipal wastes. Sectoral shares of energy consumption and carbon emissions are also based on IEA data.

Energy Overview

Organization: State Oil Company of Azerbaijani Republic (SOCAR); Azerigaz (state natural gas distribution company); Azerenergo (state electric company)

Major Oil Ports: Baku

Oil Export Pipelines: Baku-Novorossiisk (via Russia; "early oil" northern route), Baku-Supsa (via Georgia; "early oil" western route)

Major Oil Refineries (Capacities 1/1/02E): Azerineftyag (Baku) (238,978 bbl/d), and Azernettyanajag (New Baku) (202,830 bbl/d)

Major Power Plants: Yenikand (4,000 megawatts, MW) (hydro), Azerbaijan Station near Mingechaur (2,100 megawatts, MW), Ali-Bayramli (1,100 MW)

9

Bahrain

Bahrain produces only a modest quantity of oil, but it is a growing regional services center strategically located on the Persian Gulf.

General Background

A small group of islands linked by causeway to Saudi Arabia, Bahrain recently became a constitutional monarchy. Since 1999, the country has been ruled by Sheikh Hamad bin Isa Al-Khalifa, who has committed to a process of gradual democratization. In February 2001, Sheikh Hamad held a referendum on his National Action Charter, which called for constitutional reforms. The Charter received the approval of 98 per cent of the electorate, and in February 2002, Sheikh Hamad declared himself king and Bahrain a constitutional monarchy. He also called for parliamentary elections in October. The new bicameral legislature (to be made-up of a lower chamber of elected representatives and an upper house of appointed legislators) was Bahrain's first since the National Assembly was dissolved in 1975. Though the elections occurred without irregularity, turnout was only 53 per cent due to the decision by Shi'ite groups to boycott the election. The king has responded to Sunni Islamic candidates' capture of most of the seats in the lower house by appointing secular MPs to the upper house.

Since 1948, the United States Navy has made Bahrain its headquarters for activity in the Persian Gulf. At present, its base occupies more than 60 acres of land in Manama, the country's capital. This base was the site of protests against the American operation in Iraq. In March 2003, a bomb shattered windows on the base, but none was hurt.

Already a regional financial services center specializing in offshore banking, Bahrain is intent on further diversifying its generally liberalized economy. For the time being, the country remains heavily dependent on the oil sector. Petroleum revenues account for about 60 per cent of government revenue and 65 per cent of exports. Hydrocarbons also provide the foundation for Bahrain's two major non-oil

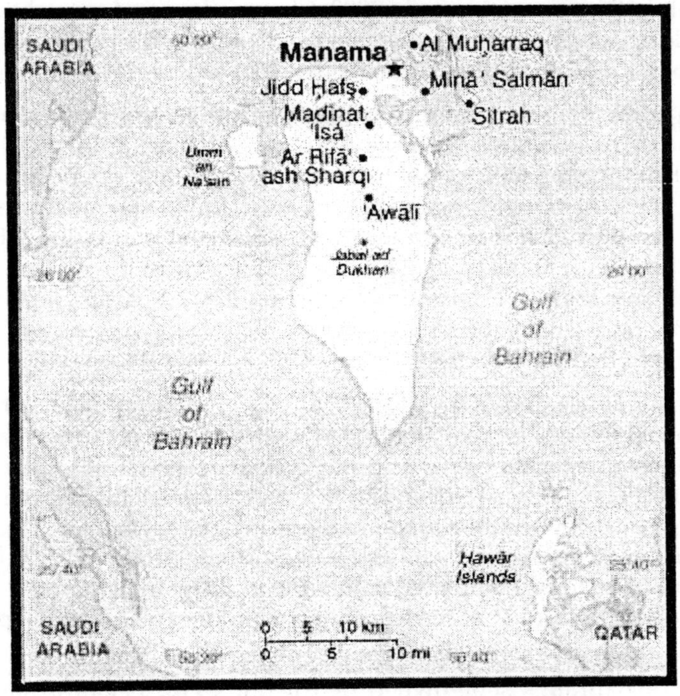

industries: refining and an aluminum smelter. Accordingly, oil prices dramatically affect Bahrain's economy. The relatively high oil prices of recent years have helped Bahrain's real gross domestic product (GDP) to grow at an estimated rate of 4.5 per cent in 2002 and 4.1 per cent in 2003. Growth for 2004 is forecasted at 3.8 per cent. These relatively high growth rates have not kept unemployment from becoming a growing economic and political problem. In December 2002, the new Minister of Social Affairs and Labor, Majid al-Alawi, stated that economic growth rates of 5-6 per cent would be required to reduce unemployment among Bahraini nationals.

Oil

All of Bahrain's 125 million barrels of onshore proven oil reserves are located in the Awali field, the first oil field to be developed in the Persian Gulf. Discovered in 1932, Awali's crude oil production peaked at more than 75,000 bbl/d in the 1970s, but has declined since then. In 2002, Awali produced only 35,000 barrels per day (bbl/d) of crude oil.

In addition to Awali, Bahrain also has potential oil reserves offshore in the Gulf of Bahrain. These areas became available for exploration and potential exploitation following the International Court of Justice's March 2001 resolution of a territorial dispute between Bahrain and Qatar over islands located between the two countries. The court awarded sovereignty over the Hawar Islands to Bahrain, allowing Qatar to retain the neighboring islands of Zubarah and Janan. Bahrain has since offered concessions to foreign investors. In November 2001, it awarded two blocks off the country's southeastern coast to Petronas (Malaysia) and another similarly located block to ChevronTexaco. Both companies have since begun to explore their concessions. In 2002, ChevronTexaco declared that it had failed to find commercially viable opportunities in another concession to the north and west of the mainland. Since then, the Calgary-based EnCana Corporation has also become involved in offshore exploration.

In addition to what is produced in its territory, Bahrain receives oil via subsea pipeline from Saudi Arabia. Some of this oil comes from the Abu Saafa offshore field, which lies in Saudi Arabian waters. As a way of supporting their neighbor's economy, the field's Saudi administrators donate all of its 140,000 bbl/d of production to Bahrain. In late 2002, Bahrain announced that the Saudi petroleum company Aramco would begin development work aimed at increasing Abu Saafa's production capacity to as much as 300,000 bbl/d by 2004. The remainder of the oil that Bahrain receives from Saudi Arabia comes from the oil fields around Dammam. The Abu Saafa pipeline passes through this area on its way to Bahrain. Unlike the Abu Saafa oil which it receives as a gift, Bahrain purchases the Dammam oil, albeit at a discounted rate.

Unlike other Gulf states, Bahrain exports refined petroleum products rather than crude. The country's only refinery, Sitra, is located south of Manama and has a current capacity of 248,900 bbl/d. Bahrain exports most of the Sitra refinery's products to India and the Far East. The Sitra refinery was built in 1936, and has been modernized several times. At present, the Bahrain Petroleum Company (Bapco) is pursuing a $600-$900 million modernization program, which will increase overall capacity. It will also introduce hydrocracking facilities to allow production of 40,000 bbl/d of low sulfphur diesel. This project is due to be completed sometime in 2004. In September 2002, the government authorized Bapco to initiate a study of the feasibility of building a $1.5 billion petrochemical complex at Sitra.

Consolidation of Bahrain's state-owned petroleum sector began in January 2000, when the upstream Bahrain National Oil Company (Banoco) began merging into Bapco. The merger was completed on June 1, 2002. The new entity, the Bahrain Petroleum Company BSC, is charged with the exploration, production, refining, marketing and distribution of Bahraini oil for domestic use and the international market.

Natural Gas

Bahrain has natural gas reserves of about 3.25 trillion cubic feet (Tcf), much of it associated gas from the Awali oil field. The country produced 314 billion cubic feet (Bcf) of natural gas in 2001, all of which was consumed locally. Gas production and processing are the responsibility of the majority state-owned Bahrain National Gas Company (Banagas). By 2001, the largest domestic consumer was Aluminium Bahrain (Alba), which uses it in its power plant. (In September 2003, the government signed an agreement with Alcoa allowing the American company to take up to a 26 per cent equity stake in the company.)

Bahrain's growing demand for electric power generation is likely to cause it to become a net natural gas importer. In January 2002, Qatar and Bahrain signed a Memorandum of Agreement indicating Bahrain's intentions to purchase natural gas from Qatar's North Field Enhanced Gas Utilization Project. The project was expected to come online in early 2003, but it now appears that it may be until 2005 or 2006 before Bahrain receives the 0.5–1 billion cubic feet per day that it is seeking.

Banagas has also pursued investment in liquefied natural gas projects. In 2002, it set up a joint-venture with the American firm Dynegy to invest in related opportunities in the Middle East and North Africa.

Electricity

The production and distribution of electricity in Bahrain is the responsibility of the Ministry of Electricity and Water. In 2001, Bahrain had an estimated electric generating capacity of 1.1 gigawatts (GW), and produced 5.8 billion kilowatt-hours (bkwh) of electricity. Installed capacity is barely able to meet current consumption, while the country's population expansion and economic development have led to 5 per cent annual growth in electricity demand. The government estimates that the country will require a new power plant every two to three years. Nevertheless, domestic power prices are expected to remain at the low level set by the government in 1992 in an attempt to help low-income groups. Like other Gulf states, Bahrain has made official appeals for voluntary conservation.

At present, Bahrain gets its power from two combined water and power production complexes and three smaller electricity-only plants. The recent focus has been on the modern Hidd complex. Phase One, which was completed in 1999, added 280 megawatts (MW). Phase Two, which is expected to add another 600-700 MW when completed in 2004, is being led by Alstom. The government is expected to announce its choice of the design consultant for Phase Three in fall 2003; as of late August 2003, the UK's Black & Veatch Consulting was the lowest bidder. In addition to the work being done at Hidd, the government awarded a contract in 2001 to Siemens Power Generation to extend the life of the Rifaa power plant. In order to meet long-term electricity needs, the government has declared that it will add a third major power and water plant by 2006. It has said this plant may be privately operated.

In addition to trying to increase supply, the Bahraini government has continued its attempts to improve the nation's transmission and distribution infrastructure. In December 2002, Bahrain awarded a contract to the Saudi Cable Co. for the supply and installation of power lines, optic cables, and other accessories.

While movement is likely to be slow, Bahrain is considering the privatization of the country's entire electricity sector. Along these lines, Bahrain retained Ernst and Young to act as financial consultants in June 2003.

Country Overview

Independence: August 15, 1971 (from United Kingdom)

Population (July 2003E): 667,238 (includes 238,108 non-nationals)

Location/Size: Persian Gulf, 257 square miles

Major Cities: Manama (capital)

Languages: Arabic, English, Farsi, Urdu

Ethnic Groups: Bahraini Arab (63 per cent), Asian (19 per cent), other Arab (10 per cent), Iranian (8 per cent)

Religion: Shi'a Muslim (70 per cent), Sunni Muslim (30 per cent)

Economic Overview

Currency: Bahraini Dinar

Fixed Exchange Rate: 1 Bahraini Dinar = $2.66 US (Fixed)

Nominal Gross Domestic Product (2002E): $8.4 billion *(2003E):* $9.0 billion *(2004F):* $9.5 billion

Real GDP Growth Rate (2002E): 4.5 per cent *(2003E):* 4.1 per cent *(2004E):* 3.8 per cent

Inflation Rate (consumer prices) (2002E): 0.7 per cent *(2003E):* 2.4 per cent *(2004F):* 2.2 per cent

Current Account Balance (2002E): -$151.2 million *(2003E):* $3.1 million *(2004F):* -$145.2 million

Major Export Products: petroleum and petroleum products, aluminum, textiles

Main Destinations of Exports (2000): India (8.4 per cent), US (3.9 per cent), Saudi Arabia (3.4 per cent), Japan (2.8 per cent), South Korea (2.1 per cent)

Major Import Products: crude oil, machinery, chemicals

Main Origins of Imports (2000): Saudi Arabia (28.7 per cent), US (12.5 per cent), UK (6.6 per cent), France (6 per cent), Japan (4 per cent)

External Debt (2002): $3.7 billion

Energy Overview

Proven Oil Reserves (1/1/03E): 125 million barrels

Oil Production (2002E): 43,000 bbl/d (of which 35,000 bbl/d was crude oil)

Oil Consumption (2002E): 34,000 bbl/d

Net Oil Exports (2002E): 9,000 bbl/d

Crude Oil Refining Capacity (1/1/03E): 248,900 bbl/d

Natural Gas Reserves (1/1/03E): 3.25 trillion cubic feet (Tcf)

Natural Gas Production (2001E): 314 billion cubic feet (Bcf)

Natural Gas Consumption (2001E): 314 Bcf

Electric Generation Capacity (1/1/2001E): 1.1 gigawatts (GW)

Electricity Production (2001E): 5.8 billion kilowatt-hours (bkwh)

Environmental Overview

Total Energy Consumption (2001E): 0.4 quadrillion Btu* (0.1 per cent of world total energy consumption)

Energy-Related Carbon Emissions (2001E): 6.2 million metric tons of carbon (0.1 per cent of world carbon emissions)

Per Capita Energy Consumption (2001E): 605.0 million Btu (vs. U.S. value of 341.8 million Btu)

Per Capita Carbon Emissions (2001E): 9.5 metric tons of carbon (vs. U.S. value of 5.5 metric tons of carbon)

Energy Intensity (2001E-PPP): 41,845 Btu/ $1995 (vs U.S. value of 10,805.4 Btu/ $1995)**

Carbon Intensity (2001E-PPP): 0.66 metric tons of carbon/thousand $1995 (vs U.S. value of 0.17 metric tons/thousand $1995)

Fuel Share of Energy Consumption (2001E): Natural Gas (85.0 per cent), Oil (15.0 per cent)

Fuel Share of Carbon Emissions (2001E): Natural Gas (77 per cent), Oil (23 per cent)

Status in Climate Change Negotiations: Non-Annex I country under the United Nations Framework Convention on Climate Change (ratified December 28, 1994). Not a signatory to the Kyoto Protocol.

Major Environmental Issues: Rapid industrialization during the 1970s and 1980s has placed increased stress on the natural environment. Limited natural fresh water resources are increasing dependence on large-scale desalination facilities. Rising sea levels also pose a long term threat.

The total energy consumption statistic includes petroleum, dry natural gas, coal, net hydro, nuclear, geothermal, solar, wind, wood and waste electric power. Fuel shares of energy consumption and carbon emissions are also based on EIA data.

Oil and Gas Industries

Organization: Bahrain Petroleum Company BSC; Bahrain Natural Gas Company (Banagas)

Major Foreign Energy Company Involvement: ChevronTexaco, Petronas, EnCana Corporation

Major Ports: Manama (Another port is under construction. The Sheikh Khalifa bin Salman Port is in the Hidd region. Begun in January 2002, it should be completed in January 2005.)

Producing Oil Fields (crude production–bbl/d)(2002E): Awali (35,000 bbl/d)

Major Refineries (capacity–bbl/d): Sitra (248,900)

10

Balkans Region

The Balkans region–here including Slovenia, Croatia, Bosnia-Herzegovina, Yugoslavia, the Former Yugoslav Republic of Macedonia (F.Y.R.O.M.), and Albania–is recovering from a decade of conflict and instability as a result of the dissolution of the former Yugoslavia in the early 1990s. As the region stabilizes, the Balkans may play an important role as a transit center for Russian and Caspian Sea region oil exports.

General Background

Following the rejection of communism in Eastern Europe in 1989 and the collapse of the Soviet Union in 1991, the Balkans region descended into internecine warfare in 1992. Yugoslavia, which was comprised of six constituent republics, was torn asunder as Slovenia, Croatia, Bosnia-Herzegovina, and Macedonia (officially known as the Former Yugoslav Republic of Macedonia, F.Y.R.O.M.) declared their independence, leading to three years of ethnic warfare and leaving a rump Yugoslav state of Serbia and Montenegro.

The war in Bosnia-Herzegovina, the site of the region's worst fighting, ended with the signing of the Dayton Peace Accords in November 1995, but hostilities erupted between Serbs and ethnic Albanians in Kosovo, an autonomous province of Serbia, in 1998. As a result, the North Atlantic Treaty Organization (NATO) systematically bombed Yugoslav targets for 78 days in the spring of 1999. In October 2000, Yugoslav President Slobodan Milosevic, whose policies many people blamed for the war in Bosnia and the conflict in Kosovo, was ousted in a popular rebellion after refusing to acknowledge his defeat in presidential elections the previous month. In June 2001, the Serbian government handed Milosevic over to the United National War Crimes Tribunal in The Hague, where he is currently imprisoned and standing trial.

Although Albania remained outside the hostilities that engulfed Yugoslavia in the 1990s, the country has suffered through its own problems over the course of the past decade. Following the demise of the country's isolationist communist regime in 1991, Albania, which was far less developed

The Balkan States
The main political divisions of the former Yugoslavia.

economically than any part of the former Yugoslavia, suffered through a period of government power struggles and economic downturns that damaged the country's already-fragile economy. After the collapse of a government-promoted investment program in 1997, millions of Albanians lost their savings, and the country slipped into near anarchy from the resulting riots and looting before order was restored.

Table 10.1: Economic and Demographic Indicators for the Balkans Region

Country	*Gross Domestic Product (Nominal GDP), 2001E (Billions of U.S. $–PPP*)*	*Real GDP Growth Rate, 2001 Estimate*	*Real GDP Growth Rate, 2003 Projection*	*Per Capita GDP, 2001E*	*Population 2001E (Millions)*
Bosnia-Herzegovina	$10.4	2.3%	3.3%	$2,422	4.3
Croatia	$20.3	4.1%	4.7%	$4,570	4.4
F.Y.R.O.M.	$7.7	-4.1%	1.6%	$3,766	2.0
Slovenia	$30.5	3.0%	3.5%	$15,319	2.0
Yugoslavia	$35.5	6.2%	6.2%	$3,331	10.7
Subtotal/weighted average	$104.4	3.7%	4.5%	$3,896	26.8
Albania	$7.3	6.5%	6.0%	$2,134	3.4
Total/weighted average	**$111.7**	**3.9%**	**4.6%**	**$3,699**	**30.2**

Following this decade of warfare, political turmoil, and economic contraction, the Balkans region has entered a period of recovery, with hopes that the decade ahead will be more peaceful, stable, and prosperous. A rebellion by ethnic Albanians in Macedonia in the spring of 2001 renewed fears of civil war in that country, but a NATO-brokered cease-fire has now been in place for over a year. In addition, plans by Montenegrin leaders to declare independence from Serb-dominated Yugoslavia have been shelved as the two republics reached an agreement on a power-sharing federal arrangement. A conference of Western nations has pledged to donate $1.3 billion for the rebuilding of Yugoslavia, including at least $100 million from the United States, and Albania has posted four consecutive years of economic growth of more than 6 per cent. Slovenia, the wealthiest country in the Balkans region, is slated to join the European Union (EU) when it expands in 2004.

Regional Energy Issues

The countries of the Balkans region are neither major energy producers or consumers. Although the region does hold some important fossil fuel deposits, these resources are not significant on a world scale, and the political and economic instability in the Balkans during the past 11 years has discouraged any substantial foreign investment in the respective countries' energy sectors. Rather, the region is becoming more important as a transit center for Russian and Caspian Sea region oil exports to Western consumers.

Table 10.2: Energy Consumption and Carbon Dioxide Emissions in the Balkans Region, 2000

Country	*Total Energy Consumption (Quadrillion Btu)*	*Petro-leum*	*Natural Gas*	*Coal*	*Nuclear*	*Hydro-Electric*	*Other Electricity*	*Net Electricity Imports*	*Carbon Dioxide Emissions (Million metric tons of carbon)*
Bosnia-Herzegovina	0.09	44.4%	11.1%	11.1%	0%	22.2%	0%	11.2%	1.16
Croatia	0.41	46.3%	24.4%	7.3%	0%	14.6%	0%	7.4%	5.74
F.Y.R.O.M.	0.13	30.8%	2.7%	53.8%	0%	7.7%	0%	5%	2.50
Slovenia	0.28	39.3%	14.3%	21.4%	17.9%	14.3%	0.1%	-7.3%	4.19
Yugoslavia	0.59	20.3%	3.4%	52.5%	0%	23.7%	0%	0.1%	10.21
Subtotal/ weighted average	**1.5**	33.3%	11.6%	32%	3.3%	18%	0.02%	1.8%	23.8
Albania	0.08	25%	0.9%	0%	0%	62.5%	0%	11.6%	0.39
Total/weighted average 24.19	**1.58**	32.9%	11%	30.4%	3.2%	20.2%	0.02%	2.3%	

Note: Percentages may not add up to 100 per cent due to rounding.

Oil Transit

Albania-Macedonia-Bulgaria Pipeline

Perhaps the most important oil pipeline project in development in the Balkans region is a proposed 750,000-barrel-per-day (bbl/d) pipeline connecting the Bulgarian Black Sea port of Burgas with the

Albanian Adriatic port of Vlore. The Albanian-Macedonian-Bulgarian Oil (AMBO) Pipeline Corporation has been established with exclusive rights to construct the pipeline, which is estimated to cost between $850 million and $1.1 billion.

Table 10.3: Energy Supply Indicators, Balkans Region

Country	*Proven Crude Oil Reserves, 1/1/02E (Million Barrels)*	*Natural Gas Reserves, 1/1/02E (Billion Cubic Feet)*	*Coal Reserves, 1/1/01E (Million Short Tons)*	*Petroleum Production, 2001E (Thousand Barrels Per Day)*	*Natural Gas Production, 2000E (Billion Cubic Feet)*	*Coal Production, 2000E (Million Short Tons)*	*Electric Generating Capacity, 2000E (Gigawatts)*	*Crude Oil Refining Capacity, 1/1/02E (Thousand Barrels Per Day)*
Bosnia-Herzegovina	minimal	minimal	minimal	minimal	minimal	1.8	3.94	0
Croatia	92.2	1,237	43	27.2	58.6	0.02	3.82	253
F.Y.R.O.M.	minimal	minimal	N.A.	minimal	minimal	7.8	1.44	51
Slovenia	minimal	minimal	303	minimal	minimal	4.9	2.64	14
Yugoslavia	77.5	1,700	17,919	19.2	19.2	37.8	10.83	158
Subtotal	**169.7**	**2,937**	**18,265**	**46.3**	**77.8**	**52.32**	**22.67**	**476**
Albania	165	100	minimal	7	0.7	0.05	1.68	26
Total	**334.7**	**3,037**	**18,265**	**53.3**	**78.5**	**52.37**	**24.35**	**502**

The AMBO pipeline, which would run via Macedonia, would allow oil that is being shipped across the Black Sea to be piped across the Balkan peninsula to the Adriatic Sea, thereby reducing tanker traffic through the congested Bosporus Straits and the Aegean Sea. With oil exports from the Caspian Sea region projected to increase rapidly in the next decade, the AMBO pipeline proposal is one of several "Bosporus bypass" oil pipeline proposals that are currently under consideration or in development.

The AMBO pipeline proposal has received letters of acceptance from the governments of Albania, Bulgaria, and Macedonia, and a $980,000 feasibility study, partially funded by the U.S. Trade and Development Agency, concluded that the 560-mile pipeline project was feasible. AMBO began to assemble financing for the project in June 2001, with construction originally scheduled to start in 2001 and completion by 2005. However, luring foreign investment to the troubled region has been difficult, and ethnic violence that erupted in Macedonia in February 2001 near the proposed route further hindered efforts to fund the pipeline's construction. Once the financing is in place, construction is expected to take three years.

Constanta-Trieste/SEEL Pipeline

The Balkans region also may play a transit role in two additional proposed oil pipelines designed to bypass the Bosporus Straits. Romanian government officials have advocated that a pipeline to transport crude oil from the Caspian Sea to European markets pass through its territory, claiming that Romania, which has sought to develop its infrastructure to increase its chances of sharing in the Caspian oil bonanza, offers the shortest route, best refining technology, and links via waterways to major ports in the West. The proposed 660,000-bbl/d Constanta-Trieste pipeline would allow crude oil from Kazakhstan to be shipped via the Novorossiisk port on the Black Sea to the Romanian port of Constanta, where it would then be piped to Italy across the Balkan Peninsula.

The pipeline, estimated to cost $900 million to construct, would be used mostly to provide oil to the countries along the route, and would incorporate existing pipelines connecting Constanta with 10 refineries. Several alternatives exist for the route, with a proposed northern route transiting southern Hungary and central Slovenia before terminating at Italy's oil terminal of Trieste. The southern route for the pipeline, sometimes known as the South-East European Line (SEEL), would transport Caspian oil from Constanta via a similar route as the northern route, but instead would pass through Yugoslavia and an intermediate transit point at Croatia's Adriatic port of Omisalj before crossing Slovenia and ending at Trieste.

Feasibility studies have shown that both proposed Constanta routes are viable, but neither pipeline has moved forward as potential investors await a political accord providing security guarantees for the lines. Representatives of Romanian, Yugoslav, and Croatian oil companies have agreed that an inter-governmental accord likely would boost the pipeline's prospects and help to secure financial resources to construct pipeline, which would provide a significant amount of revenue in the form of transit tariffs.

Druzhba-Adria Integration

Another key project is the reversal of the 400,000 bbl/d-capacity Croatian Adriatic Oil (Adria) Pipeline and the integration of the Adria pipeline with the southern Druzhba pipeline. Currently, oil that arrives by tanker at the Croatian Adriatic port of Omisalj is piped to Sisak in the interior of the country. At Sisak, the Adria pipeline splits, with one branch going to Hungary and the other branch going to Yugoslavia (Vojvodina), touching the border with Bosnia at the refinery at Srpski (Bosanski) Brod, before heading on to a connection with a pipeline to Novi Sad.October 2000, Russian oil major Yukos announced plans to integrate the Druzhba southern pipeline, which runs through Belarus, Ukraine, Slovakia, and Hungary, with the Adria pipeline. Yukos signed a $20 million agreement with Croatian oil transport company Jadranski Naftovod (JANAF) to modernize the Adria pipeline to help integrate the two pipelines. By reversing the flows of the 110-mile pipeline between Omisalj and Sisak, the integration of the Druzhba and Adria pipelines will allow Russian oil exports to flow directly to the deepwater port of Omisalj on the Adriatic Sea, thereby bypassing the Bosporus Straits. Omisalj can accommodate tankers of up to 500,000 dead weight tons.

According to Yukos, Russian Urals Blend crude oil, which includes a mixture of Caspian-origin oils, should be flowing the 1,987-mile route to the deepwater Omisalj port by the end of 2002. The entire Druzhba-Adria pipeline route would handle 100,000 bbl/d in 2003, the first full year of operation. Transneft and JANAF have said that exports via the pipeline would rise to 200,000 bbl/d after five years, and to 300,000 bbl/d after 10 years.

Oil

The Balkans region has total proven oil reserves of approximately 335 million barrels, with 165 million barrels of that located in Albania. Croatia has proven reserves of 92 million barrels, while Yugoslavia holds the remaining 77 million barrels. Although a number of companies are active in the region and exploration is picking up, especially in Albania and offshore areas in the Adriatic Sea, oil production in the Balkans remains low: the entire region produced a total of 53,300 bbl/d in 2001.

Croatia led the region with 27,200 bbl/d, followed by Yugoslavia (19,200 bbl/d) and Albania (7,000 bbl/d).

As a result, each of the countries in the Balkans remains a net oil importer. Overall oil consumption in the Balkans in 2000 was 247,000 bbl/d, leaving the region dependent on imports for nearly 80 per cent of its oil needs. Slovenia, which, along with Yugoslavia, consumed 55,000 bbl/d in 2000, is entirely reliant on imports, as are Bosnia-Herzegovina and Macedonia. Most of the region's oil imports come overland from Russia and from tankers at Adriatic ports, but in July 2002, the Skopje-Thessaloniki oil pipeline was completed, allowing Macedonia to import oil via Greece. The 133-mile, $110 million pipeline has the capacity to deliver 50,200 bbl/d to Macedonia, which is nearly triple Macedonia's current oil supply needs. Yugoslavia and Bosnia-Herzegovina, which recently settled a dispute with Croatia over an oil products ban, may look to impo. the additional oil from the pipeline.

Natural Gas

Total natural gas reserves in the Balkans region are just over 3 trillion cubic feet (Tcf). Yugoslavia, with 1.7 Tcf in proven natural gas reserves, holds nearly 60 per cent of the region's total, while Croatia contains a further 1.2 Tcf and Albania has just 100 billion cubic feet (Bcf). In 2000, Croatia produced 58.6 Bcf of natural gas, while Yugoslavia's natural gas output stood at 19.2 Bcf. Albania, the only other natural-gas producing country in the region, produced just 0.7 Bcf to fulfill its domestic demand.

Overall, the Balkans region consumed 169 Bcf of natural gas in 2000, meaning that just over 50 per cent of the region's natural gas came from imports. Croatia, the region's largest consumer of natural gas (88 Bcf), imported its 40 Bcf supply shortfall from Russia via Austria, while Slovenia, which consumed 36 Bcf and is fully dependent on imports, received 23 Bcf from Russia as well as 13 Bcf from Algeria (via Italy). Both Macedonia (4 Bcf) and Bosnia-Herzegovina (11 Bcf) received all of their natural gas supplies from imports, while Yugoslavia supplemented its own production by re-starting imports from Russia. Under former President Slobodan Milosevic, from 1994 to 2000 Yugoslavia ran up a $255 million natural gas debt to Gazprom, Russia's natural gas monopoly, but since 2000 the country has been paying for its supplies regularly.

Coal

Yugoslavia holds the bulk of the Balkans' coal reserves, with 17,919 million short tons (Mmst) of the region's 18,265 Mmst, according to EIA estimates. Yugoslavia's coal reserves are found in five basins: Kostolac, Kolubara, Kosovo, Metohija, and Pljevlja. Serbian officials believe that nearly one-third of Yugoslavia's coal resources are located in Kosovo. Macedonia also is thought to hold large coal deposits, with some estimates stating the country's reserves as high as 1,000 Mmst. Slovenia has 303 Mmst in proven coal reserves as well. Bosnia-Herzegovina, Croatia, and Albania have minimal coal reserves and correspondingly low rates of coal production.

In 2000, the Balkans region as a whole produced 52.5 Mmst of coal. Yugoslavia accounted for 37.8 Mmst of that total, while Macedonia (7.8 Mmst) and Slovenia (4.9 Mmst) made up much of the remainder. Coal consumption in the Balkans region reached 54.5 Mmst in 2000, with Yugoslavia (38 Mmst) and Macedonia (8.2 Mmst) the top two consumers. Coal accounts for over 50 per cent of total energy consumption in both Yugoslavia and Macedonia whereas it makes up less than 12 per cent of the total energy consumption in Bosnia-Herzegovina, Croatia, and Albania. Slovenia uses coal for approximately one-fifth of its energy consumption needs.

Electricity

The Balkans region had a total installed electricity-generating capacity of 24.35 gigawatts (GW) in 2000. Yugoslavia, with 10.83 GW, holds over 40 per cent of the region's installed capacity, although much of its power infrastructure was damaged during the Kosovo conflict in 1999. Similarly, many of the region's electricity grids, power lines, and power plants were damaged by warfare in the 1990s, with the notable exception of Slovenia. Aside from Slovenia, an overall lack of investment, combined with low levels of customer payments (and, in the case of Yugoslavia, international economic sanctions that only ended in the fall of 2000), has added to the deterioration of the power sector in the Balkans region.

The aggregate effect has been that the region barely produces enough power for its needs. Bosnia-Herzegovina, Macedonia, Slovenia, and Yugoslavia generated more electricity than they consumed in 2000, but Croatia and Albania imported additional power to meet their domestic needs. Yugoslavia generated nearly half of the region's electricity output, with 33 billion kilowatt-hours (Bkwh) of power produced in 2000. Overall, the Balkans region produced 70.14 Bkwh of electricity in 2000 while consuming 68.6 Bkwh.

Prior to its dissolution in the early 1990s, the former Yugoslavia had a single electricity grid, the Union for the Coordination of the Transmission of Electricity (UCTE) network. UCTE was connected to the Western European power grid. Only Croatia, Slovenia, and the Federation of Bosnia and Herzegovina (along with the Republika Srpska, one of two autonomous entities that make up Bosnia-Herzegovina) have been reconnected to the UCTE. In June 2001, energy ministers from Albania, Bosnia-Herzegovina, Bulgaria, Greece, Macedonia, and Romania signed a memorandum for the creation of a competitive energy market in the Balkans. In addition, the Regional Association of Energy Regulators (ERRA) was established in December 2000, in order to create a common power market in southeastern Europe and the former Soviet Union.

Nuclear

Slovenia has the region's sole nuclear power plant (NPP) at Krsko. The plant, which has a 675-megawatt (MW) capacity, was built jointly with the Croatian Republic when both countries were still part of Yugoslavia. Construction on the Krsko NPP began in 1975, and the plant was connected to Yugoslavia's electricity grid in 1981, with commercial operations commencing in 1983. Nuclear power from the Krsko NPP supplied nearly 18 per cent of Slovenia's total energy demand in 2000.

Since Slovenia and Croatia became independent countries, however, ownership and responsibility for the nuclear power plant have been contentious issues between the two countries. In the mid-1990s, Slovenia made the Krsko NPP into a public company, and then the country stopped supplying Croatia with power from the plant in 1998. In June 2001, the countries resolved the dispute by making Croatia co-owner of the plant, including partial responsibility for the nuclear waste produced. Croatia was to begin receiving electricity from the plant in July 2002.

Nevertheless, on April 29, 2002, Slovene Agriculture Minister Franci But proposed that Slovenia withdraw from the deal with Croatia, saying that the agreement did not provide for nuclear waste management on an equal basis. In July 2002, Croatian officials demanded that Elektro-Slovenija, the Slovene electric company, pay nearly $70 million for a bill for electricity that was never transmitted from Krsko to Croatia.

Hydroelectric

With a lack of significant indigenous fossil fuel resources, the countries of the Balkans region have turned to hydroelectricity to meet their power needs. Fully 40 per cent of all installed electricity-generating capacity in the region is hydroelectric, and over 20 per cent of the region's total energy consumption is from hydropower. Albania, in particular, is extremely dependent on hydropower, as it accounts for 86 per cent of the country's installed generating capacity, and hydroelectricity made up 62.5 per cent of Albania's total energy consumption in 2000. Croatia (54 per cent of installed capacity) and Bosnia-Herzegovina (50 per cent of installed capacity) are also highly reliant on hydropower.

11

Baltic Sea Region

The Baltic states–including Estonia, Latvia, and Lithuania–occupy a strategic location as transit centers for Russia's northern oil exports. In addition, Belarus is a major transit center for Russian natural gas exports to Europe.

General Background

Alone among the former Soviet republics, the Baltic Sea region states of Estonia, Latvia, and Lithuania were quick to adopt market economies and to implement democratic reforms. As a result, they largely have avoided the economic and political crises that have beset other regions in transition from a centrally planned economy, including the Balkan region and southeastern Europe. Privatization in the Baltics is nearly complete, and in 2001, despite the downturn in the global economy, the three countries posted an average 6.3 per cent increase in their real gross domestic product (GDP).

Table 11.1: Economic and Demographic Indicators for Selected Baltic Sea Region Countries

Country	*Gross Domestic Product (Nominal GDP), 2001E (Billions of U.S. $ Market Exchange Rate)*	*Real GDP Growth Rate, 2001 Estimate*	*Real GDP Growth Rate, 2002 Projection*	*Per Capita GDP, 2001E (Market Exchange Rate)*	*Population 2002E (Millions)*
Estonia	$5.4	5.4%	5.3%	$4,000	1.4
Latvia	$7.5	7.6%	4.8%	$3,190	2.3
Lithuania	$12.0	5.9%	5.9%	$3,444	3.5
Subtotal/weighted average	**$24.9**	**6.3%**	**5.4%**	**$3,458**	**7.2**
Belarus	$12.2	4.1%	4.6%	$1,228	9.9
Total/weighted average	**$37.1**	**5.6%**	**5.2%**	**$2,170**	**17.1**

With a combined population of only 7.2 million people, Estonia, Latvia, and Lithuania have achieved greater presence in the international community by joining forces in a number of political and economic arenas. In November 2002, Estonia, Latvia, and Lithuania received invitations to join the North Atlantic Treaty Organization (NATO) in 2004. In addition, efforts to integrate with European institutions will soon pay off as the Baltic states are expected to join the European Union (EU) in mid-2004. Membership in NATO and the EU has been a stated foreign policy goal in each of the three countries since they became independent.

Belarus, on the other hand, has refused to implement political and economic reforms since the collapse of the Soviet Union. Rather than integrating with Europe, Belarussian President Alyaksandr Lukashenko has isolated the country from the West following his election in 1994 by clamping down on the political opposition and by returning the Belarussian economy to a form of market socialism. Belarus continues to suffer from 60 per cent inflation, and although the official unemployment rate is approximately 2.3 per cent, the actual figure is believed to be much higher. Nevertheless, the country's real GDP grew by 4.1 per cent in 2001, with a projected 4.6 per cent increase in 2002.

Regional Energy Issues

With little in the way of indigenous natural resources, Estonia, Latvia, Lithuania, and Belarus are dependent on energy imports in order to satisfy their domestic energy needs. Just as during Soviet times, the Baltic states and Belarus today are heavily reliant on Russian oil and natural gas, although Estonia, Latvia, and Lithuania have made efforts to diversify their energy suppliers. Belarus has maintained a closer relationship with Russia, and as such it has benefited by receiving cheaper energy supplies from its eastern neighbor, making Belarus even more dependent on Russia.

Although the Baltic states are not important energy consumers or producers, together they occupy a key transit location for Russian oil exports. Belarus is an important transit country for Russian natural gas exports to Europe, and Belarus also hosts a section of the Druzhba oil export pipeline originating in Russia. As the Baltic states seek to coordinate their energy sector legislation with the EU, energy sector privatization and energy market liberalization are becoming important regional issues as well.

Table 11.2: Energy Consumption and Carbon Dioxide Emissions in Selected Baltic Sea Region Countries, 2000

Country	*Total Energy Consumption (Quadrillion Btu)*	*Petro-leum*	*Natural Gas*	*Coal*	*Nuclear*	*Hydro-Electric*	*Other Electricity*	*Net Electricity Imports*	*Carbon Dioxide Emissions (Million metric tons of carbon)*
Estonia	0.09	55.5%	44.4%	11.1%	0%	0.04%	0.1%	-11.1%	1.89
Latvia	0.16	31.3%	37.5%	1.9%	0%	12.5%	0%	16.8%	1.92
Lithuania	0.27	40.7%	33.3%	1.5%	33.3%	1.1%	0%	-9.9%	3.56
Subtotal/ weighted average	**0.52**	**40.4%**	**36.5%**	**3.3%**	**17.3%**	**4.4%**	**0.02%**	**-4.2%**	**7.37**
Belarus	1.08	28.7%	66.6%	1.9%	0%	0.02%	0.1%	2.7%	16.47
Total/ weighted average	**1.6**	**32.5%**	**56.9%**	**2.3%**	**5.6%**	**1.5%**	**0.07%**	**-1.5%**	**23.84**

Note: percentages may not add up to 100 per cent due to rounding.

Oil

Estonia has no proven crude oil reserves, but *polevkivi* (oil shale) is abundant in the northeastern part of the country. Oil shale provides over 75 per cent of Estonia's total energy supply, making Estonia the only country in the world where oil shale is the primary source of energy. Oil shale is produced by majority state-owned Eesti Polevkivi (Estonian Oil Shale) near Kohtla-Jarve. Oil shale is consumed for power generation by the Eesti Energia and Kohtla-Jarve Soojus electric companies and for shale-to-oil processing by Kiviter AS, which processes the oil shale to produce about 4,400 barrels per day (bbl/d) of distillate liquid fuels. Estonia's indigenous oil shale production, however, is not sufficient to meet the country's demand for oil, which, at 25,000 bbl/d in 2001, has remained relatively steady throughout the past decade of transition. Estonia has no refineries, so it must import all petroleum products, either by rail or by pipeline.

Table 11.3: Energy Supply Indicators, Selected Baltic Sea Region Countries

Country	*Proven Crude Oil Reserves, 1/1/02 (Million Barrels)*	*Natural Gas Reserves, 1/1/02 (Billion Cubic Feet)*	*Coal Reserves, 1/1/01 (Million Short Tons)*	*Petroleum Production, 2001 (Thousand Barrels Per Day)*	*Natural Gas Production, 2000 (Billion Cubic Feet)*	*Coal Production, 2000 (Million Short Tons)*	*Electric Generating Capacity, 2000 (Gigawatts)*	*Crude Oil Refining Capacity, 1/1/02 (Thousand Barrels Per Day)*
Estonia	4 million metric tons of oil shale	minimal	minimal	4.4 (oil shale)	0	0	3.4	0
Latvia	minimal	minimal	minimal	0	0	0	2.1	0
Lithuania	12	minimal	minimal	4.2	0	0	5.8	263
Subtotal	**12 + oil shale**	**minimal**	**minimal**	**8.6**	**0**	**0**	**11.3**	**263**
Belarus	198	100	minimal	37	7.4	0	7.5	493
Total	**210 + oil shale**	**100**	**minimal**	**45.6**	**7.4**	**0**	**18.8**	**756**

Oil shale production is heavily polluting, and as such Estonia is under heavy pressure from the EU to cut back significantly on its oil shale output. Eesti Polevkivi has indicated that it expects the oil shale industry to continue for another 40 years, but no new mines are scheduled to be built, and as Estonia tries to curb pollution from the oil shale industry in an effort to meet EU environmental regulations, Eesti Polevkivi is forecasting its production target will shift downward by 2006, from 12 million tons per year to 10.5 million tons per year. On July 30, 2002, the EU granted Estonia's request for oil shale operations to be approached the same way that the EU does coal, since the problems of the two natural resources are similar. The agreement, in which Estonia was given a long transition period

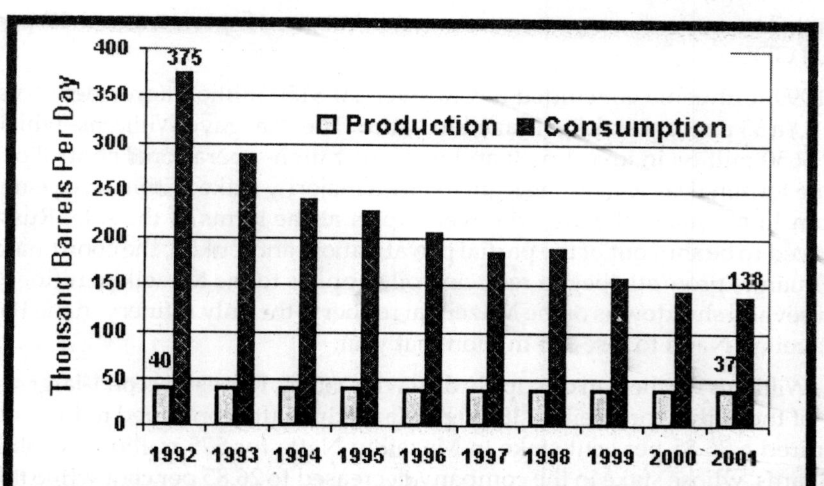

BELARUSSIAN OIL CONSUMPTION AND PRODUCTION, 1992-2001

in exchange for phasing out oil shale operations, allowed Estonia to close the energy chapter of its accession negotiations with the EU.

Latvia has no domestic oil production or refineries, so it is entirely dependent on imports of petroleum products to meet its consumption needs. Since 1992, when Latvia consumed an average of 52,200 bbl/d of oil products, domestic consumption has been on the decline, dropping to approximately 25,000 bbl/d in 2000. In 2001, Latvia imported around 21,000 bbl/d of oil products, mostly from Belarus, Russia, and Lithuania, with less than 3 per cent coming from the EU. Of these imports, diesel fuel accounted for 43 per cent of the total, while gasoline made up 29 per cent and fuel oil 6 per cent.

Latvia's territorial waters in the Baltic Sea are thought to contain up to 300 million barrels of oil, and on April 3, 2002, Latvia awarded TGS-Nopec, a U.S.-Norwegian joint venture, a non-exclusive oil exploration license for five years covering over 2,000 square kilometers in the Baltic Sea. In October 2002, Latvia's Economy Ministry said that it received only one bid, from Odin Energy, for a tender it announced in May 2002 for a 30-year exclusive oil exploration and drilling license in all of the Latvian waters of the Baltic Sea.

Lithuania has 12 million barrels of proven oil reserves, but the country's estimated total onshore oil resources could reach 337 million barrels, and possible reserves in the Lithuanian shelf of the Baltic Sea are estimated between 220 million and 440 million barrels. Geonafta, Lithuania's oil exploration company, and several joint ventures are undertaking onshore drilling projects in western Lithuania. Minijos Nafta, a Lithuanian-Danish joint venture, maintained its status as the country's biggest oil producer in 2001, producing approximately 63 per cent of the country's oil in 2001, when Lithuania produced 4,600 bbl/d. Lithuania's oil production is projected to rise to 5,000 bbl/d in 2002, but with approximately 60,000 bbl/d of oil consumption in 2001, the country will remain a net oil importer. Russia is Lithuania's main supplier of crude oil.

After winning its independence from the Soviet Union, Lithuania reorganized and unified much of its oil industry, creating Mazeikiu Nafta by merging the Lithuania's only refinery (the 263,000-bbl/d-capacity Mazeikiai refinery), Butinge Nafta (which operates a new 160,000-bbl/d-capacity oil terminal at Butinge that is connected by pipeline to the Mazeikiai refinery), and Naftotiekis of Birzu (which operates the Birzu oil pipeline bringing Russian crude oil into Lithuania via the Russian Druzhba pipeline). The unified company accounts for between 5 per cent and 10 per cent of the country's nominal GDP.

In October 1999, Lithuania concluded a controversial $150 million agreement to sell Williams International (U.S.) a 33 per cent stake in Mazeikiu Naftu. The deal gave Williams–which committed to make another $650 million in investment and modernization–operational control of the refinery, pipeline and crude terminal, as well as the right to buy a majority stake within five years. In addition to opposition from Lithuania's citizens, who were upset at the terms of the sale, Russian oil giant Lukoil was dismayed to be shut out of the partial privatization, and Lukoil, the coordinator of Russian oil exports to Lithuania, promptly began reducing oil supplies to the Mazeikiai refinery. Oil supply problems caused several shutdowns of the Mazeikiai refinery–the only refinery in the Baltic states–in 2000, causing Mazeikiu Nafta to lose $45 million that year.

In June 2001, Williams reached an oil supply deal with Yukos, Russia's second-largest oil company, alleviating some of the supply problems, although Mazeikiu Nafta continued to lose money. In June 2002, Yukos acquired a 26.85 per cent stake in Mazeikiu Nafta for $75 million, becoming an equal partner with Williams, whose stake in the company decreased to 26.85 per cent while the Lithuanian government's stake in Mazeikiu Nafta decreased from 59 per cent to 40.66 per cent. Yukos pledged to provide Mazeikiu Nafta with at least 4.8 million tons of oil per year (96,400 bbl/d) for 10 years. In September 2002, Yukos increased its shareholding in Mazeikiu Nafta to 53.7 per cent, buying out Williams' stake for $85 million and taking over management rights and operational control in the Lithuanian oil complex. Yukos is boosting crude supplies to the Mazeikiai refinery to 600,000 tons per month (145,000 bbl/d) during the last quarter of 2002, citing a 25 per cent increase in refining margins. In the first nine months of 2002, the Mazeikiai refinery processed 4.725 million tons of crude (126,500 bbl/d), a 6.8 per cent decline from the same time period in 2001.

Belarus has a small oil industry, and the country produced 37,000 bbl/d of oil in 2001. The country has 198 million barrels of oil in proven reserves, but the lack of political and economic reform in the past decade has hindered the entrance of any foreign investment to the sector in order to help boost production. Belarusnafta, the state-owned oil production monopoly, estimates that active oil deposits may last for another 17 years, with more difficult deposits (e.g. those with a water content of over 80 per cent or with high viscosity) lasting for 34 years, taking into account the company's plan for oil extraction will remain around 40,000 bbl/d. On December 6, 2002, the Belarussian government sold its 10.83 per cent stake in Slavneft, the Belarussian-Russian state-owned oil company operating in Russia, to Sibneft (Russia) for $207 million. Slavneft produced an average of 300,000 bbl/d of crude oil in 2001, a 22 per cent increase over 2000.

Oil consumption in Belarus has fallen by half in the past decade, from 375,000 bbl/d in 1992 to approximately 138,000 bbl/d in 2001, yet Belarus still must import nearly 75 per cent of its oil from Russia. According to a intergovernmental agreement, Russia will supply Belarus with 13 million tons of oil (261,000 bbl/d) in 2002. Although some excess capacity has been shuttered, Belarus has two refineries, the Naftan refinery in Navapolatsk Vitsebsk Region and the Mazyr refinery in the Homel Region, with a combined refining capacity of 493,000 bbl/d–still far higher than the country's oil consumption needs. In 2001, Belarussian refineries processed 266,500 bbl/d of oil, 146,500 bbl/d of which was refined by Naftan and 120,500 bbl/d by Mazyr. According to the Belarussian government, In the first ten months of 2002 Belarus refined 12.6 million tons of oil (an average of 303,000 bbl/d).

Oil Transit and Export

Latvia, Lithuania, and Estonia each have important ports for Russian crude oil and petroleum product exports. Transit fees for these oil and petroleum products that are destined for export are an important source of revenue in the Baltic states. Until recently, Latvia's Ventspils port was Russia's primary northern crude oil export terminal. However, in an effort to avoid these transit fees, Russia constructed its own oil export terminal at Primorsk, part of the country's new Baltic Pipeline System (BPS). The Primorsk terminal, with an initial capacity of 240,000 barrels per day (bbl/d), opened in December 2001. According to various estimates, ports in the Baltic states could lose between 10 per cent and 50 per cent of their current Russian oil export volumes. Nevertheless, projected increases in Russian oil exports, along with increased oil exports from the Caspian Sea region, especially Kazakhstan, appear to ensure that the ports in Estonia, Latvia, and Lithuania will remain important oil export terminals in the future.

Belarus is also an important transit country for Russian oil exports. The northern branch of Russia's 1-million-bb/d-capacity Druzhba oil export pipeline runs through Belarus on its way to the oil terminal in Ventspils, as well as to Poland and Germany. Oil exported from Russia via Belarus (approximately 50 per cent of Russia's net oil exports go through Belarus) is not subject to export duties due to the Russian-Belarussian Union agreement, which, along with relatively high world oil prices and a 20 per cent increase in Russian oil production since 1998, has contributed to the Druzhba pipeline running at or near capacity for the past several years.

Estonian Ports

In recent years, Estonia's ports at Tallinn and nearby Muuga have become major terminals for the export of Russian petroleum products, rivaling Ventspils as the largest transshipment center in the Baltics. Estonia's ports, which only export petroleum products such as heavy fuel oil, are not reliant on pipelines to deliver supplies. Instead, Estonian transit companies–such as Estonia Oil Service (E.O.S.), Eurodek, and Pakterminal–use rail cars to transport oil products from Russia to Estonian sea ports.

The Port of Tallinn, comprised of four separate harbors, handles the vast majority of Estonia's oil product exports. In 2001, port officials said that oil throughput jumped 17.8 per cent, to 20.98 million tons (approximately 421,300 bbl/d), compared to 2000. Estonia's two largest oil transit companies, Pakterminal and Eurodek, handled 8.7 million tons (174,700 bbl/d) and 7.4 million tons (148,600 bbl/d), respectively, in 2001, while E.O.S. handled 5.5 million tons (110,500 bbl/d) during 2001. Through the first three months of 2002, the Port of Tallinn reported handling 5.68 million tons (456,000 bbl/d), up 25 per cent year-on-year, as Russian oil companies increased petroleum product exports due to government-imposed crude export cuts.

Latvian Ports and Pipelines

Latvia's main oil and oil product pipelines stretch from Polotsk, Belarus across Lithuania to Ventspils. These pipelines are managed by the Latvian-Russian joint venture LatRosTrans, owned by the Ventspils Nafta oil terminal from Latvia and Transneft, Russia's pipeline monopoly. Latvia's Ventspils port, which is ice-free year-round, is the largest oil export terminal in the Baltics, and both Riga and Liepaja also have ports for exporting oil and oil products. The Ventspils oil terminal can handle about 500,000 bbl/d of crude, and Latvian officials have stated that its throughput capacity could be increased to 1.8 million bbl/d (although this would require about $30 million in investment). In 2001, Ventspils Nafta handled 22.3 million tons (447,800 bbl/d) of oil and oil products, up 18.6 per cent from 2000.

Oil exports from Ventspils have been far below capacity, however, in 2002, mainly due to the opening of Russia's own export terminal at Primorsk in December 2001, but also due to competition from other Baltic ports. Despite several reduction in reloading tariffs over the past few years, Ventspils remains the highest-cost route for Russian crude oil exports in the Baltics, and the Primorsk terminal has taken away a a part of Ventspils' share of the oil export market. In the first ten months of 2002, Ventspils Nafta reloaded just 12.5 million tons (301,200 bbl/d) of oil and oil products, a 34.2 per cent drop year-on-year. Latvia, which holds a 38.6 per cent stake in Ventspils Nafta and another 5 per cent already reserved for the terminal's current largest private shareholder, Latvijas Naftas Tranzits (which currently holds 47 per cent), may be forced to sell its stake in the terminal as profits dry up and Russian oil companies send their exports elsewhere. Several Russian oil companies, as well as Transneft have expressed interest in Latvia's stake in Ventspils Nafta.

Lithuanian Ports and Pipelines

Lithuania hosts 140 miles of the 280,000-bbl/d-capacity Polotsk-Birzai-Mazeikiai pipeline, with oil transported to Lithuania via two main pipelines from Novopolotsk, Belarus to Birzai, from where one arm turns to Mazeikiu Nafta's oil refinery, and the other to Ventspils. Lithuanian recently completed a $120 million upgrade of the port at Klaipeda, expanding petroleum product export capacity to from

90,000 bbl/d to 160,000 bbl/d. In addition, in July 1999 Lithuanian launched the $267 million crude oil export/import terminal at Butinge near the Latvian border, with capacity of 160,000 bbl/d for crude oil and 50,000 bbl/d of oil products.

The Butinge terminal has been beset by problems since it was launched. After the controversial privatization of Mazeikiu Nafta, the company that operates the Butinge terminal, in 1999, Lithuania was unable to secure enough oil supplies to load at the terminal; Butinge exported just 60,000 bbl/d of crude oil and petroleum products in 2000, followed by 102,000 bbl/d in 2001. Furthermore, several accidents and oil spills at the Butinge terminal in 2001 caused tensions between Latvia and Lithuania as winds blew the spills into Latvian waters. The Butinge terminal was out of operation from January 2002 to March 2002 following a November 2001 oil spill, but almost 153,000 bbl/d were loaded at the terminal in the second quarter of 2002. For the first nine months of 2002, the Butinge terminal handled an average of 103,000 bbl/d. Since Yukos took over operational control of Mazeikiu Nafta in September 2002, the terminal has been operating at near capacity.

Natural Gas

Estonia has no natural gas reserves and therefore must import all of its natural gas for domestic consumption. Currently, Estonia imports all of its supplies via the country's 250-mile pipeline network from Russia, but Estonia is keen to diversify, and Norway is a potential supplier. Estonia's natural gas consumption collapsed from 53 billion cubic feet (Bcf) in 1992 to just 21 Bcf in 1993 as Estonia attempted to reorient its economy to the West, but the country's consumption of Russian natural gas has crept slowly upwards in ensuing years.

In 2000, Estonia consumed 39.6 Bcf of natural gas, while Eesti Gaas (Estonian Gas), the former state-owned natural gas company that was fully privatized in 1997, reported that Estonia consumed 31.1 Bcf of natural gas in 2001. Eesti Gaas is the largest natural gas supplier in Estonia. In November 1999, Eesti Gaas, which is owned by Ruhrgas (Germany, 32 per cent), Gazprom (Russia, 31 per cent), Fortum (Finland, 10 per cent), and Itera (9.5 per cent), signed a long-term agreement with Gazprom to

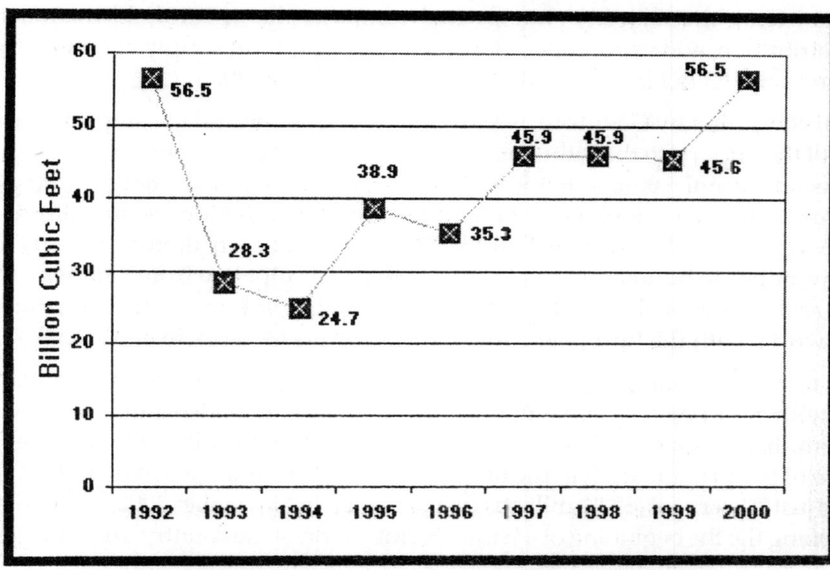

LATVIAN NATUAL GAS CONSUMPTION, 1992-2000

supply Estonia with natural gas from 2000 to -2005. Over 90 per cent of Estonian district heating stations use natural gas, and natural gas is the primary fuel of the Viru Power Station, which produces both heat and power.

Since Latvia has no domestic natural gas reserves, all of the country's natural gas for domestic consumption is imported, mainly from Russia. After fluctuating wildly in the first few years after independence, Latvia's domestic consumption of natural gas remained steady at approximately 45 Bcf in the late 1990s before jumping to 56.5 Bcf in 2000. Of this total, approximately 50 per cent is consumed by the country's main electric utility, Latvenergo.

Latvijas Gaze (Latvia Gas) controls the country's natural gas distribution system and its huge underground storage facility near Riga at Incukalns, the only natural gas storage facility in the Baltics and the third-largest storage facility in Europe. The former state-owned company has been restructured as a joint-stock company and has been substantially privatized. Latvijas Gaze's largest shareholders are Gazprom, Germany's Ruhrgas and EON Energie, and Itera Latvija, a subsidiary of Itera.

In October 2002, Latvijas Gaze reached an agreement with Russia's Gazprom on natural gas supplies to Latvia for the period from 2003 to 2005, providing for a 15 per cent to 20 per cent increase of supply price. Analyst have suggested the Latvijas Gaze agreed to the supply price increase in order to force the Latvian government to allow the company to hike its prices–unchanged since 1997–to consumers. Latvijas Gaze and the Latvian government have been at odds over liberalizing natural gas prices since May 2001 when the government rejected amendments to the country's energy law, proposed by Latvijas Gaze's shareholders, that would have stopped natural gas price regulation for industrial consumers.

Lithuania has minimal natural gas reserves and no natural gas production, making the country completely reliant on imports. Lithuania consumed 92 Bcf of natural gas in 2000, up from 76 Bcf in 1999 but still 35 per cent less than the country consumed in 1992 just after its independence. According to Lithuanian government data, the country consumed 95 Bcf of natural gas in 2001, and the country is planning to increase its natural gas consumption in 2002 to 113 Bcf. In December 1999, Lietuvos Dujos (Lithuanian Gas), the majority state-owned company that controls Lithuania's natural gas transmission, distribution, and export operations, signed a long-term supply agreement with Russia's Gazprom, starting with 53 Bcf in 2000 and increasing to 88 Bcf in 2005.

Dujotekana, controlled by Gazprom, is Lithuania's largest natural gas importer, accounting for over 40 per cent of the imports to the Lithuanian market. Itera also has begun supplying the Lithuanian market with Russian natural gas, and in line with the Lithuania's national energy strategy to diversify its supply sources, Lithuania is looking to Poland to supply it with Norwegian natural gas. Poland, which signed a 6-year, $11 billion deal in 2001 to import a total of 2.6 trillion cubic feet (Tcf) of natural gas from Norway, hopes to build another part of a natural gas pipeline from Gdansk to Lithuania by 2004. Lithuania and Poland will present a project study on the link-up of the two countries' natural gas pipeline networks with the European Union's natural gas network in early 2003.

After reorganizing the company, the Lithuanian government has begun to privatize Lietuvos Dujos, starting with a 34 per cent stake that it sold to a German consortium of EON Energie and Ruhrgas for $33 million in June 2002. Russia's Gazprom was the sole bidder for a 34 per cent stake in the second phase of the privatization in the fall of 2002, but as Lithuania's main natural gas supplier, Gazprom offered just 70 per cent ($23.3 million) for the tender. In November 2002, Lithuanian authorities proposed postponing the the beginning of a wider liberalization of the country's natural gas distribution sector, hoping to delay deregulation in order to boost the value of Lietuvos Dujos and negotiate a higher sale price from Gazprom for its stake in the Lithuanian company. The Lithuanian government is planing to keep a 24.36 per cent stake in Lietuvos Dujos and potentially sell it later on the stock exchange.

With just 7.4 Bcf in indigenous natural gas production in 2002, Belarus is heavily reliant on natural gas imports from Russia. Natural gas consumption in Belarus rose to 692 Bcf in 2000, surpassing levels not seen since before the country's independence. However, due to Belarus' inadequate natural gas pricing structure and payment recovery from consumers, the country has built up large arrears to Russia's Gazprom, despite the fact that Gazprom, which accounts for 62 per cent of Belarussian natural gas imports, only charges Belarus $24 per 1,000 cubic meters (35,300 cubic feet) for natural gas supplies. (Itera, which accounts for 38 per cent of Belarus' natural gas imports, charges $36 per 1,000 cubic meters.) In April 2001, Gazprom and Beltransgaz, the Belarussian state-run natural gas distributor, signed an agreement on restructuring debt for the natural gas supplied to Belarus between 1997 and 1999, but Belarus has continued to rack up payment arrears; as of November 22, 2002, Belarus also owed Russia $80 million for natural gas supplies in 2002.

On November 1, 2002, Gazprom announced that Belarus was nearing its contracted supply of natural gas for all of 2002, and consequently the Russian natural gas supplier cut shipments to Belarus by 50 per cent. Belarussian officials protested, but Belarussian Prime Minister Gennady Novitsky later agreed that Minsk would pay Gazprom $202 million of the debt and speed up privatization of state-owned natural gas monopoly Beltransgaz, ending the dispute with Moscow. Under the deal, Gazprom increased it s supplies to 75 per cent of their former levels while Belarus agreed to buy natural gas from the independent supplier Itera for the remainder of 2002. In addition, Gazprom expects to get 25 per cent to 30 per cent in Beltransgaz, which operates 4,100 miles of natural gas pipelines in Belarus, as well as eight compressor stations, 250 natural gas distribution stations and two natural gas storage reservoirs, as compensation for Belarussian natural gas arrears. Belarus pledged to transfer the stake to Gazprom before June 1, 2003.

Natural Gas Transit

Belarus hosts a 350-mile stretch of the 2,500-mile, trans-continental Yamal-Europe natural gas pipeline, as well as a section of the Northern Lights pipeline, both of which carry Russian natural gas exports to European consumers. Although Belarus does not transit nearly as much Russian natural gas as Ukraine, its importance as a transit state is growing. The first strand of the Belarussian section of the Yamal-Europe pipeline, which runs from Russia's Yamal peninsula through Belarus and Poland to Germany's eastern border, was completed in September 1999, substantially increasing Belarus' transit capacity. Nearly 1.5 Tcf of Russian natural gas transited Belarus via the Northern Lights and Yamal-Europe I pipelines to customers in Europe in 2001.

In July 2002, an additional 108-mile stretch of the Belarussian section of the Yamal pipeline was commissioned, increasing the country's transit capacity from 1.77 Tcf to 2.01 Tcf. By 2005, analysts say, natural gas transit to Europe via the transcontinental Yamal pipeline alone may exceed 1 Tcf. Russia charges Belarus lower prices for natural gas since Russia receives reduced tariff rates for its natural gas that transits Belarus to customers in Western Europe. Beltransgaz charges Russia a transit tariff that is approximately one-third of the tariff in other countries. However, Russia's dispute with Belarus over natural gas supplies for Belarus' domestic consumption, as well as Belarus' mounting debts to Russia for supplies already received, calls into question the reliability of Belarus as a transit state for Russian natural gas exports to Europe.

Gazprom's proposed North TransGas project, viewed by many as a way for Russian natural gas to bypass Belarus en route to European consumers, could result in the Baltic states becoming transit centers for natural gas. The project, which would be carried out by Gazprom in conjunction with Finland's Fortum and Germany's Wintershall, would pump natural gas to Scandinavian and German customers via a pipeline beneath the Baltic Sea. Three options for the pipeline route have been identified: Russia-Finland-Gulf of Bothnia-Sweden-Baltic Sea-Germany; Finland-Baltic Sea-Gotland-Germany; or St. Petersburg-Germany via a pipeline on the bottom of the Baltic Sea. Although none of these options call for the pipeline to transit any of the Baltic states, a pipeline extension to Estonia may be possible and, according to Estonian officials, it may even be desirable for the project's participants, since it could cut down on the costs of part of the subsea pipeline.

In addition, Latvia's huge natural gas storage facilities could play an important role in the project. An extension from the North TransGas pipeline to Latvia would allow Russia to supply the pipeline with natural gas stored in Latvia. Latvijas Gaze typically pumps natural gas into the 141-Bcf-capacity Incukalns storage facility during the spring and summer, drawing it down as needed during heating season in the winter. Latvia has a number of other large natural gas storage facilities, and Gazprom exports a portion of the natural gas stored in these facilities to Estonian, Lithuanian, and Russian consumers, while Latvijas Gaze is charged for the natural gas which is distributed within Latvia. For the purposes of the 2002-2003 heating season, Latvia stored a total of 1.832 billion cubic meters (64 Bcf) of natural gas in the Incukalns. Latvijas Gaze is working to expand the capacity at Incukalns to 177 Bcf by 2005.

Coal

The Baltic Sea region countries do not contain any significant coal deposits, so there is no regional coal production. Since the collapse of the Soviet Union, the Baltic states and Belarus have reduced their coal consumption, shifting to more domestic sources of energy for power generation and heat. Mindful of EU environmental standards, Estonia, Latvia, and Lithuania have also attempted to shift their energy balance towards cleaner fuels in an effort to mirror European norms and join the EU. Nevertheless, each country still imports a small amount of coal–mainly from Poland and Russia–for domestic energy purposes.

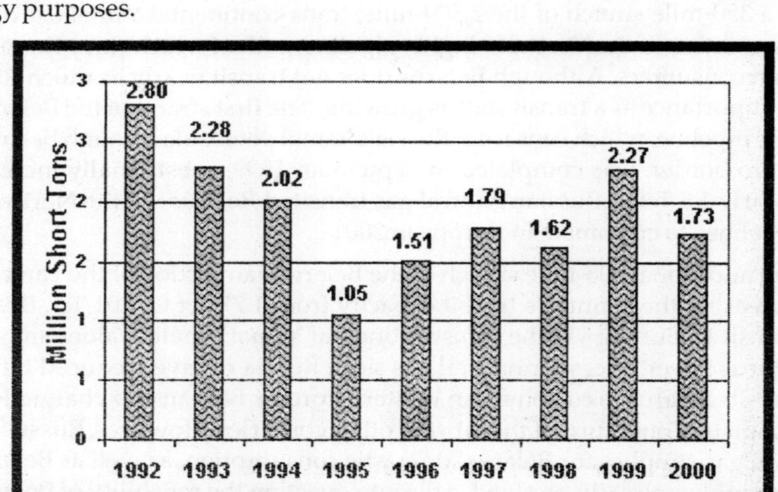

ESTONIAN COAL CONSUMPTION, 1992-2000

Estonia imports a small amount of lignite coal, mainly for district heating. Throughout much of the 1990s, Estonia's coal consumption was on the decline, with consumption falling from 2.08 million short tons (Mmst) in 1992 to 1.62 Mmst in 1998. Estonia's coal consumption rose to 2.27 mmst in 1999 before falling to 1.73 Mmst in 2000.

Latvia imports a small amount of coal as well, mostly from Poland. The country's coal consumption declined 80 per cent from 1992 to 2000, from 0.74 Mmst to 0.15 Mmst. Although Latvia has no coal production, it does produce about 500,000 metric tons of peat each year. Peat covers approximately 10 per cent of Latvia's territory, with the heaviest concentration in the eastern plains near Riga. All peat production companies in Latvia have been privatized, but most of the peat deposits are still owned by municipal governments, which rent the deposits for extraction.

Coal consumption has been substantially reduced in both Lithuania and Belarus as well. Lietuvos Kuras, a joint-stock company that operates a number of Lithuania's service stations, imports a small amount of coal, but in 2000, Lithuania consumed just 0.18 Mmst of coal, down 75 per cent from its consumption level of 0.73 Mmst in 1992. Likewise, Belarus consumed 2.1 Mmst of coal in 1992, but since the fall of the Soviet Union, coal imports have been cut by two-thirds, and in 2000 Belarus imported only 0.7 Mmst of coal for domestic consumption.

Electricity

Estonia has 3.4 gigawatts (GW) of electric-generating capacity, providing the country with ample electricity to meet domestic demand. The majority of this generating capacity comes from Estonia's two oil-shale-fired power plants in the northeast of the country, the 1,610-MW Eesti plant and the 1,390-MW Balti plant. The two power stations, the Narva Power Plants, supply approximately 95 per cent of Estonia's electricity. Although Estonia has a small amount of hydropower and other renewable energy capacity, the power produced at these facilities costs twice as much as the electricity generated at the oil-shale-fired power plants.

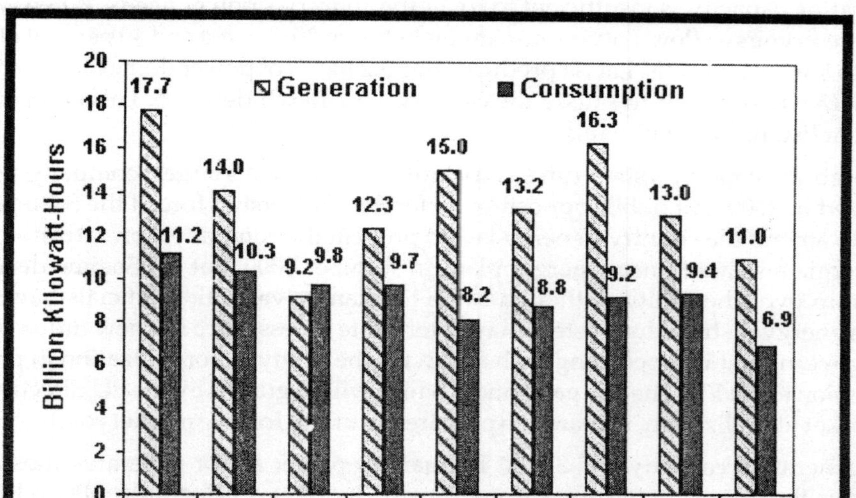

LITHUANIAN ELECTRICTIY GENERATION AND CONSUMPTION, 1992-2000

In 2000, Estonia generated 7.1 billion kilowatt-hours (Bkwh) of electricity, which was more than enough to cover the country's overall consumption of 5.4 Bkwh of power. Estonia exports its excess power to Latvia and to northwestern Russia. In Estonia's negotiations with the EU with regard to energy, the EU recognized the priority position of oil shale-fired power plants, granting the country a transition period to phase out the polluting power plants. In order to shift to cleaner and less polluting energy sources, Eesti Energia, the state-owned electric company, is hoping to boost hydropower production to 10 per cent of the country's overall electricity generation.

Prior to 1995, all of Estonia's electricity had been produced entirely by Eesti Energia, but in 1996, the first steps towards privatization were taken with the establishment of the joint-stock company Kohtla-Jarve Soojus comprising Estonia's two smaller oil shale-fired power plants, the Ahtme and Kohtla-Jarve Power Stations. In August 2000, the Estonian government finalized a controversial $70.5-million deal to sell a 49 per cent stake in the Narva Power Plants to NRG Energy (U.S.), which agreed to invest nearly $200 million more to refurbish the Soviet-era power plants. In January 2002, prior to the actual privatization, Estonia canceled the sale, citing a deadline at the end of December 2001 that NRG Energy missed to secure financing for renovation of the facilities. NRG Energy has filed a $100-million lawsuit against Eesti Energia, which is now the sole owner of the Narva Power Plants, demanding compensation for the abortive privatization.

Hydroelectricity is the main generating source of electricity in Latvia, as approximately 73 per cent of the country's 2.1-GW power-generating capacity is hydroelectric. Three hydropower plants, Kegums-Plavinas-Riga, constitute the Daugava cascade located on the Daugava River. All three hydroelectric plants have recently been modernized, with the Kegums hydropower station, built in 1939, officially re-opened in August 2001 after $21 million worth of renovation to extend its service life for another 40 years.

Hydroelectric power plants on the Daugava River are Latvia's main power producers, but their output, along with the TEC-1 and TEC-2 power plants that constitute 97 per cent of Latvia's thermal power-generating capacity, is insufficient to cover the country's power needs. When water levels in hydroelectric reservoirs are low, Latvia must import between 30 per cent and 40 per cent of its electricity. In 2000, due to low water levels, Latvia produced just 3.2 Bkwh of power domestically and was forced to import 1.9 Bkwh of power to satisfy the country's 5.2 Bkwh demand. Latvia regularly imports electricity from Estonia and Lithuania.

Latvenergo, the country's state-run electric utility and Latvia's largest company, was scheduled to be privatized in 2000, but public opposition to foreign ownership forced the Saeima (the Latvian parliament) to amend the country's energy law to prevent the company's privatization. Latvenergo and Eesti Energia also announced merger plans in summer 2000, but the Saeima decided to retain state ownership over the utility. Although some Latvian privatization officials have warned that Latvenergo's energy distribution system may deteriorate unless there is a new influx of investment, the Latvian government is proceeding with a plan for the utility's reorganization in preparation for Latvia's accession to the EU. The reorganization, which will be guided by the EU directive with regard to energy market liberalization, is geared to prepare the utility for free market conditions.

With a generation capacity of 5.8 GW, Lithuania's power sector generates substantially more electricity than the country consumes domestically. In 2000, Lithuania produced 11.0 Bkwh of electricity–far more than the 6.9 Bkwh that the country consumed that year, leaving Lithuania with excess power to export to Latvia, to the Kaliningrad region of Russia, and to Belarus. However, a dispute with Belarus over payment for $50 million worth of power supplied in 1998 and 1999 continues to linger, prompting Lithuania to shut off power exports to its neighbor periodically. Nuclear power from the massive 2,370-MW Ignalina Nuclear Power Plant (NPP), the sole nuclear plant in Lithuania, accounts for over 70 per cent of the electricity generated in Lithuania, earning it the distinction as the most nuclear-dependent country in the world.

Lietuvos Energija (Lithuanian Energy), a joint-stock company formed by the reorganization of the Lithuanian state power system in 1995, is the largest electric power company in Lithuania. The Lithuanian government holds an 86.5 per cent share in the company. Besides transmitting and distributing all electricity generated in Lithuania, Lietuvos Energija owns the Ignalina NPP, as well as all the major conventional fuel power plants in Lithuania, including Elektrenai, Kruonis Hydro Power Plant, Kaunas Hydro Power Storage Plant, and Mazeikiai Combined Heat and Power Plant. In 1998, the Vilnius Power Station became independent of Lietuvos Energij, and in July 2002, Lithuania updated its national energy strategy with plans for selling at least 50.1 per cent of Mazeikiu Elektrine (Mazeikiai Power Plant) and two electricity distribution network operators, Rytu Skirstomieji Tinklai (RST) and Vakaru Skirstomieji Tinklai (VST), by the spring of 2003. The state would retain 34 per cent shares in the companies, as well as Lietuvos Energija and the power grid.

The EU has expressed safety concerns over the Ignalina NPP, which is of the same Soviet-era design as Ukraine's ill-fated Chornobyl NPP. Despite a number of safety measures introduced at Ignalina over the past decade, the EU considers the plant's two power units to be dangerous, and as a result the EU has made closure of the Ignalina plant a pre-condition for Lithuanian membership in that organization. In exchange for the right to begin EU membership negotiations, in 1999 Lithuania's parliament pledged to take Ignalina-1 out of operation by January 1, 2005, while a decommissioning date for the second reactor is still undecided. The EU is trying to convince Lithuania to shut down the second reactor before 2009.

To facilitate the closure and to develop alternative sources of power, the EU announced at the end of 1999 that it would provide additional aid to Lithuania through its Phare program. In June 2000, representatives from countries around the world pledged nearly $195 million to help Lithuania shut Ignalina-1. However, the Lithuanian Economics Ministry predicts costs to close the nuclear plant down will reach $2.4 billion by 2020 and will eventually top $3 billion. The country also needs an estimated $910 million to modernize its non-nuclear power plants and transmission lines to ensure sufficient generation after Ignalina is decommissioned.

The most critical project is modernization of the 1,800-MW combined heat and power Lithuanian Power Plant (LPP). The LPP, which can run on both oil and natural gas, was built between 1962 and 1972 and operated at full capacity until 1992. Although the LPP has used over 80 per cent of its technical lifetime, a Lithuanian government study has shown that it is less costly to modernize the plant than to build a modern plant of similar size. Once Ignalina shuts down, the LPP will be the country's primary source of power generation. Approximately $13 million will be needed to renovate the LPP before 2005, with a further $264 million needed between 2006 and 2010. A second project involves renovation of the Kaunas combined heat and power plant, a 170-MW unit built in 1960. Another $331 million will be needed for modernizing Lithuania's transmission grid, a two-stage project planned for 2001-2005 and 2006-2010.

Belarus has a power-generating capacity of 7.5 GW. Oil- and natural gas-fired power plants make up 99.9 per cent of Belarus' power generation, with hydroelectric accounting for just 0.1 per cent. Both electricity consumption and generation have declined in the decade since independence, but the country's decaying infrastructure and a lack of investment in maintenance and upgrades has resulted in power generation decreasing faster than consumption, meaning Belarus is still a net importer of electricity for about 20 per cent of its annual power demand. In 2000, Belarus produced 24.7 Bkwh of power but consumed 26.8 Bkwh.

The Belarussian government has attempted to stifle domestic power consumption and cover production costs by implementing incremental price increases, including a 25 per cent price rise for residential consumers in February 2002. However, the country's accelerating inflation consistently has outpaced tariff increases, and nonpayment by consumers, as well as the Belarussian government's unwillingness to cut off debtors, has forced Belenerha into a dire financial position, unable to pay for imported electricity supplies. With Lithuania periodically suspending electricity supplies to Belarus in order to force debt payment, Belarus has turned to Russia as its main source for its power imports. Belenerha, the state-owned energy utility, imported over 5 Bkwh of Russian electricity in 2001, with an additional 0.9 Bkwh of power imported from Lithuania. Russia's electricity monopoly, Unified Energy Systems (UES), agreed to supply Belarus with up to 5.5 Bkwh of electricity in 2002.

12

Bangladesh

Bangladesh is important to world energy markets because of its large potential natural gas reserves.

General Background

Bangladesh has received more than $30 billion in disbursed grant aid and loans from foreign donors (including the World Bank, the Asian Development Bank, the U.N. Development Program, the United States, Japan, Saudi Arabia, and Western Europe) since its independence in 1971, but remains one of the poorest and most densely populated countries in the world. Bangladesh historically has run a large trade deficit, which it finances largely through foreign aid and remittances from the many Bangladeshi workers abroad (largely in the Persian Gulf region). Overall, foreign aid provides Bangladesh with around 40 per cent of government revenues and 50 per cent of foreign exchange. The World Trade Organization (WTO) has stated that Bangladesh's main problems include civil unrest and political instability, natural disasters, and inadequate infrastructure. Bangladesh's real GDP grew at an estimated rate of 4.9 per cent in 2002, and growth is projected to increase slightly to 5.4 per cent in 2003.

In November 2002, Germany agreed to provide Bangladesh with $27 million to be used for the improvement of feeder roads and market facilities in rural areas, solar energy systems, and business development services. Previously undisbursed funds from Germany's other commitments to the country will now be earmarked for improvements in the energy sector.

Bangladesh is primarily agricultural (around two-thirds of the labor force and 35 per cent of the gross domestic product–GDP), although urbanization is proceeding rapidly. This heavy reliance on agriculture makes Bangladesh vulnerable to natural disasters such as cyclones, floods, and droughts, as well as to world commodity prices. Over the past several years, however, Bangladesh has experienced bumper crops and strong growth in the agricultural sector.

Bangladesh has moved increasingly towards a market-oriented economy since the mid-1970s, although the majority of enterprises remain under state control. Bangladesh is attempting to diversify its economy away from agriculture, and has made industrial development a priority. Exports are increasing at around 8 per cent annually. Currently, cotton textiles and garments account for around 80 per cent of Bangladeshi exports. Bangladesh is attempting to attract foreign investment, and has established export processing zones (EPZs) in Chittagong (the country's major port) and Dhaka (the capital), with plans for more such zones. Most investment is coming in the natural gas, electricity, and physical infrastructure areas. Exports of natural gas could provide a major additional revenue source for Bangladesh, but the issue remains controversial, and no final decision has been made on whether to allow exports.

A new government under the leadership of Prime Minister Khaleda Zia took office in October 2001 after her Bangladesh National Party (BNP) won the majority of seats in parliament in a national election. Political tensions remain high as the opposition Awami League party continues to stage strikes.

Bangladesh (along with Bhutan, India, Maldives, Nepal, Pakistan and Sri Lanka) is a member of the South Asian Association for Regional Cooperation (SAARC), created in 1985 to help promote regional economic cooperation, plus economic and social development in general in the South Asian region.

Energy

Bangladesh has small reserves of oil and coal, but potentially very large natural gas resources. Commercial energy consumption is around 67 per cent natural gas, with the remainder mostly oil (plus limited amounts of hydropower and coal). Only around 18 per cent of the population (25 per cent in urban areas and 10 per cent in rural areas) has access to electricity, and per capita commercial energy consumption is among the lowest in the world. Noncommercial energy sources, such as wood, animal wastes, and crop residues, are estimated to account for over half of the country's energy consumption. Consumption of wood for fuel has contributed to deforestation and other environmental problems in Bangladesh. The World Bank has estimated that Bangladesh loses around $1 billion per year due to power outages and unreliable energy supplies.

Bangladesh's Ministry of Energy and Mineral Resources (MEMR) has overall responsibility for the country's energy sector, with policy formulation and investment decisions under its control. Within MEMR, the "Power Cell" acts as a single point of contact to facilitate the electricity reform and restructuring process, such as development of Independent Power Projects (IPPs).

Oil

Bangladesh contains small proven oil reserves of 56.9 million barrels and produces around 6,200 barrels per day (bbl/d), of which 6,000 bbl/d is crude oil. Until the beginning of the 1990s, state oil and gas company Petrobangla, along with its eight operating companies (OCs), was the sole player in the Bangladeshi oil and gas sectors. Over the past few years, however, Bangladesh has encouraged foreign oil companies to do business in the country. At present, Shell, Texaco, Tullow Oil; Holland Sea Search, Unocal, and UMC Bangladesh Corporation are active in exploration under six Production Sharing Contracts (PSCs) partnership with Petrobangla. To date, oil exploration has proven largely unsuccessful, although hopes continue, especially onshore. Petrobangla regulates the activities of foreign companies under PSCs, and serves as the sole purchaser of oil and gas from the companies. Around 65 per cent of Petrobangla's gross revenues are paid to the government in the form of taxes and compulsory dividends. Petrobangla has been characterized in recent years by a low level of investment and a lack of sufficient financing.

Refining/Downstream

Bangladesh has one refinery, a 33,000-bbl/d unit at Chittagong. In December 2000, TotalFinaElf said that it would set up a $16 million plant to bottle liquefied petroleum gas (LPG), in a joint venture with Bangladesh's Premier LP Gas Ltd. LPG is used in Bangladesh mainly for domestic cooking, as well as in some industries and vehicles.

In July 1999, Bangladesh decided to remove lead from gasoline sold in the country. The decision was taken mainly due to health and environmental concerns, particularly in Dhaka, the capital. In January 2003, the Bangladeshi government announced that it had granted approval for price increases on retail sales of petroleum products by the Bangladesh Petroleum Corporation (BPC). The move reduced consumption subsidies and has helped to reduce border smuggling, which had existed due to the difference in retail petroleum product prices between Bangladesh and India. The government also has reportedly considered allowing firms other than BPC to enter the downstream market, but none has yet done so.

Natural Gas

Natural gas is Bangladesh's only significant source of commercial energy, with 2001 production of 349.6 billion cubic feet (Bcf). Bangladeshi natural gas production began in 1960 from the Chattak

Field. There is much uncertainty and debate about the level of natural gas reserves in Bangladesh. Current Bangladeshi government estimates, revised upward in May 2003, put net proven reserves at 18.0 Tcf. The US Geological Survey has estimated that Bangladesh contains an additional 32.1 Tcf in additional "undiscovered reserves." Bangladesh may have the potential to become a major gas producer (as well as supplier to the vast potential market in neighboring India) at some point. Bangladesh also could use its natural gas resources to power vehicles (the government already has announced plans to convert government vehicles to compressed natural gas to help alleviate pollution problems in Dhaka, and also in response to high oil prices), to produce electricity, petrochemicals, and fertilizers, which it also could use both within the country as well as for export. Natural gas exports are controversial within Bangladesh, with many people feeling that Bangladeshi gas resources first should be used for domestic purposes (*i.e.*, electric power generation, fertilizer production, transportation), and also that the size of the country's gas reserves remains highly uncertain, particularly in relation to future domestic demand projections. Both major political parties are officially committed to considering natural gas exports only if Bangladesh has proven reserves sufficient to cover 50 years of domestic demand.

Unocal had submitted a formal proposal for an export pipeline to India in November 2001, which would link the Bibiyana field to India's main natural gas backbone, the HBJ Pipeline. In April 2003, it was reported by Agence France Presse that the Bangladeshi government had decided to put off its decision on natural gas exports due to the conflict in Iraq. Energy ministry sources were quoted as saying that the situation had made the climate "unfavorable" for a decision.

Petrobangla has approximately 20 natural gas fields nationwide, half of which are active. The main fields include: Bibiyana (discovered by Unocal in Block 12), Titas (the country's second largest natural gas field), Habiganj, Kailashtilla, Rashidpur, and Jalalabad, nearly all of which are located in the eastern part of the country, plus the Sangu offshore natural gas field (being developed by Cairn Energy, Shell, and Halliburton) in Block 16 of the Bay of Bengal, 30 miles southwest of Chittagong. Production from Sangu, Bangladesh's first offshore field (with estimated reserves of around 850 Bcf), began in June 1998. Sangu is one of Bangladesh's most important natural gas discoveries to date, and the first foreign-run natural gas field. In January 2000, Shell Bangladesh Exploration and Development–SBED–along with partners Cairn Energy and HBR Energy reportedly discovered a new natural gas field near Sangu (South Sangu-1). In August 2000, SBED announced that it had invested $40-$50 million in new offshore natural gas exploration projects in Bangladesh, including the Sandwip East 1 well in Block 15 (Bay of Bengal). Other possible natural gas fields include Shaldanadi (estimated reserves of 500-1,000 Bcf), Fenchuganj, Feni, Kumta, and Shahbajpur.

Major foreign energy companies active in natural gas exploration and development in Bangladesh include Shell, and Unocal, which operates in Bangladesh through its wholly owned subsidiary, Unocal Bangladesh, Ltd. In early 1997, Unocal acquired 50 per cent interest in Occidental blocks 12, 13, and 14. Unocal also is involved in two PSCs with Petrobangla covering Blocks 12, 13, and 14. In 1998, Occidental-Unocal discovered an estimated 4-5 Tcf of gas-in-place on Block 12. In May 1999, Unocal took over the assets and operations in Bangladesh of Occidental, which had experienced a major explosion and fire at one of its wells in the Sylhet area in 1997. Unocal announced in March 2003 that it would begin development of the Moulavi Bazar field in Block 14, which is to begin deliveries of 100 million cubic feet per day (Mmcf/d) in 2004, for use in the domestic market. Unocal has said that it hopes the additional domestic supply from Moulavi Bazar will help pave the way for exports to India from its Bibiyana field.

Besides foreign energy companies, natural gas in Bangladesh is being produced by two subsidiaries of state energy company Petrobangla–Sylhet Gas Fields Ltd. and Bangladesh Gas Fields Co. Ltd. These two companies produce natural gas for domestic consumption. More than 80 per cent of the natural gas is consumed for power and fertilizer production, and the remainder by industry and households.

Bangladesh's natural gas demand is expected by some independent analysts to grow by around 6 per cent annually over the next two decades. Potential uses for natural gas in Bangladesh include: petrochemicals, compressed natural gas (CNG) for vehicles, power generation, and fertilizer. Bangladesh also contains around 55 million barrels of natural gas liquids (NGLs), which could be used for petrochemicals production or as a cooking fuel to help reduce deforestation and pollution. Production of NGLs is currently only about 200 bbl/d.

Shahbazpur, discovered by Petrobangla subsidiary Bapex (Bangladesh Petroleum Exploration Company) in 1995, is estimated to contain 330-400 Bcf of recoverable natural gas. In September 1998, Unocal and Petrobangla initialed a PSC for development of Shahbazpur.

In July 1998, Cairn Energy reportedly made a large natural gas discovery in the Halda valley. Meanwhile, Unocal, along with Petrobangla, has developed the 1.6-Tcf Jalalabad field, which came onstream in 1999 and is currently producing 80 Mmcf/d. In late September 1998, Shell and Cairn said they had agreed to an alliance over natural gas development in Bangladesh (including the Sangu and Semutang fields), plus northeastern India.

In late November 1998, Bangladesh raised the price of natural gas by 15 per cent as part of an effort to reduce government subsidies as recommended by international lending institutions and countries. Bangladesh has had relatively low natural gas prices by international standards, with electricity consumers, plus fertilizer plants and households, receiving around $600 million a year in direct subsidies and savings associated with their gas consumption.

In addition to possible exports of its own natural gas, Bangladesh may eventually also become a transit corridor for natural gas from the Tripura state in eastern India to heavily-populated West Bengal. Unocal has proposed that natural gas from Tripura be linked into the pipeline it hopes to build to export Bangladeshi natural gas to India.

Coal

Bangladesh began its first significant coal production in April 2003 with the opening of the Barapukuria Coal Mine in the Dinjapur area of northwest Bangladesh. The project, backed by Chinese investors, will produce about one million short tons of coal per year, to be used primarily for electricity generation. Another possible coal mining project, at Khalashpir, is under consideration.

Electricity

Bangladesh's installed electric generating capacity in 2001 was 3.6 gigawatts (GW), of which 94 per cent was thermal (mainly natural-gas-fired), and the remainder hydroelectric, at 18 power stations. Only around two-thirds of Bangladesh's total electric generating capacity is considered to be "available," however. Problems in the Bangladeshi electric power sector include high system losses (up to 40 per cent), delays in completion of new plants, low plant efficiencies, natural gas availability problems, erratic power supply, electricity theft, and blackouts, shortages of funds for needed maintenance at the country's power plants and other power infrastructure, and unwillingness of customers to pay bills. Overall, the country's generation plants have been chronically unable to meet system demand over the past decade. With only around 18 per cent of the population connected to the electricity grid, and with power demand growing rapidly, Bangladesh's Power System Master Plan (PSMP) projects a required doubling of electric generating capacity by 2010. Total investment required for this increased capacity is estimated at $4.4 billion through 2005. In addition, Bangladesh also may need to replace 30 per cent -40 per cent of its current generating capacity, due to aging infrastructure.

The Padma-Jamuna-Meghna river system divides Bangladesh into two zones, East and West. The East contains nearly all of the country's electric generating capacity, while the West, with almost no natural resources, must import power from the East. Electricity interconnection from the East to the West was accomplished in 1982 by a new, 230-kilovolt (kV) power transmission line. The vast majority of Bangladesh's electricity consumption takes place in the East, with the entire region west of the Jamuna River accounting for only 22 per cent of the total. Greater Dhaka alone consumes around half of Bangladeshi electricity.

Through MEMR, the Bangladeshi government owns and supervises the Bangladesh Power Development Board (BPDB). BPDB is an integrated utility distributing electricity directly to retail consumers, as well as to two other distribution utilities–the Dhaka Electric Supply Authority (DESA, established in 1991), and the Rural Electrification Board (REB, established in 1977).

Given Bangladesh's electricity supply shortage, the government decided in October 1996 to issue a "Private Sector Power Generation Policy of Bangladesh." As part of this plan (and also following the Power Systems Master Plan developed by Acres International Ltd. in 1995), the government decided to solicit proposals from international companies for IPPs. This has resulted in solicitations for a number of fast-track barge-mounted plants, plus two large-capacity gas-fired, combined-cycle plants (a 360-MW plant at Haripur and a 450-MW plant at Meghnaghat), and a 124-MW gas-fired plant at Baghabari. The Haripur plant began operation in October 2001, and the Meghnaghat plant began operation in November 2002. Both projects were built by AES Corporation of the United States, but were sold to the British firm CDC Globeleq in April 2003. Bharat Heavy Electricals of India completed the gas-fired Baghabari generating unit in November 2001. A consortium of Chinese firms concluded an agreement with Bangladesh in June 2001 for the country's first coal-fired power plant. It is to be built by 2004 at Barapukria in northern Bangladesh, near the country's main coal deposit, and will have a capacity of 250 MW. In addition to large IPP projects, in April 1998, Bangladesh adopted a "Small Power Generation Policy," which encourages development of small local generation projects of up to 10-MW in capacity in underserved areas. The country also has an active rural electrification program. All of these initiatives aim to increase power generation and to reduce the country's power shortage significantly in coming years, with a goal of achieving universal electrification by 2020.

Bangladesh has several barge-mounted power stations under construction. Westont Power's 130-MW, $37 million Baghabari project began delivering power to BPDB in July 1999. A $103 million, 110-MW, fuel-oil-fired power plant at Khulna, built on two barges, was commissioned in October 1998 and is being operated by Wartsila NSD of Finland in a joint venture with two local companies (plus financing from the International Finance Corporation). Power from the plant supplies western Bangladesh.

Discussions have been underway for several years about the possibility of Bangladesh connection its electric grid to those of India, Nepal, and Bhutan. Nepal and Bhutan have substantial untapped hydroelectricity potential. This power could be consumed in those two countries and also exported to India, Pakistan, and Bangladesh. In March 1999, it was reported that India's Power Grid Corporation had completed a feasibility study on possible exchange of 150 MW of power between Bangladesh and

India. Interconnection points would be Ishwardi, Bangladesh-Farakka, India and Shahjibazar, Bangladesh-Kurnarghat, India.

Country Overview

Independence: December 16, 1971 (from Pakistan)

Population (July 2002E): 133 million

Location/Size: Southern Asia, bordering Bay of Bengal, between India and Burma/55,813 square miles (about the size of Wisconsin)

Major Cities: Dhaka (capital–population, 10 million), Chittagong (2.8 million), Khulna (1.8 million), Rajshahi (1 million)

Languages: Bangla (official, also known as Bengali), English

Ethnic Groups: Bengali (98 per cent), tribal groups, non-Bengali Muslims

Religions: Muslim (88 per cent), Hindu (11 per cent), Christian, Buddhist, others (1 per cent)

Defense (8/98): Total manpower 121,000 (Army 101,000; Navy 10,500; Air Force 9,500); Paramilitary (49,700)

Economic Overview

Currency: Taka (Tk)

Market Exchange Rate (5/1/03): US$1 = 58.2 Tk

Gross Domestic Product (GDP) (2002E, market exchange rates): $48.3 billion

Per Capita GDP (market exchange rate, 2002E): $356

Real GDP Growth Rate (2002E): 4.9 per cent *(2003F):* 5.4 per cent

Inflation Rate (consumer prices) (2002E): 3.6 per cent

Current Account Balance (2002E): $240 million (Bangladeshi government figure)

Merchandise Exports (2002E): $6.8 billion

Merchandise Imports (2002E): $7.5 billion

Merchandise Trade Balance (2002E): -$1.3 billion

Major Trading Partners (2002): United States, India, China, Japan, United Kingdom, Germany, France

Major Export Products: Garments and knitwear, frozen fish, jute and jute goods, leather and leather products, tea, urea fertilizer, ceramic tableware

Major Import Products: Capital goods, foodgrains, petroleum, textiles, chemicals, vegetable oils

Total Foreign Debt (2002E): $16.7 billion

Energy Overview

Proven Oil Reserves (1/1/03E): 56.9 million barrels

Oil Production (2002E): 6,200 bbl/d, of which 6,000 bbl/d was crude oil

Oil Consumption (2002E): 75,000 bbl/d

Net Oil Imports (2002E): 68,800 bbl/d

Crude Oil Refining Capacity (1/1/03E): 33,000 bbl/d

Natural Gas Reserves (May 2003E): 18.0 trillion cubic feet (Tcf) (current estimate of the Bangladeshi government. Other estimates vary widely. The US Geological Survey has estimated that Bangladesh has an additional 32.1 Tcf in "undiscovered reserves.")

Natural Gas Production/Consumption (2001E): 349.6 billion cubic feet (Bcf)

Coal Production (2001): None

Coal Consumption (2001): 0.55 Mmst

Electric Generation Capacity (2001E): 3.6 gigawatts

Electricity Production (2001E): 15.4 billion kilowatt hours (88 per cent natural gas, 6 per cent oil, 6.3 per cent hydro)

Environmental Overview

Total Energy Consumption (2001E): 0.51 quadrillion Btu* (0.1 per cent of world total energy consumption)

Energy-Related Carbon Emissions (2001E): 8.1 million metric tons of carbon (0.1 per cent of world carbon emissions)

Per Capita Energy Consumption (2001E): 3.7 million Btu (vs U.S. value of 341.8 million Btu)

Per Capita Carbon Emissions (2001E): 0.06 metric tons of carbon (vs U.S. value of 5.5 metric tons of carbon)

Energy Intensity (2001E): 12,743 Btu/ $1995 (vs U.S. value of 10,736 Btu/ $1995)**

Carbon Intensity (2001E): 0.20 metric tons of carbon/thousand $1995 (vs U.S. value of 0.17 metric tons/thousand $1995)

Fuel Share of Energy Consumption (2001E): Natural Gas (66.8 per cent), Oil (29.2 per cent), Coal (2.0 per cent)

Fuel Share of Carbon Emissions (2001E): Natural Gas (61.2 per cent), Oil (35.6 per cent), Coal (3.2 per cent)

Status in Climate Change Negotiations: Non-Annex I country under the United Nations Framework Convention on Climate Change (ratified on April 15, 1994). Not a signatory to the Kyoto Protocol.

Major Environmental Issues: Many people are landless and forced to live on and cultivate flood-prone land; limited access to potable water; water-borne diseases prevalent; water pollution especially of fishing areas results from the use of commercial pesticides; intermittent water shortages because of falling water tables in the northern and central parts of the country; soil degradation; deforestation; severe overpopulation.

Major International Environmental Agreements: A party to the Conventions on Biodiversity, Climate Change, Desertification, Endangered Species, Environmental Modification, Hazardous Wastes, Nuclear Test Ban, Ozone Layer Protection and Wetlands. Has signed, but not ratified, the Law of the Sea.

The total energy consumption statistic includes petroleum, dry natural gas, coal, net hydro, nuclear, geothermal, solar, wind, wood and waste electric power. The renewable energy consumption statistic is based on International Energy Agency (IEA) data and includes hydropower, solar, wind, tide, geothermal, solid biomass and animal products, biomass gas and liquids, industrial and municipal wastes. Sectoral shares of energy consumption and carbon emissions are also based on IEA data.

Oil and Gas Industries

Organizations: Bangladesh Oil, Gas, and Minerals Corp. (also known as *Petrobangla*), formed in 1974, is the state company responsible for oil and gas exploration, production, and distribution. Petrobangla also is involved in exploration and production for minerals, including coal. Petrobangla has 10 operating companies, including *Bangladesh Petroleum Corporation*, formed in 1976 and a separate corporate entity, handles oil imports, refining, and marketing. *Bangladesh Petroleum Exploration Company (Bapex)* is the exploration subsidiary of Petrobangla. Besides Bapex, Petrobangla has 7 other subsidiaries: *Bangladesh Gas Fields Company Ltd.* (gas development and production, mainly in central gas fields); *Sylhet Gas Fields Ltd.* (responsible for northern gas fields operation) Gas Transmission Company Limited (national gas transmission system); *Rupantarita Prakritik Gas Company Ltd.* (natural gas liquids and liquefied petroleum gas); *Titas Gas Transmission and Distribution Company* (regional gas distribution, with 73 per cent of the market); *Bakhrabad Gas Systems Ltd.* (regional gas distribution, with 21 per cent of the market); *Jalalabad Gas Transmission and Distribution System Ltd.* (regional gas distribution, with 6 per cent of the market).

Refinery: Chittagong (33,000 bbl/d)

Foreign Energy Company Involvement: Cairn, Halliburton, Occidental, Rexwood-Okland, Shell, Texaco, Unocal

Gas Fields: Bakhrabad, Beani Bazar, Chattak, Feni, Habiganj, Jalalabad, Kailashtilla, Narshingdi, Rashidpur, Sangu, Shahbazpur, Sylhet, and Titas

Ports: Chittagong, Mongla (Khulna)

13

Bolivia

Bolivia is on its way to becoming a natural gas hub for the Southern Cone, as well as a major exporter of liquefied natural gas to Mexico and the United States. The country's oil and natural gas production potential has not yet been reached.

General Background

Bolivia, already one of the poorest countries in South America, is enduring a period of serious economic difficulties. Slow economic growth in the United States, Bolivia's major trading partner, as well as regional economic problems (*i.e.*, in Argentina and Brazil), plus low prices for Bolivia's commodity exports (e.g., tin), have not helped matters. In general, the weak global economy has stifled both Bolivian exports as well as foreign investment flows into the country. Making matters even worse, Bolivia's capital, La Paz, was hit by the worst flooding in 25 years in late February 2002. Approximately 10 per cent of Bolivia's population is unemployed, another 13 per cent or so are underemployed, and three-quarters live in poverty. For 2002, real economic growth of only 1.6 per cent is expected.

Bolivia's economic problems have contributed to social unrest. In 2000 and 2001, for instance, protestors demanded more public investment in rural areas and an end to the country's coca eradication program. In spring 2001, blockades were regularly halting traffic on many of the country's major highways around the capital, La Paz.

On August 6, 2002, Gonzalo "Goni" Sanchez de Lozada was sworn in as Bolivia's new president, succeeding Jorge Quiroga, who himself became president a year earlier when Hugo Banzer resigned due to ill health (cancer). Sanchez de Lozada, who also served as president from 1993 to 1997, was elected with only 22.5 per cent of the June 2002 vote, and was selected by Congress over the second-place finisher, Evo Morales, leader of Bolivia's coca growers. Sanchez de Lozada quickly announced emergency economic measures to pull Bolivia out of its economic problems, which include high

unemployment and low economic growth. Bolivia's options remain limited, however, due to a high debt load, spending constraints due to tight fiscal spending restrictions imposed by International Monetary Fund (IMF) loan requirements. President Sanchez de Lozada generally is considered to be in favor of free-market reforms.

Bolivia's main hope for future economic growth hinges on increasing natural gas exports and becoming a major energy hub for South America's Southern Cone in coming years. Possible markets for Bolivian natural gas include Brazil, Mexico, Argentina, and the United States. Progress has been slow to date, however. In August 2002, the US Congress renewed the Andean Trade Preferences Act, which expired in December 2001. This Act lowers tariffs on 6,000 products from Bolivia, Colombia, Ecuador and Peru.

Bolivia was the second country in South America (after Chile) to implement International Monetary Fund (IMF) "structural adjustments" in the 1980s, and the 1990s saw the privatization of the five largest state-held companies. After difficult years at the end of the 1990s, strict economic policies that reigned in government spending were expected to result in strong gross domestic product growth for 2000-2002, but so far have not lived up to expectations. Also, in the past few years, there have been collaborative efforts between the United States and Bolivia to control growth of coca plants, which are the raw material for cocaine. Coca eradication efforts have been highly contentious within Bolivia, and have led at various times to bloody protests by indigenous coca growers. Bolivia is one of the world's largest producers of coca, which is the sole source of income for many poor Bolivian farmers.

Bolivia has been without access to the Pacific Ocean since 1884, after it lost a four-year war to Chile. Bolivia and Chile severed diplomatic relations over the issue in 1978, but rapprochement now appears possible, as Chile reportedly is considering an arrangement that would allow Bolivia access to one of its Pacific ports, particularly for expanded Bolivian exports of liquefied natural gas (LNG). However, many Bolivians are opposed to using Chile's port facilities due to the sovereignty question, and many prefer a southern Peruvian port option (Ilo) instead. On July 12, 2002, for instance, thousands of Bolivians took to the streets to oppose the Chilean port option. Bolivia's government is promising to put the issue to a vote before making a decision.

Oil

As of January 1, 2002, Bolivia's total proven oil reserves amounted to 440.5 million barrels. Bolivia is largely self sufficient in crude oil, consuming an estimated 43,000 barrels per day (bbl/d), about what it produced during 2001. Almost all crude oil produced in Bolivia is for domestic consumption, except for a small amount that is exported to Chile through the Sica Sica-Arica pipeline.

Industry Organization

The privatization of *Yacimientos Petroliferos Fiscales Bolivianos* (YPFB), the Bolivian state oil company, has been in progress since passage of the 1994 Capitalization Law. The law paved the way for partial privatization of several national industries. YPFB was divided into two upstream units, a transport company, a refining company, and several service companies. The two upstream units were privatized in 1996/1997, with sales to U.S.-based Amoco (now BP) and a consortium of Argentine YPF (now Repsol YPF), Perez Companc, and Pluspetrol. The oil and gas transport company, Transredes, was sold to a consortium of U.S.-based Enron and Royal Dutch/Shell. By December 2000, most of YPFB's assets had been sold off. In April 2002, no bids were submitted for YPFB's natural gas distribution networks.

In 1994, a Superintendency of Hydrocarbons was established to regulate the oil and gas industries. Its responsibilities include implementing the relevant laws, granting concessions and licenses, and ensuring a competitive industry. Once YPFB has sold its remaining assets, the Superintendency of Hydrocarbons is expected to take on a new role as the administrator of international negotiations and contracts with foreign oil companies.

Bid Rounds

Bolivia has held annual oil and natural gas exploration and production licensing rounds since 1997. In 1997, 16 blocks were awarded. Only six blocks were awarded in 1998, and five in 1999. In 1999, Pluspetrol was the only bidder. One block each was awarded to Petrogas Energy and Matpetrol in 2000. In 2001, Bolivia awarded four exploration blocks (Rio Hondo, Aquio, Ingre, Chaco).

Downstream

Of Bolivia's 63,000 bbl/d in crude oil refinery capacity, YPFB owns and operates 3,000 bbl/d. The other 60,000 bbl/d was sold to Brazil's Petrobras and Perez Companc as part of the wider privatization scheme in November 1999. Oil storage facilities, some domestic pipelines, and airport jet fuel terminals have also been sold, leaving only retail gasoline and diesel stations to be privatized. Dates for this final step have not been announced.

In July 2000, Bolivia introduced a "price collar" mechanism to prevent retail gasoline prices from fluctuating with world oil prices. The retail price is pegged to an international reference price, so higher prices lead to refiners accruing debt, while lower prices then reduce that debt. However, crude prices have averaged higher than expected, resulting in unsustainably high debt for producers and refiners. The government hopes to decide how to respond to this situation in early August.

In January 2002, Peru's President Alejandro Toledo, on his first official trip to Bolivia, proposed that the two countries form a "strategic alliance" in order to become the petrochemical centers of South America. Toledo emphasized the importance of adding value to raw materials, in this natural gas, by processing them into higher-end commodities.

Natural Gas

According to the *Oil and Gas Journal*, Bolivia's proven natural gas reserves were 24 trillion cubic feet (Tcf) as of January 1, 2002. However, other estimates are much higher–52 Tcf or more–making Bolivia a potentially major world natural gas exporter with the second-largest gas reserves (after Venezuela) in South America. Bolivia's proven natural gas reserves have increased enormously over the past few years, from only about 4.3 Tcf in 1999.

Bolivia has been producing natural gas since the 1960s. Exports have primarily been to Brazil (under a 20-year contract) since December 1998, but small volumes began flowing to Argentina as well in June 2002. The domestic Bolivian gas market is small, and while it is expected to grow, export markets will remain the key destination for Bolivian gas. In February 2002, Brazil and Bolivia pledged to strengthen bilateral ties in the energy field. One possibility is a $5 billion gas pipeline, known as Gafin, connecting southern Bolivia with southern Paraguay, northern Argentina, the southern Brazilian state of Santa Catarina, and Brazil's capital city, Brasilia.

Aggressive exploration efforts in recent years have led to many large discoveries, with several fields over 5 Tcf. Blocks containing significant gas reserves include Block 20 (Tarija West), San Antonio and San Alberto, El Dorado, and Caipipendi. Caipipendi contains what is currently the largest field, Margarita, which holds over 13 Tcf, according to some estimates. Growth in supply has outpaced growth in demand, and exploration has slowed considerably since 1999. In February 2002, Petrobras said that it expected natural gas production at the southern San Alberto field to double by 2003 to

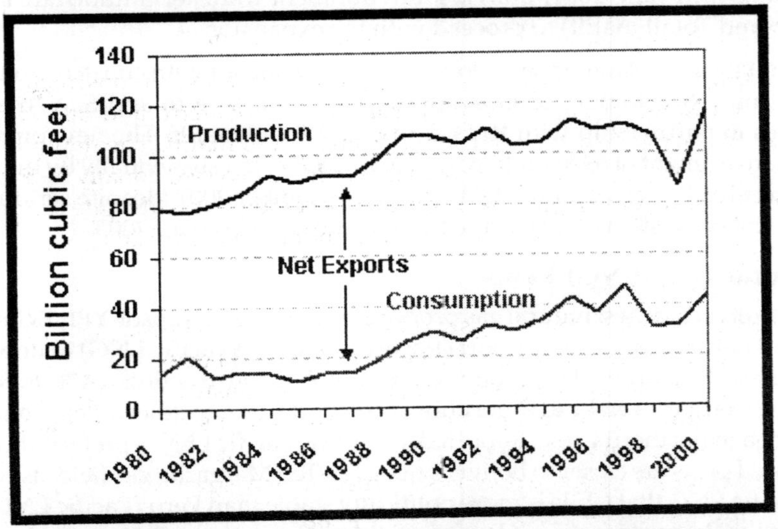

BOLIVIA'S NATUAL GAS PRODUCTION AND CONSUMPTION, 1980-2000

nearly 500 million cubic feet per day (MMcf/d). The adjacent San Antonio gas field is being developed by the same consortium.

Natural Gas Pipelines

The Bolivia-Brazil pipeline, tapping Bolivia's Rio Grande sources, came onstream in July 1999. It is the largest private sector infrastructure project in South America, and partners include Enron (although Enron is planning to sell its share), Shell, Petrobras, BG Group (British Gas), TotalFinaElf, and El Paso. The project began in 1996 and cost $2.1 billion. It serves Sao Paulo, Brazil, with an extension southward to Porto Alegre. The completed pipeline covers almost 2,000 miles. The pipeline is currently being utilized at only about one-third its full capacity (around 1 Bcf per day). This subject was discussed at a meeting between the Bolivian and Brazilian presidents in June 2001, where the two leaders signed a letter of intent agreeing to make more Bolivian gas available to Brazil, sooner rather than later. In March 2002, Brazil's National Petroleum Agency (ANP) announced a suspension of bidding on expansion of the Bolivia-Brazil gas pipeline due to uncertainties surrounding Brazil's power sector, which has been experiencing shortages and other difficulties in recent years. Regulatory hurdles in Brazil, for instance, reportedly have slowed construction of several dozen natural-gas-fired power plants there.

In addition, there has been controversy regarding Petrobras's preferential access to the Brazilian section of the Bolivia-Brazil pipeline, as well as disputes over pricing of Bolivian gas. Among other problems, the depreciation of Brazil's currency and high transportation tariffs have raised the price of Bolivian gas, making it less competitive with fuel oil and liquefied petroleum gas (LPG). Currently, Bolivia sells gas to Brazil for around $1.00-$1.20 pei million Btu (MMBtu), but transportation costs raise this to around $3.20 per MMBtu Brazilian energy companies have complained that this is too high, but Bolivia had not lowered its price as of early August 2002.

Bolivia's San Alberto and San Antonio fields began exports to Brazil in January 2001. The gas travels through an existing, 30-year-old pipeline (Yabog, owned by Bolivia's Transredes), connecting to the Bolivia-to-Brazil pipeline. Plans to expand Yabog over the next seven years or so would increase capacity substantially. In June 2001, Bolivia's Environment Minister authorized Petrobras (along with Repsol/YPF and TotalFinaElf) to proceed with the expansion.

Besides the Yabog expansion project, Bolivia is moving ahead with construction of another new gas pipeline (Yacuiba-Rio Grande, or Gasyrg) from the San Alberto and San Antonio fields that Petrobras operates in partnership with Repsol-YPF and TotalFinaElf. The new pipeline reportedly would require an investment of $300 million, which would be invested equally by the three firms. The pipeline project received an environmental license in December 2001 (despite serious concerns over damage to dry forest ecosystems), and is due to come online in January 2003.

Liquefied Natural Gas (LNG) Plans

In July 2001, Bolivia's largest natural gas producing companies–Repsol-YPF, BG, and BP–decided to form a liquefied natural gas (LNG) export consortium (called Pacific LNG) in hopes of delivering Bolivian gas to the U.S. market. The project could involve constructing a 430-mile pipeline from Bolivia's southern Margarita field to a port (most likely ILO) in southern Peru, or alternatively to Mejillones or Cobija in northern Chile. Reportedly, Chile is Pacific LNG's preferred choice due to the fact that its proposed ports are closer to Bolivia's huge (13-Tcf) Margarita gas field, as well as due to the perception by Pacific LNG that Chile is more politically stable than Peru (Pacific LNG has stated that it would consider the Peruvian option if Peru paid the extra cost–possibly as much as $600 million – entailed in building a pipeline to ILO).

In July 2002, Bolivia's newspaper *La Prensa* reported that Bolivia had reached an agreement to export through Chile, but this has not been officially confirmed. In either case, a liquefaction plant would be built at the port of choice, and the LNG would then be shipped to a regasification plant in northern Mexico (to be constructed by US firms Sempra Energy and CMS Energy). From there, the gas–possibly 800 MMcf/d–would be piped to northern Mexican (*i.e.*, Baja California), to southern California, and to the US Southwest. The entire project would cost an estimated $5-$8 billion, and earn Bolivia as much as $1 billion per year by 2007. In December 2001, Sempra and Pacific LNG signed a memorandum of understanding on gas deliveries beginning in 2004 at the earliest and lasting for 20 years.

Electricity

Approximately 71 per cent of Bolivia's electri . power generating capacity is thermal (oil and natural gas), with the remaining 29 per cent hydroelectricity. In 2000, the country generated about 3.9 billion kilowatthours, approximately half of which was thermal and half hydropower.

In 1994, Bolivia passed an electricity law that established the Superintendency of Electricity as the regulatory body for the Bolivian electricity sector. Its mandate includes protecting consumer rights; granting and amending concessions and licenses; approving international interconnections and stipulating the quantity of electricity exports and imports (should interconnections be established); supervising the activities of the National Committee of Load Dispatch, the body responsible for the coordination and administration of transactions on the national grid, the *Sistema Interconnectado Nacional* (SIN); and approving and setting prices and tariffs for the electricity industry.

The 1994 law divided the assets of ENDE, the national electric company, and separated the responsibilities for electric generation, transmission and distribution. Any generating company operating on the SIN is prohibited from owning more than 35 per cent of national installed generating capacity. The law also stipulated that companies on the SIN are only allowed to participate in one activity: generation, transmission, or distribution. However, power companies that do not operate on the SIN are allowed to be vertically integrated.

Generation

There are four major power generating companies in Bolivia–Guaracachi, Valle Hermosa, Corani and Cobee. The first three companies are regional utilities that resulted when ENDE was split up. They are now 50 per cent privately-owned (by U.S.-based Energy Initiatives, Constellation Energy, and Duke Energy) and the remaining 50 per cent of each of the company's shares have been placed in private pension funds for Bolivians. The fourth generation company, Cobee, is owned by U.S.-based NRG Energy, and it operates in the La Paz and Oruro departments. Cobee has always been privately owned and is subject to slightly different regulations than the other three generators.

Eight hydroelectric plants power the SIN, with Corani and Cobee supplying the bulk of the power. The country's significant hydropower potential has not yet been fully exploited. Several new hydro plants are in various stages of the planning process. There are six thermal power plants on the SIN. Guaracachi and Valle Hermosa are the largest thermal generators. Thermal capacity has risen considerably in the last several years, and electricity supply now outstrips demand.

Transmission and Distribution

The national grid, the SIN, accounts for the vast majority of Bolivia's installed electric capacity and power generation, reaching five of the country's nine "departments". The grid was privatized in 1997 and is owned and operated by Transportadora de Electricidad (TDE), itself owned by Red

Electrica Espana following its purchase of Union Fenosa's 69 per cent share in July 2002. A plan to expand the SIN received regulatory approval in March 2001. In May 2002, Bolivia's Electricity Superintendency launched a tender for construction and operation of three power transmission lines: Santivanez-Sucre, Sucre-Punutuma, and Carrasco-Urubo.

Although the SIN connects Bolivia's major cities, an estimated 80 per cent of the country's rural population does not have access to electricity. The National Rural Electrification Program (Proner) aims to increase the rural electrification rate in coming years. Bolivia has been developing alternative electricity sources for its isolated communities. Renewable energy sources such as solar and biomass fuel small power generators in the highlands. Also, grid connections are expanding to include areas previously not accessed by the grid.

As part of the restructuring of the electricity sector, Cobee and ENDE sold their distribution assets in 1996. Eléctropaz, CRE, and Elfec are the three largest distributors in Bolivia, together accounting for more than three-quarters of power sales. The government plans to offer several regional distribution companies for sale in the near future. In late 2000, the sales process for provincial electricity system Servicios Electricos Tarija SA (SETAR) was restarted by the solicitation of bids for the advisory contract to manage the privatization.

Country Overview

Independence: August 6, 1825 (from Spain)

Population (July 2001E): 8.3 million

Location/Size: Central South America; 1,098,580 sq. km (424,164 sq. mi.), slightly less than three times the size of Montana/Central South America, southwest of Brazil

Major Cities: La Paz (executive capital), Sucre (judicial capital), Santa Cruz de la Sierra, Cochabamba

Languages: Spanish (official), Quechua (official), Aymara (official)

Ethnic Groups: Quechua (30 per cent), Aymara (25 per cent), Mestizo (mixed European and Indian ancestry- 25-30 per cent), European (5-15 per cent)

Religions: Roman Catholic (95 per cent), Protestant and other (5 per cent)

Defense (8/98): Army (25,000), Navy (4,500), Air Force (4,000)

Economic Overview

Currency: 1 bolivianos ($B) = 100 centavos

Market Exchange Rate (8/13/01): US$1 = $B7.51

Gross Domestic Product (GDP at market exchange rates, 2002E): $7.4 billion

Real GDP Growth Rate (2001E): 1.2 per cent *(2002E):* 1.6 per cent

Inflation Rate (consumer prices, 2001E): 1.8 per cent *(2002E):* 1.0 per cent

Current Account Balance, as per cent of GDP (2001E): -$0.5 per cent

Merchandise Exports (2002E): $1.3 billion

Merchandise Imports (2002E): $1.6 billion

Net Merchandise Trade Deficit (2002E): -$0.3 billion

Major Export Products: Processed soya, zinc, natural gas, gold

Major Import Products: Capital goods, raw materials, consumer goods

Major Trading Partners: United States, Argentina, Brazil, Peru

Energy Overview

Proven Oil Reserves (1/1/02E): 440.5 million barrels

Oil Production (2001E): 43,841 barrels per day (bbl/d), of which 34,337 bbl/d was crude oil

Oil Consumption (2001E): 43,000 bbl/d

Net Oil Exports (2001E): 841 bbl/d

Crude Oil Refining Capacity (1/1/02E): 63,000 bbl/d

Proven Natural Gas Reserves (1/1/02E): 24 trillion cubic feet–Tcf (some estimates are as high as 52 Tcf)

Natural Gas Production (2000E): 117 billion cubic feet (Bcf)

Natural Gas Consumption (2000E): 44 Bcf

Net Natural Gas Exports (2000E): 73 Bcf

Electric Generation Capacity (2000E): 1.34 million kilowatts (of which 71 per cent is thermal, 29 per cent is hydro)

Electricity Generation (2000E): 3.9 billion kilowatthours

Environmental Overview

Total Energy Consumption (2000E): 0.15 quadrillion Btu* (<0.1 per cent of world total energy consumption)

Energy-Related Carbon Emissions (2000E): 2.45 million metric tons of carbon (<0.1 per cent of world carbon emissions)

Per Capita Energy Consumption (2000E): 18.4 million Btu (vs. U.S. value of 351.0 million Btu)

Per Capita Carbon Emissions (2000E): 0.29 metric tons of carbon (vs U.S. value of 5.6 metric tons of carbon)

Energy Intensity (2000E): 19,215 Btu/ $1995 (vs U.S. value of 10,918 Btu/ $1995)**

Carbon Intensity (2000E): 0.31 metric tons of carbon/thousand $1995 (vs U.S. value of 0.17 metric tons/thousand $1995)

Sectoral Share of Energy Consumption (1998E): Industrial (36.5 per cent), Transportation (28.9 per cent), Residential (29.5 per cent), Commercial (5.1 per cent)

Sectoral Share of Carbon Emissions (1998E): Industrial (31.8 per cent), Transportation (45.2 per cent), Residential (19.5 per cent), Commercial (3.5 per cent)

Fuel Share of Energy Consumption (2000E): Oil (56.2 per cent), Natural Gas (30.1 per cent), Hydroelectricity (13.1 per cent)

Fuel Share of Carbon Emissions (2000E): Oil (68 per cent), Natural Gas (32 per cent)

Number of People per Motor Vehicle (1998): 19.2 (vs U.S. value of 1.3)

Status in Climate Change Negotiations: Non-Annex I country under the United Nations Framework

Convention on Climate Change (ratified October 3rd, 1994). Bolivia ratified the Kyoto Protocol on November 30th, 1999.

Major Environmental Issues: Clearing of land for agricultural purposes and the international demand for tropical timber are contributing to deforestation; soil erosion from overgrazing and poor cultivation methods (including slash-and-burn agriculture); desertification; loss of biodiversity; industrial pollution of water supplies used for drinking and irrigation.

Major International Environmental Agreements: A party to Conventions on Biodiversity, Climate Change, Desertification, Endangered Species, Hazardous Wastes, Law of the Sea, Nuclear Test Ban, Tropical Timber 83, Tropical Timber 94 and Wetlands. Has signed, but not ratified, Environmental Modification, Marine Dumping, Marine Life Conservation and Ozone Layer Protection.

The total energy consumption statistic includes petroleum, dry natural gas, coal, net hydro, nuclear, geothermal, solar, wind, wood and waste electric power. The renewable energy consumption statistic is based on International Energy Agency (IEA) data and includes hydropower, solar, wind, tide, geothermal, solid biomass and animal products, biomass gas and liquids, industrial and municipal wastes. Sectoral shares of energy consumption and carbon emissions are also based on IEA data.

Oil and Gas Industries

Organization: Yacimientos Petroliferos Fiscales Bolivianos (YPFB), foreign companies

Major Ports: None (landlocked)

Foreign Energy Company Involvement: BG, BHP, BP, Orca Petroleum, Perez Companc, Petrobras, Pluspetrol, Red Electrica Espana, Repsol-YPF, Royal Dutch/Shell, TotalFinaElf

Major Oil and Gas Blocks: Tarija, Tarija West, San Antonio, San Alberto, El Dorado, Caipipendi

Major Pipelines: Bolivia-to-Brazil

Major Refineries (crude oil capacity, 2002): Cochabamba (40,000 bbl/d), Santa Cruz (20,000 bbl/d), Sucre (3,000 bbl/d)

14

Brazil

Brazil is the largest country in South America and has experienced rapidly expanding oil, natural gas, and electricity markets in recent years. The country is in the process of recovering from an energy crisis in 2001.

General Background

In October 2002, Luiz Inacio Lula da Silva was elected president, replacing Fernando Henrique Cardoso. The election marked a return of the left to power, 38 years after a military coup overthrew President Joao Goulart of the Brazilian Labor Party (PTB). President da Silva was elected with the support of an alliance composed of his own leftist Workers' Party, the PT, the center right Liberal Party, (PL), the National Mobilization Party (PMN), the Popular Socialist Party (PPS, formerly the PCB), and the Communist Party of Brazil (PCdoB). During his election campaign, Mr. da Silva promised to lead campaigns against hunger (Zero Hunger Initiative) and for increased social spending in the areas of health, education and public security. The new government continues to work on reforming the pension system and the tax code after submitting two bills to the Congress in late spring.

In June 2003, Brazil signed two energy cooperation agreements with the United States. As part of the agreements, the two countries will begin discussion on collaborating on hydrogen and fuel cells, making Brazil the first Latin American country to join the Bush administration's proposed International Partnership for the Hydrogen Economy. Brazil and the United Stated also signed a bilateral agreement on electricity deregulation, offshore safety, and the development of clean energy technologies, as well as began negotiations for Brazil to become a charter member of the Carbon Sequestration Leadership Forum.

Oil

Brazil contains the second largest oil reserves in South America (after Venezuela), at 8.3 billion barrels. The country continues to strive for self-sufficiency in oil production by 2006 and has made positive steps towards reaching this goal. In May 2003, Petrobrás announced that it plans to invest

$34 billion over the next four years, mainly in exploration. The state-owned oil company also expects its average domestic oil production (not including production from other companies) to reach 1.82 million barrels per day (bbl/d) by 2005.

The drivers behind the increase are a number of new fields coming onstream and improved production performance, according to Petrobrás. However, new government regulations have delayed the construction of new floating production platforms, which resulted in Petrobrás downgrading its initial production estimate of 1.9 milliuon bbl/d by 2005.

Overall, total oil production has been rising steadily since the early 1990s, averaging nearly 1.6 bbl/d in 2002. The offshore Campos Basin, north of Rio de Janeiro, is the country's most prolific production area (see map below). Brazil's oil consumption for 2002 was estimated at almost 2.2 million bbl/d. Brazil's oil imports come mostly from Africa and the Middle East and, to a lesser extent, from Venezuela and Argentina, according to 2000 data.

Petrobrás is seeking to expand its oil and natural gas operations outside of Brazil. In October 2002, The company acquired a majority stake in Argentina's Perez Companc, increasing its oil and natural gas reserves significantly.

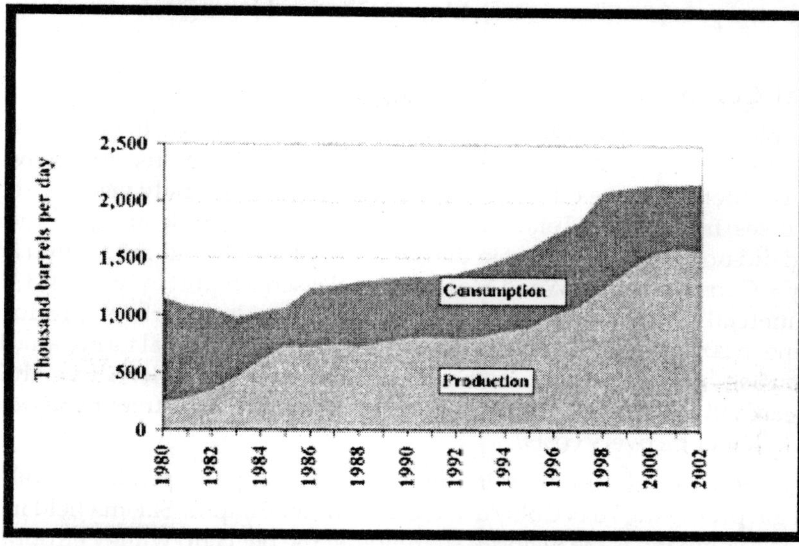

BRAZILIAN TOTAL OIL PRODUCTION AND CONSUMPTION, 1980-2002

Exploration and Production

Petrobrás is the only company to have made commercially viable discoveries in Brazil in recent years. Most of these are located in the northern Campos Basin, along the coast of Espirito Santo State, at block BC-60. In June 2003, Petrobrás announced that it had made three new discoveries in the block.

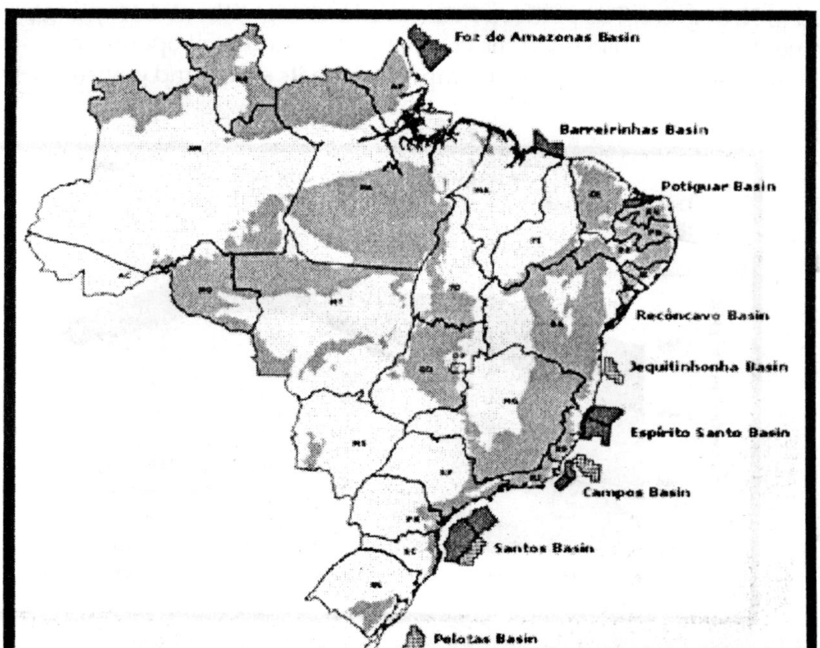

These finds follow a previous discovery in May 2003 and in March 2003, at the Sergipe offshore SEAL 100 block.

International Companies

Despite exploration efforts, Total, BP and ConocoPhillips have yet to find an oil field in Brazil that is large enough to be commercialized. According to these companies, recent discoveries are small and located in deep waters, and the oil contained in the fields is extremely viscous, requiring expensive recovery processes. In July 2003, Total announced plans to scale back on exploration activities for the time being and did not plan to take part in the next oil exploration and production licensing round, set for August 2003. Consulting firm Wood MacKenzie released a report in June 2003 stating that the lack of recent commercial discoveries will impact Brazil's mid-to-long-term oil production, with decline possibly starting in 2007. Wood MacKenzie added that a change in fiscal terms, attached to exploration and production activities, could make some fields commercially viable. Although Brazil has large oil reserves, increases in production will be depend on attracting investment and on new technologies that would help lower recovery costs.

Repsol-YPF is currently the only international oil company producing oil in Brazil. Shell is expected to begin producing 70,000 bbl/d at the deepwater Bijupira-Salema field in the Campos Basin by mid-2003. The company is the majority shareholder (80 per cent) while Petrobrás holds the rest. In May 2003, Shell also received permission to export oil.

Accidents and the Environment

Over the years, *poor safety and environmental records* have tarnished Petrobrás' image. In March 2001, explosions on the deepwater platform P-36, located at the Roncador field, caused the world's largest platform to sink into the Atlantic Ocean. Production at the Roncador stopped but was brought back onstream in December 2002. In October 2002, platform P-34 had to be evacuated when the platform tilted to 32 degrees. The platform was saved but not without disrupting production. Brazil's state environmental agency also fined the company for running oil operations without the required environmental licenses. Petrobrás plans to improve both its safety and environmental records.

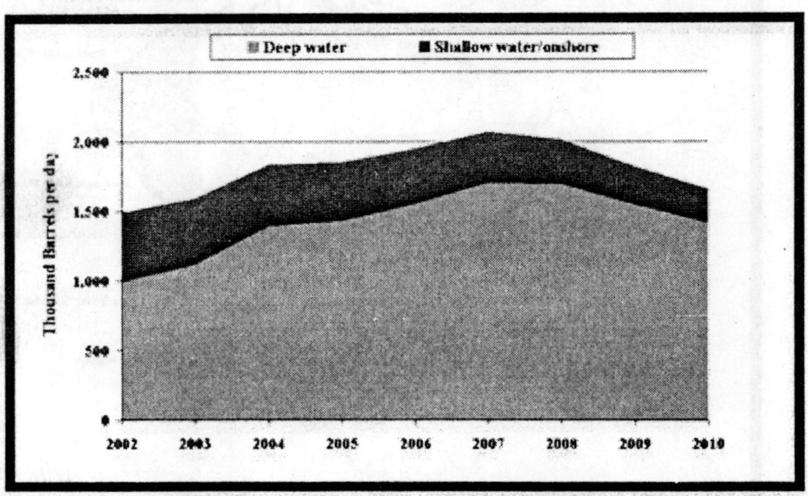

Sector Organization

Prior to 1997, Petrobrás, which was created in 1953, monopolized rights to explore, produce, refine, and distribute oil in Brazil. Prices for Petrobrás oil were also fixed. In 1997, the former President Cardoso signed the Petroleum Investment Law. The law permitted joint ventures between foreign oil companies and Petrobrás. The law also opened up Brazil's sedimentary basins to drilling operations by foreign and Brazilian firms. The new law also created National Petroleum Agency (ANP), charged with issuing tenders, granting concessions for domestic and foreign companies, and monitoring the activities of the sector, including establishing rights to explore for and develop oil and natural gas in Brazil. One of the main drivers behind opening the oil sector was to increase oil production in order to reduce dependence on oil imports and eventually achieve self-sufficiency. Most of Brazil's reserves were and remain located offshore in deep water, requiring extensive capital investment to develop.

In July 1998, ANP announced that more than 92 per cent of the nation's sedimentary basins were to be put up for bidding by other oil companies. Also in 1998, oil prices in Brazil became linked to world oil prices. In August 2000, the government sold a 28.5 per cent stake in Petrobrás. In January 2002, the company lost its monopoly on importing oil and refined products.

New Government

After taking office, President da Silva began showing signs of reversing some of the privatization policies of the former administration. For instance, the government now requires companies which are awarded oil and natural gas leases to devote a percentage of their investments to *purchase goods and services provided by Brazilian firms*. This requirement fulfills one of President da Silva's electoral promises to increase the role of the domestic industry in developing the country's oil and natural gas reserves. During the presidential campaign, Mr. da Silva stated that the oil platforms P-51 and P-52 should be built domestically in order to create economic growth.

In July 2003, Petrobrás launched a bidding round for the building of P-54 production platform. According to the new regulations, the project will require between 60 per cent and 75 per cent of the parts to be constructed in Brazil

Tax in Rio de Janeiro State

On June 30, 2003, the governor of Rio de Janeiro signed a bill that imposes an 18 per cent tax on oil production, but then suspended its implementation in the face of strong opposition from oil companies (Shell and ChevronTexaco) operating in the country. If the tax is eventually approved, it will be applied at the wellhead, but not to exports, sparing foreign companies producing oil in Brazil. Nonetheless, some analysts suggest that the new tax would further discourage interest from foreign companies, as they have already expressed concern about recent small oil finds. In addition, these finds are mainly heavy crude and at great depths, which only increases production costs.

International Cooperation

Brazil and Venezuela have held talks discussing potential oil and natural gas partnerships between Petrobrás and Petroleos de Venezuela (PDVSA). Activities include heavy crude oil refining, exploration, production, as well as joint ventures in the petrochemicals and natural gas sectors. The first project could be a partnership in constructing a new heavy crude refinery, which Petrobrás intends to build.

Bid Rounds

Since opening up the oil industry, ANP has held four bidding rounds. The fifth round, planned for August 19-20, 2003, differs in several ways from previous rounds. The government decided (as mentioned above) to require increased inclusion of local equipment and services related to oil operations. Second, a majority of the blocks up for bid are small and located in shallow waters, whereas previous rounds offered larger blocks. ANP hopes to attract smaller oil companies to bid on these blocks. In all, 1,075 blocks will be up for bidding, of which 262 are located onshore, 742 in shallow waters, and 71 in deep waters, according to ANP. Finally, analysts have suggested that this round is less attractive for larger companies than previous rounds, citing uncertainties about taxes, lack of recent commercial discoveries, the geological challenge of oil recovery in deep waters, and new opportunities for oil companies in the Middle East.

Previous Rounds

Foreign companies first became involved in the Brazilian oil sector in 1997, through joint-ventures with Petrobrás. The first bidding round that allowed foreign companies to compete against Petrobrás occurred in 1999. The blocks for offer in the round, which were extremely large (the average size of each area was 1,800 square miles, equivalent to 225 blocks in the Gulf of Mexico), were sold to 10 foreign firms, as well as to Petrobrás, which won 5 of the 12 blocks. Foreign firms, such as Agip, Unocal, Texaco, Amerada Hess, Repsol-YPF, and ExxonMobil made successful exploration bids, with some acting in alliance with Petrobrás. Most of the bids were for relatively unexplored but highly coveted areas in over 6,560 feet (2,000 meters) of water off Brazil's Atlantic coast.

In June 2000, Brazil's second licensing round offered 23 blocks in nine sedimentary basins, including Campos (the Campos Basin is responsible for roughly 70 per cent of the country's current total crude oil output), Amazonas, Camamu-Almada, Parana, Para-Maranhao, Potiguar, Reconcavo, Sergipe-Alagoas, and Santos. The second round was hailed as more successful than anticipated, with Petrobrás winning many bids in partnerships with foreign companies. The Campos and Santos Basin blocks generated the most interest, receiving as many as four bids. Only two blocks received no bids.

Brazil's third bidding round, held in June 2001, offered 53 blocks, 43 offshore, and mostly in deep and ultra-deep water areas. Although about a third of the offered blocks received no bids, the round was considered a success because of the wide range of aggressive bidders that participated. The bidding round attracted major international oil companies, such as ExxonMobil, Royal Dutch/Shell, Total, and Statoil, as well as some smaller companies that were new to the Brazilian oil sector, such as U.S.-based Ocean Energy and Germany's Wintershall. As in the previous rounds, Petrobrás was the largest benefactor, winning 13 blocks and joint venture partner in two others (with operators ExxonMobil and Total).

The fourth round offered 39 offshore and 15 onshore blocks. Eight blocks were in "frontier areas" and six were in the Campos Basin. The average block size was considerably larger than those in the third round. The auction was held in June 2002, with Petrobrás once again the largest winner. Statoil, El Paso, and several other companies increased their investments in Brazil. (Complete round results are available from brasil.gov)

Downstream

Petrobrás plans to invest $5.5 billion by 2007 to increase domestic refining capacity by 300,000 bbl/d, to 1.9 million bbl/d, and to upgrade many of its refineries. One of the goals of the da Silva administration is to increase Brazil's ability to refine heavy crude oil, which makes up a large portion of the country's crude oil production. In April 2003, Petrobrás revealed plans to build a 150,000-bbl/d refinery, which would be the first one in Brazil to process heavy crude. It remains unclear where the refinery will be built and whether it will be built in conjunction with Venezuela's PDVSA).

There are 13 crude oil refineries in Brazil, 11 of which belong to Petrobrás and comprise almost 99 per cent of total refinery capacity.

Natural Gas

Brazil's natural gas consumption and production in Brazil rose steadily throughout the 1990s, with imports beginning in 1999 as demand outstripped supply. Natural gas production reached its highest level in 2000 at 257 billion cubic feet (Bcf), but declined in 2001 by 18 per cent mainly due to an economic downturn. Consumption, however, continues to increase, reaching an estimated 339 Bcf in 2001. As of January 2003, natural gas reserves stood at 8.1 trillion cubic feet (Tcf), the fifth largest in South America behind Venezuela, Argentina, Bolivia, and Peru.

Industry Organization

Natural gas exploration and production historically have been carried out by Petrobrás. Upstream privatization has not progressed as quickly with natural gas as it has with oil. Distribution, however, traditionly has been handled at the state level. In accordance with Brazil's national privatization agenda, plans are underway to privatize parts of the country's natural gas sector. Because Brazil maintains a highly federalized government system, many state governments are now shouldering heavy fiscal responsibilities. In an effort to raise necessary working capital, state governments have begun to sell their state natural gas distribution companies. Natural gas distribution companies in Rio de Janeiro and Sao Paulo are now in private hands.

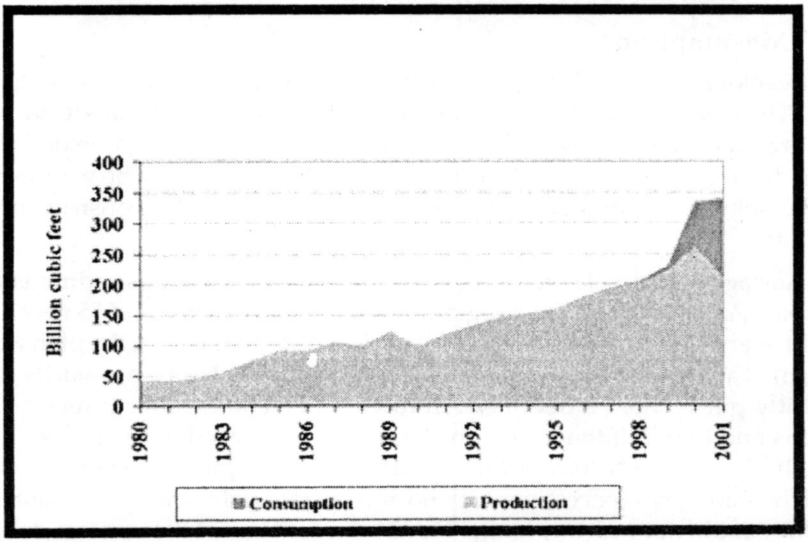

BRAZILIAN NATURAL GAS CONSUMPTION AND PRODUCTION, 1980-2001

Exploration

The Campos and Santos basins hold the largest Brazilian natural gas fields. Other natural gas basins include Foz do Amazonas, Ceara e Patiguar, Pernambuco e Paraíba, Sergipe/Alagoos, Espírito and Amazonas (onshore). Offshore southeast Brazil is hydrocarbon rich yet under explored, leaving potential for significant increases in reserves and production. In May 2003, Petrobrás confirmed Brazil's largest natural gas discovery to date, an estimated 2.45 Tcf, on Block BS-400 in the Santos Basin.

Promotion of Natural Gas

Unlike the previous administration, the da Silva government wants hydroelectricity to remain Brazil's main source of power. In the late-1990s, the Brazilian government under former President Cardoso had been encouraging the use of natural gas in an attempt to reduce the country's dependence on hydropower. In the past, hydroelectric plants were able to keep pace with Brazil's growing demand for electricity, but rapid electricity demand growth in the 1990s resulted in power cuts, blackouts and rationing programs.

In 1997, the government set out to increase natural gas' share of total energy consumption from 2.5 per cent to 12 per cent by 2002. Along with encouraging the construction of natural gas-fired power plants, Petrobrás signed a 20-year take-or-pay commitment with Bolivia to import natural gas through the long-awaited Brazil-Bolivia (Gasbol) pipeline, which came onstream in June 1999. At the original signing of the contract, the supply volume was tied to Brazil building 16 natural-gas-fired power plants. Out of the 16 plants, Petrobrás has pulled out of six planned projects and is contemplating its role in two others. On a positive note, EdF announced in July 2003 that it plans to restart construction of the $400 million Paracambi thermoelectric power plant. EdF also plans to begin operation later this year of its 720MW Norte Fluminense thermoelectric plant.

Natural Gas Consumption

Economic developments hindered the Cardoso administration's efforts to enlarge Brazil's market for natural gas. The devaluation of *the real* in January 1999, for example, made natural gas imports from Bolivia more expensive, as power producers had to pay for natural gas in dollars. In addition, a severe recession deterred banks from making loans to power developers, who were expected to absorb nearly half of the Bolivian imports. All of these effectively disrupted the construction of natural gas-fired power plants.

In 2001, a shortage of hydroelectricity due to a drought led the Cardoso administration to launch the *Priority Thermal Power Program*. The program envisaged development of 55 new natural gas-fired plants over eight years with a combined capacity of 23,000 MW. However, the plan was discontinued when rains returned along with the availability of cheap hydroelectricity. A majority of the new plants were subsequently put on hold indefinitely. In July 2002, metal manufacturers, such as Canadian Alcan, U.S. Alcoa and BHP Billiton in conjunction with power producers Tractebel, increased their commitment to hydroelectricity to support expanding their output. The Brazilian energy regulator Aneel awarded the companies concessions to build and operate eight new hydro plants. The companies indicated that they could not wait any longer for natural gas to meet their energy requirements.

New Developments: Bolivian Imports

Since signing an agreement with Bolivian state natural gas exporter YPFB in 1997, Brazil has taken only a fraction of the natural gas, to which it had agreed. Government delegations from Bolivia and Brazil have met repeatedly over the last six months in attempt to renegotiate the contract. Brazil wants to import less and pay less for natural gas from Bolivia, suggesting a more flexible agreement. Bolivia indicated after a meeting held in May 2003 that it would not change the contract unless Brazil would guarantee that exports would increase in the future.

In response, the Brazilian government revealed in June 2003 its new *Natural Gas Expansion Project*. The objective of the plan is to expand the country's natural gas pipeline network (see pipelines below), bringing natural gas to residential, industrial, and thermal power plants, particularly to southeastern and northeastern regions of the country where no pipelines currently exist, and which now rely on fuel oil. One of the drivers of the project is Brazil's commitment to expanding the domestic natural gas market as part of its ongoing negotiations with Bolivia.

Pipelines

Brazil has two existing international pipeline connections, with several more under construction. The first pipeline to connect Brazil to foreign natural gas sources was the *Bolivia-to-Brazil pipeline (Gasbol)*, tapping Bolivia's Rio Grande sources. The completed covers almost 2,000 miles, and came onstream in June 1999, with service to Sao Paulo and a terminus in Porto Alegre. The pipeline continues to operate well below its 1.05 billion cubic feet per day capacity, due to the high price of dollar-denominated Bolivian natural gas and the minimal development of natural gas-fired generating capacity in Brazil. As a result, the Brazilian government has suspended plans to expand the Brazil-Bolivia pipeline.

MAIN BRAZILIAN GAS PIPILINES: INTERCONNECTION OF SOUTH AMERICAN PIPELINES

In 2002, BG signed nine-year agreement with Petrobrás, guaranteeing access to 22.8 million cubic feet per day from the pipeline from January 1, 2003 to December 31, 2011. Enron and El Paso Energy announced in February 2003 that they plan to sell their stakes in the pipeline as the expected revenue from the project has yet to materialize.

Brazil's second operational international natural gas pipeline links Paraná, Argentina, to Uruguaiana, Brazil. The pipeline, *Transportadora de Gas del Mercosur,* is 270 miles long. The pipeline supplies natural gas to a $350-million, 600-megawatt (MW) AES power plant in Uruguaiana. Service began in July 2000. The Argentinean government is currently promoting a 384-mile pipeline extension, the *Transportadora Sulbrasileira de Gas (TSB)*, which would connect *Uruguaiana to Porte Alegre*. However, the project most likely will not proceed until there is a decision on constructing two new natural gas-fired power plants in Porto Alegre.

New Pipeline Developments

Petrobrás recently released a statement that it is conducting a feasibility study on a 745-mile pipeline that would link Brazil's southeast and northeast networks, allowing it to transport natural gas from neighboring Bolivia and from Petrobrás' fields in the Santos Basin to the northeast.

Petrobrás also is negotiating with the states of Amazonas and Rondonia to build a 342-mile *Urucu to Porto Velho* pipeline. The pipeline is expected to supply 80.5 million cubic feet per day, mainly to power a 404 MW thermoelectric plant owned by El Paso and CS Participacoes. Environmental licensing problems have delayed the construction of the pipeline. In the same region, Petrobrás is proceeding with plans to build a 261-mile *Coari to Manaus* pipeline to be able to supply natural gas from the Urucu field to regional industry.

In January 2003, Argentinean natural gas began to flow to Montevideo, Uruguay, through the *Gasoducto Cruz del Sur* (GCDS, Southern Cross pipeline). The pipeline is significant because the GCDS project includes a concession covering a possible extension from Uruguay to Porto Alegre in southern Brazil. Partners in the project are British Gas (40 per cent), Pan American Energy (30 per cent), Uruguayan state company ANCAP (20 per cent) and Wintershall (10 per cent). (For more details on natural gas distributors in Brazil).

Liquefied Natural Gas (LNG)

In January 2003, Petrobrás announced the development of a plan to bring stranded natural gas reserves in the Amazon to the market. The *Green LNG* project would pipe natural gas from the Jurua, Uruco, and Polo Arara fields to a floating liquefaction plant on the Amazon River near the existing Solimoes refinery. At the refinery, the natural gas would be liquefied and loaded onto a specially designed LNG transport vessel, then shipped to a smaller 1.5-million-ton-per-year capacity regasification terminal located in Suapeto. The vessel would have a capacity of 3.5 million cubic feet in order to be able to navigate the Amazon River. According to Petrobrás, the next steps of the project include a more rigorous feasibility study.

Coal

Brazil's recoverable coal reserves are estimated at approximately 13.1 billion short tons of lignite and sub-bituminous coal, the largest coal reserves in Latin America. Much of Brazil's coal is characterized by high ash and sulfur contents, as well as low caloric values. Overall, Brazil's coal production in 2001 was approximately 4.53 million short tons (Mmst), while consumption was an estimated 20.8 Mmst, with net imports of 16.3 Mmst.

Coal is used predominantly for Brazil's domestic steel industry, with a small portion burned to generate electricity. A 1996 study by the International Energy Agency (IEA) concluded that Brazil will continue to be one of the world's major coal importers during the next 10 to 15 years, with imports possibly doubling by 2010. The country's steel industry is expected to remain the largest domestic coal consumer for the foreseeable future. Brazil plans to increase its use of steam coal (both domestic and imported) for electric power generation.

Electricity

Brazil has installed electric capacity of 73.4 million kilowatts, 85 per cent of which is hydropower (as of January 1, 2001). Of the 321.2 billion kilowatthours (bkwh) generated in Brazil in 2001, 83 per cent was from hydropower (down from 91 per cent in 1999). Brazil ranks consistently as one the world's top hydropower producers. Together with Paraguay, Brazil maintains the world's largest operational hydroelectric complex, the Itaipu facility on the Paraná River, with a capacity of 12,600 megawatts (MW). Brazil's remaining electricity generation capacity comes from coal and an increasing amount from natural gas. Brazil's small northern and larger southern electrical grids were joined in January 1999 into one grid that serves 98 per cent of the country. Brazil's domestic supply is augmented by imports from neighboring Argentina.

Electricity Shortage

In 2001, Brazil faced a critical electricity shortage. The shortage was a result of insufficient rainfall, as well as under investment in the industry. Several years of below-average rainfall left reservoirs 70 per cent depleted. During the 1990s, energy demand, on account of a growing economy and a rising standard of living, had been consistently outstripping increments ingeneration capacity. In 2000, consumption was about 58 per cent higher than it was in 1990, while installed generation capacity

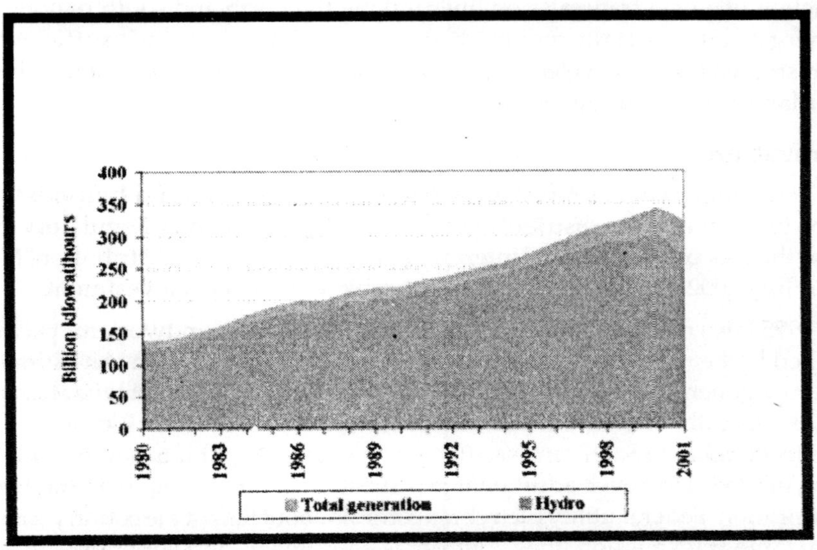

BRAZILIAN ELECTRICITY GENERATION, 1980-2001

grew about 32 per cent during the same period. Analysts had long predicted that this demand growth, if not supported by capacity growth, had the potential to lead to shortages.

The government implemented an energy-rationing program in June 2001, which prevented rolling blackouts by requiring customers to reduce consumption by 20 per cent. Restrictions were most severe in the heavily populated southeast and central-west regions, and northeastern states had three mandatory holidays declared in the autumn of 2001 in order to reduce demand. Restrictions ending on March 1, 2002.

New Developments

Once the government lifted the rationing program, the country found itself facing a power glut. This pushed many electricity companies deep into debt, such as Cemar, Eletropaulo Metropolitana, Cesp, Cemig and AES. The power glut has reduced incentives for companies to begin construction of thermoelectric power plants. In July 2003, the Energy and Mines Minister Dilma Roussef warned that country could face another power crisis by 2007, particularly if the economy begins to rebound. A report from the Brazilian Basic Infrastructure and Industry Association reported that the country would need to invest around $4.8 billion annually through 2020 to sustain moderate growth and avoid another crisis.

In July 2003, Brazilian Energy Minister Dilma Rousseff revealed a new model for the electricity sector, with the goals of ensuring reliable supply, stabilizing prices for consumers and attracting long-term investment to the sector. One of the proposed reforms included pooling cheaper hydroelectricity with more expensive thermoelectric plants (natural gas). By pooling the various sources, the ministry hopes to reduce the electricity tariffs and to ensure power is purchased from the newly constructed thermal plants, which have yet fully amortized. The Brazilian government plans to implement the program by January 2004.

As a first step in ensuring a more reliable supply, the government has put out tenders (July 2003) for the construction of a new transmission lines linking the north and south regions of the country. During the energy crisis, certain regions had excess supply but an insufficient transmission infrastructure restricted the flow of electricity, particularly to the southeast region, where are majority of Brazil's population and industry resides.

Sector Organization

Although generation remains mostly under government control and transmission is not slated for privatization in the near term, distribution is mostly in private hands. Regulatory difficulties have been blamed for the lack of international interest in Brazilian electricity distribution. New regulations in February and July 2002 aimed to address these problems and boost investment.

After two 1995 laws restructured the industry and laid the groundwork for private investment, the process stalled in the wake of the country's 1999 currency devaluation. An estimated 80 per cent of Brazilian electric generation remains in public hands. As part of the privatization program, three remaining major federally owned utilities, Eletronorte, Furnas Centrais Eletricas (Furnas) and the Companhia Hidroeletrica do Sao Francisco (Chesf), have been split into several smaller generating and distributing utilities. However, additional privatization of these companies has been ruled out by the Lula administration. Federal utility Eletrobrás controls about half of the country's installed capacity and most of the large transmission lines. Eletrobrás coordinates and supervises the expansion and operation of the generation, transmission and distribution systems. Private capital flows resulting from privatization had been expected to play a key role in bolstering the industry, especially as state-owned generators have not had investment capital available.

Nuclear Power

Brazil has two operational nuclear plants, Angra-1 and Angra-2, and one under construction, Angra-3, although construction has stalled. Angra-1 was bought from the U.S. company Westinghouse in 1969. The Angra-2 plant came online in 2000, 23 years and $10 billion after construction began. The nuclear program historically came under the jurisdiction of the Ministry of Defense rather than the Ministry of Mines and Energy, and decreased funding for the military translated to delays in nuclear power plant construction. A government company, Eletronuclear, now has been created to assume responsibility for the plants.

New Developments

In 2003, the government plans to invest $35 million in the Angra 1 nuclear power plant. The money will be used to acquire two new steam generators for the plant. The partly completed Angra 3 nuclear power plant still requires a reportedly $1.7 billion to begin the next construction steps.

Electrobrás has repeatedly appealed to the Brazil energy regulator ANEEL to authorize tariff hikes for nuclear power in order to raise the necessary funding for maintenance of the two operating power plants. Electronuclear, which runs the plants, is in debt and continues to have difficulty in raising the necessary revenues to conduct regular maintenance.

Environment

Brazil is a major player in discussions regarding the environment. Brazil's Amazon rainforest comprises 30 per cent of the world's remaining tropical forests, and in addition to providing shelter to at least one tenth of the world's plant and animal species, the rainforest acts as a mechanism for absorbing carbon dioxide from the atmosphere.

Brazil is the largest energy consumer in South America (consuming 8.78 quadrillion Btu of commercial energy in 2001), and the third largest in the Western Hemisphere, behind the United States and Canada. While total energy consumption statistics place the country as prominent in the region, Brazil's per capita energy consumption is comparable to the average per capita energy consumption for all of Central and South America. Brazil also is the largest emitter of carbon dioxide in the region, releasing 95.8 million metric tons of carbon into the atmosphere in 2001. Although Brazil's carbon emissions are fairly significant in the region, carbon intensity, the amount of carbon emitted per dollar of GDP, is comparatively low.

One reason for the comparatively lower carbon intensity in Brazil is the significant use of hydropower in the energy mix, as well as the use of biofuels and other forms of renewable energy. One prominent biofuel in the Brazilian economy is ethanol. The ethanol program was initiated partially in response to the oil shock of 1973, and partly as an alternative to oil to promote self-sufficiency. The ethanol program also has been one of Brazil's strategies to mitigate the environmental effects of rapid urbanization.

In January 1999, Brazil's Congress implemented an environmental pollution law that fines polluters for environmental standards violations. Industrial polluters have five years to come into compliance with the new law, after which they can be fined between $50 and $50 million dollars for violations.

Country Overview

Independence: September 7, 1822 (from Portugal)

Population (2002E): 174.6 million

Location/Size: Eastern South America/3.3 million square miles, slightly smaller than the United States

Major Cities: Sao Paulo, Rio de Janeiro, Belo Horizonte, Brasilia (capital)

Languages: Portuguese (official), Spanish, English, French

Ethnic Groups: white (55 per cent–includes Portuguese, German, Italian, Spanish and Polish), mixed (38 per cent), black (6 per cent), other (1 per cent–includes Japanese, Arab, and Amerindian)

Religions: Roman Catholic (80 per cent)

Defense (8/98): Army (195,000), Navy (68,250), Air Force (50,000), Public Security Forces (385,600)

Economic Overview

Currency: 1 Real (R) = 100 centavos

Exchange Rate (7/22/03): US$1 = R 2.96

Gross Domestic Product (GDP, 2002E): $451.7 billion

Real GDP Growth Rate (2002E): 1.5 per cent *(2003F):* 2.4 per cent

Inflation Rate (consumer prices, 2002E): 8.5 per cent *(2003F):* 15.3 per cent

Current Account Deficit (2002E): -1.72 per cent of GDP

Merchandise Exports (2002E, $U.S.): $60 billion

Merchandise Imports (2002E, $U.S.): $47 billion

Net Merchandise Trade Surplus (2002E, $U.S.): 13

Major Trading Partners: United States, Argentina, Japan, Germany, Italy

Unemployment Rate (2002E): 7.1 per cent *(2003F):* 7.4 per cent

Total External Debt (2002E): 208.2

Foreign Exchange Reserves (2002E): 37.7

Energy Overview

Proven Oil Reserves (1/1/03E): 8.3 billion barrels

Oil Production (2002E): 1.57 million barrels per day (bbl/d), of which 1.3 million bbl/d was crude.

Oil Consumption (2002E): 2.17 million bbl/d

Net Oil Imports (2002E): 0.6 million bbl/d

Crude Oil Refining Capacity (1/1/03E): 1.87 million bbl/d

Natural Gas Reserves (1/1/03E): 8.1trillion cubic feet (tcf)

Natural Gas Production (2001E): 210 billion cubic feet (bcf)

Natural Gas Consumption (2001E): 339 bcf

Coal Reserves (2001): 13.1 billion short tons

Coal Production (2001E): 4.53 million short tons (Mmst)

Coal Consumption (2001E): 20.8 Mmst

Electric Generation Capacity (2001E): 73.4 gigawatts (85 per cent hydroelectric)

Net Electricity Generation (2001E): 321.2 billion kilowatthours (bkwh)

Net Electricity Consumption (2001E): 336 bkwh

Environmental Overview

Total Energy Consumption (2001E): 8.78 quadrillion Btu*

Energy-Related Carbon Emissions (2001E): 95.8 million metric tons of carbon

Per Capita Energy Consumption (2001E): 50.9 million Btu (vs. U.S. value of 341.8 million Btu)

Per Capita Carbon Emissions (2001E): 0.56 metric tons of carbon (vs U.S. value of 5.5 metric tons of carbon)

Energy Intensity (2001E): 11,384 Btu/ $1995 (vs U.S. value of 10,736 Btu/ $1995)**

Carbon Intensity (2001E): 0.12 metric tons of carbon/thousand $1995 (vs U.S. value of 0.17 metric tons/thousand $1995)

Fuel Share of Energy Consumption (2001E): Oil (51 per cent), Hydro (31 per cent), Coal (6 per cent), Other Renewables (6 per cent), Natural Gas (4 per cent), Nuclear (2 per cent)

Fuel Share of Carbon Emissions (2001E): Oil (80 per cent), Coal (13 per cent), Natural Gas (7 per cent)

Status in Climate Change Negotiations: Non-Annex I country under the United Nations Framework Convention on Climate Change (ratified June 4th, 1992). Signatory to the Kyoto Protocol (April 29th, 1998 and ratified on August 23, 2002).

Major Environmental Issues: Deforestation in Amazon Basin destroys the habitat and endangers the existence of a multitude of plant and animal species indigenous to the area; air and water pollution in Rio de Janeiro, Sao Paulo, and several other large cities; land degradation and water pollution caused by improper mining activities

Major International Environmental Agreements: A party to the Antarctic-Environmental Protocol, Antarctic Treaty, Biodiversity, Climate Change, Desertification, Endangered Species, Environmental Modification, Law of the Sea, Marine Dumping, Nuclear Test Ban, Ozone Layer Protection, Ship Pollution, Tropical Timber 83, Tropical Timber 94, Wetlands and Whaling.

The total energy consumption statistic includes petroleum, dry natural gas, coal, net hydro, nuclear, geothermal, solar, wind, wood and waste electric power. The renewable energy consumption statistic is based on International Energy Agency (IEA) data and includes hydropower, solar, wind, tide, geothermal, solid biomass and animal products, biomass gas and liquids, industrial and municipal wastes. Sectoral shares of energy consumption and carbon emissions are also based on IEA data.

Energy Industry Overview

Organization: Petroleo Brasileiro (Petrobrás)–government-owned (majority shareholder) oil and natural gas company; *Centrais Eletricas Brasileira (Eletrobrás)*–federal holding company for planning and coordination of generation, transmission and distribution of electrical power

Major Petroleum Terminals: Sao Sebastiao, Paranagua, Salvador, Tramandai, Sao Francisco do Sul, Aracaju, Maceio, Recidfe, Natal, Fortaleza, and Belem

Major Ports: Santos, Rio de Janeiro, Praia Mole, Vitoria, Rio Grande

Major Oil and Gas Fields: Campos Basin (includes Marlim, Albacora, and Barracuda fields), Santos Basin

Major Refineries (January 2003 capacity, all belonging to Petrobrás): Paulinia–Sao Paulo (337,764 bbl/d), Mataripe-Bahia (283,734 bbl/d), Duque de Caxias–Rio de Janeiro (232,213 bbl/d), Sao Jose dos Campos–Sao Paulo (217,134 bbl/d), Canoas, Rio Grande do Sul (180,945), Araucaria–Parana (180,945 bbl/d), Cubatao-Sao Paulo (162,851 bbl/d), Betim-Minas Gerais (144,756 bbl/d)

15

Brunei

Brunei is important to world energy markets because it exports nearly 180,000 barrels per day of oil, is a large liquefied natural gas producer (third largest in Asia), and because it is located close to vital sea lanes through the South China Sea linking the Indian and Pacific Oceans.

General Background

Brunei's small, wealthy economy is based heavily upon proceeds from exports of crude oil and natural gas, with revenues from the hydrocarbons sector accounting for over 50 per cent of gross domestic product (GDP), around 80 per cent -90 per cent of exports, and 75 per cent -90 per cent of government revenues. Per capita GDP is far above most other developing countries (although sharply down from its high point around 1980), and substantial income from overseas investment supplements income from domestic sources. The manufacturing sector is very small. The government provides for all medical services and subsidizes food and housing. Real GDP growth in 2003 is projected at 3.0 per cent, about the same as in 2002, but sharply higher than in 1999, largely due to higher oil prices.

Brunei's main economic problems include lack of diversity in the economy, heavy reliance on the sometimes-volatile oil and natural gas sectors, huge state subsidies, a civil service which employs around 75 per cent of Brunei's workforce, extensive state economic controls, a small tax base (the country has no personal income tax and a low tariff regime), and only slow movement towards privatization. In 1998, the collapse of Amedeo Development Corporation, run by Prince Jefri, and the loss of many billions of dollars–possibly half the country's foreign exchange reserves–as a result, caused a financial crisis which is still being resolved. Former Amedeo assests reportedly have been transferred to the Brunei Investment Agency, and most of the debt apparently has been settled.

Brunei would like to diversify away from hydrocarbons into areas like tourism, financial services, rubber, rice farming, halal (Muslim dietary law) food, and forestry services, plus energy-intensive industries like petrochemicals, oil refining, and aluminum smelting. Brunei also would like to turn

itself into a major shipping hub. In January 2003, Brunei unveiled plans aimed at attracting $4.5 billion in foreign investment by 2008 and at diversifying the economy. The Brunei Economic Development Board (BEDB) reportedly is looking at building a 500-megawatt power plant, a new jetty, and a container port in the Sungai Liang area both to tap into the country's natural gas resources as well as to help establish new industries.

Oil

Brunei contains proven crude oil reserves of 1.35 billion barrels and during 2002 produced 163,000 barrels per day (bbl/d) of mainly low-sulfur crude oil, plus around 27,000 bbl/d of natural gas liquids. Oil production began in Brunei in 1929, with the discovery of the giant onshore Seria Field. Production from Seria peaked at around 115,000 bbl/d in the 1950s, but has now fallen to around 27,000 bbl/d. Brunei's oil production peaked in 1979 at about 240,000 bbl/d, but was cut back deliberately to extend life of the fields and to improve recovery rates. Overall, the country's upstream oil sector requires around $300 million annually, according to the *Petroleum Economist*, in order simply to maintain current production capacity from mature fields.

Brunei has seven offshore oil fields, including Champion (which contains about 40 per cent of total oil reserves and produces around 50,000 bbl/d), Southwest Ampa (the oldest field, with more than half of Brunei's natural gas reserves and production), Fairley, Fairley-Baram, Magpie (producing since 1977, now at 10,000 bbl/d), Gannet, and Iron Duke, plus two more fields onshore. Major customers for Brunei's oil include Japan, South Korea, the United States, Australia, New Zealand, China, and

India. In November 2000, Brunei Shell Petroleum signed an agreement with Sinopec of China to export 10,000 bbl/d, the first time Brunei has exported to mainland China.

Brunei Shell Petroleum (BSP), a 50-50 joint venture between Royal Dutch/Shell and the government of Brunei, for years has been the only oil producer in the country and also operates the country's only oil refinery. This now appears to be changing, however. In early 2002, for instance, Total (which has been active in Brunei since the 1980s) was awarded an exploration license in Brunei's deepwater Block J, located around 60 miles offshore, along with partners BHP Billiton Petroleum (25 per cent share) and Amerada Hess (15 per cent).

In June 2003, however, Total suspended exploration work in Block J, following an April 2003 incident in which several naval patrol boats from Malaysia, which disputes the potentially oil- and natural gas-rich area, and which itself recently discovered a large (700-million-barrel) oil field–called "Kikeh" (awarded to Murphy Oil and Petronas)–which may extend into Block J, chased away a Total ship. Brunei claims that Block J, and the adjacent Block K (awarded to a joint venture of Shell, Conoco and Mitsubishi) are completely within its Exclusive Economic Zone (EEZ), but will now need to negotiate rights to the area with its much larger neighbor, Malaysia. Brunei is counting on Blocks J and K to maintain the country's oil and gas output another decade or more, and without these Blocks, Brunei may be forced to move more rapidly away from an economy based on hydrocarbons. A possible compromise with Malaysia could involve some sort of joint-development zone, according to Malaysia's Foreign Minister.

In November 2001, Sultan Hasan Bolkiah signed an order creating Brunei's first national oil company, wholly owned by the government, in a move towards further diversification. The company would "consolidate and mobilize the petroleum industry of Brunei Darussalam and play a more active role in both petroleum exploration and development" according to the Prime Minister's office. Oil and gas production from the offshore Maharajah/Lela/Jamaludin field, operated by TotalFinaElf (along with New Zealand's Fletcher Energy), came online in early 1999.

The country is located close to vital sea lanes through the South China Sea, linking the Indian and Pacific Oceans.

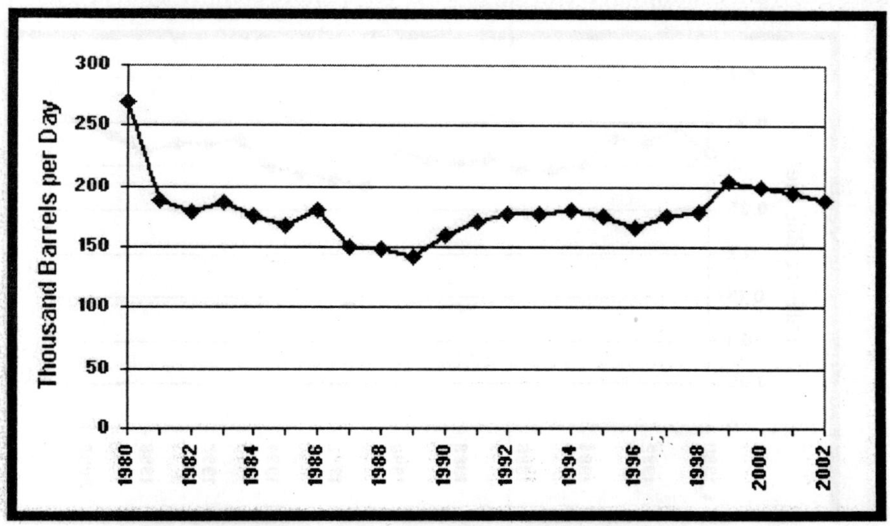

OIL PRODUCTION IN BRUNEI, 1980-2002

Downstream

Brunei has one refinery, part of BSP, with a capacity of about 8,600 bbl/d. Around 5,000-6,000 bbl/d of the refinery's output is used for local consumption. The remainder of Brunei's crude oil is exported and refined elsewhere.

Natural Gas

Brunei produced around 366 billion cubic feet (Bcf) of natural gas in 2001. Around 90 per cent is produced by BSP and 10 per cent by TotalFinaElf/Fletcher. In November 2001, BSP announced that phase 1 of the development of Egret field will produce its first natural gas by August 2003. The gas will be gathered at the platform and transported by a new pipeline to the Ampa 6 platform. Phase 2 will target oil and remaining natural gas in the field. Brunei began the first Asian liquefied natural gas (LNG) exports–to Japan–in 1972. Brunei is the fourth-largest producer of LNG in the world and the third-largest natural gas producer in Southeast Asia. Today, major customers for Brunei's LNG exports include Japan, which takes around 90 per cent of Brunei's LNG exports under a 20-year contract renewed in 1993, and South Korea. The Brunei LNG (BLNG) joint venture between Mitsubishi (25 per cent), Shell (25 per cent), and the Brunei government (50 per cent) produces Brunei's LNG, with LNG beingliquefied at BLNG's Lamut plant. In 2001, BLNG shipped 6.7 million tons of LNG, the highest volume to date, and in 2002 reportedly shipped 203 cargoes. The vast majority of BLNG sales are on a long-term contract basis, although there have been a few spot sales.

Long-term prospects for natural gas and LNG development in Brunei are excellent. BLNG has a master plan to expand its capacity with an additional 4 -million-metric-ton-per-year train by 2008. Key to maintaining or expanding capacity is finding new natural gas reserves, possibly 7 Tcf through 2033. The plan also includes the rejuvenation of the existing plant, such as the $2.4 billion modernization program carried out over the past decade. Besides exports, Brunei would like to use its natural gas to develop domestic petrochemicals and energy-intensive industries (like aluminum smelting).

In 2002, BSP completed the East High-Pressure Gas Expansion Project at the Champion field. High pressure gas from Champion is carried to a BLNG onshore compression facility. In addition, the

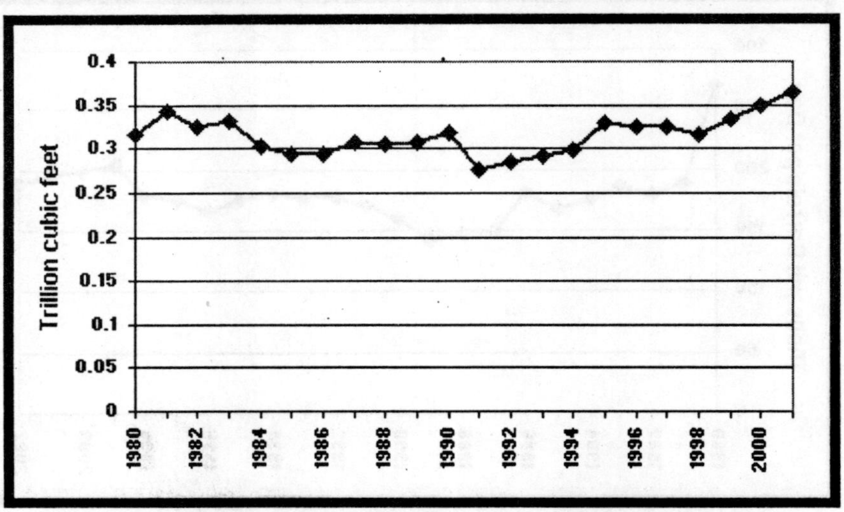

NATURAL GAS PRODUCTION IN BRUNEI, 1980-2000

$350 million Ampa-Fairley Rationalization Progject (AFRP) aims to modernize and revitalize existing BSP natural gas producing fields, such as Southwest Ampa, Fairley, and Gannet.

Electricity

Brunei's installed electric generating capacity in 2001 was 0.76 gigawatts (GW), of which all was natural gas-fired. Brunei's power demand is growing at a rapid rate of around 7 per cent -10 per cent annually. Brunei's power plants all operate single natural gas turbines, with the exception of the Lumut cogeneration facilities and the new Belingus Power Station. In October 2001, two, 3-megawatt (MW) diesel engine generators were inaugurated at the Belingus Power Station in Temburong. In January 2003, Brunei announced plans for a new 500-MW, natural-gas-fired power plant as part of its economic diversification strategy.

Country Overview

Independence: January 1, 1984 (from the United Kingdom)

Population (2001E): 342,000

Location/Size: Southeastern Asia, bordering the South China Sea and Malaysia on the north-west corner of the island of Borneo/2,228 square miles (slightly larger than Delaware)

Major Cities: Bandar Seri Begawan (capital)

Languages: Malay (official), English, Chinese, Iban, and other indigenous dialects

Ethnic Groups: Malay (62 per cent), Chinese (15 per cent), indigenous (6 per cent)

Religions: Muslim (67 per cent), Buddhist (13 per cent), Christian (10 per cent), indigenous beliefs and other (10 per cent)

Defense (2000E): Total manpower available 104,000 (Land Forces; Navy; Air Force; Royal Brunei Police)

Economic Overview

Currency: Brunei Dollar (BND)

Market Exchange Rate (7/10/03): US$1 = 1.76 BND

Gross Domestic Product (GDP, Market Exchange Rates) (2002E): $5.6 billion

Per Capita GDP (2002E, PPP): $19,044

Real GDP Growth Rate (2002E): 3.1 per cent *(2003F):* 3.0 per cent

Inflation Rate (consumer prices) (2002E): 1.8 per cent *(2003F):* 1.5 per cent

Unemployment Rate (unofficial estimate, 2000E): 10 per cent (higher for recent graduates)

Merchandise Exports (2000E): $3 billion

Merchandise Imports (2000E): $1.4 billion

Merchandise Trade Balance (2000E): $1.6 billion

Major Trading Partners (2000): Japan, Singapore, United Kingdom, United States, South Korea, Malaysia

Major Export Products: Natural gas and crude oil, refined oil products

Major Import Products: Machinery and transport equipment, manufactured goods, food

Energy Overview

Proven Oil Reserves (1/1/03E): 1.35 billion barrels

Oil Production (2002E): 189,000 bbl/d, of which 163,000 bbl/d was crude oil

Oil Consumption (2002E): 13,000 bbl/d

Net Oil Exports (2002E): 176,000 bbl/d

Crude Oil Refining Capacity (1/1/03E): 8,600 bbl/d

Natural Gas Reserves (1/1/03E): 13.8 trillion cubic feet (Tcf)

Natural Gas Production (2001E): 366 billion cubic feet (Bcf)

Natural Gas Consumption (2001E): 48 billion cubic feet (Bcf)

Electric Generation Capacity (1/1/01E): 0.76 gigawatts

Net Electricity Generation (2001E): 2.5 billion kilowatthours (all natural gas-fired)

Environmental Overview

Total Energy Consumption (2001E): 0.08 quadrillion Btu* (<0.1 per cent of world total energy consumption)

Energy-Related Carbon Emissions (2001E): 1.3 million metric tons of carbon (<0.1 per cent of world carbon emissions)

Per Capita Energy Consumption (2001E): 231.9 million Btu (vs. U.S. value of 341.8 million Btu)

Per Capita Carbon Emissions (2001E): 3.6 metric tons of carbon (vs U.S. value of 5.5 metric tons of carbon)

Energy Intensity (2000E): 11,705 Btu/ $1995 (vs U.S. value of 11,014 Btu/ $1995)**

Carbon Intensity (2000E): 0.18 metric tons of carbon/thousand $1995 (vs U.S. value of 0.18 metric tons/thousand $1995)

Fuel Share of Energy Consumption (2001E): Natural Gas (67.7 per cent), Oil (32.0 per cent)

Fuel Share of Carbon Emissions (2001E): Natural Gas (62.2 per cent), Oil (37.8 per cent)

Status in Climate Change Negotiations: Not a party to or signatory of the Kyoto Protocol.

Major Environmental Issues: Seasonal smoke/haze resulting from forest fires in Indonesia.

Major International Environmental Agreements: Party to the Endangered Species, Law of the the Sea, Ozone Layer Protection, and Ship Pollution treaties.

The total energy consumption statistic includes petroleum, dry natural gas, coal, net hydro, nuclear, geothermal, solar, wind, wood and waste electric power.

Oil and Gas Industries

Refinery: Brunei Shell Petroelum Co. Sdn. Bhd.–Seria (8,600 bbl/d)

Foreign Energy Company Involvement: Shell, in a joint venture called Brunei Shell Petroleum, is the country's main oil and gas production company, and operates Brunei's only refinery. Mitsubishi is active in the Brunei LNG joint venture, and TotalFinaElf (along with New Zealand's Fletcher Energy) is the operator at a new offshore oil and gas field, Maharajah/Lela/Jamaludin.

Oil and Gas Fields: Champion, Southwest Ampa, Fairley, Fairley-Baram, Magpie, Gannet, and Iron Duke

16

Cameroon

While Cameroon's oil production is predicted to steadily decline, its refinery capacity and its position as the terminus of the 225,000-barrels-per-day crude oil export pipeline from Chad could transform the country into a significant oil transport center.

General Background

The Republic of Cameroon achieved independence from France on January 1, 1960. Paul Biya was elected president in 1984, and since has been re-elected for three additional terms. In December 1995, a new constitution was approved, with provisions included to increase the president's term from five to seven years with a maximum presidential tenure of two terms. Biya began a seven-year term in office upon his re-election in October 1997.

Cameroon is richly endowed with natural resources and has a diversified commodity-based economy. In recent years the services sector has grown and in 2001 accounted for just over 40 per cent of Cameroon's GDP. Industry accounts for 31 per cent of GDP while agriculture only contributes 28.8 per cent of GDP. The sale of petroleum products brought in $878 million during 2001, or 60.4 per cent of export receipts. Timber, coffee, and cocoa are the country's principal non-oil exports.

Since 1996, Cameroon's GDP growth has improved and remained strong, although slower-than-expected governmental reforms such as privatization may have a negative impact on future GDP growth. In 2001, real GDP grew by an estimated 5.5 per cent due to increased construction activity, notably on the Chad-Cameroon pipeline. Inflation has been more volatile, with 2001 inflation measured at 2.8 per cent, up from 0.8 per cent in 2000. Cameroon's foreign debt is extremely high; the total level of external debt in 2001 was $9.5 billion, or 95 per cent of GDP.

In September 2002, the International Monetary Fund (IMF) completed the last in a series of three reviews for the Poverty Reduction and Growth Facility arrangement and allowed Cameroon to

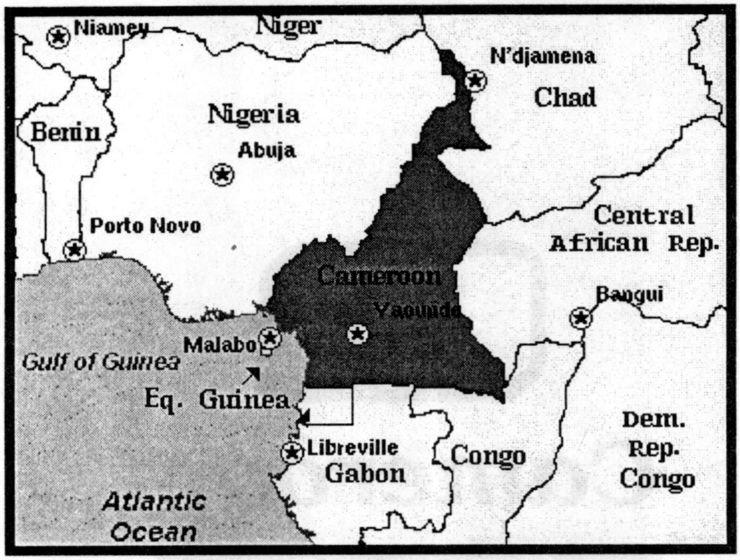

withdraw $21 million. In 2000, the (IMF) and the World Bank agreed to a comprehensive debt reduction package for Cameroon under the Enhanced Heavily Indebted Poor Countries (HIPC) Initiative. Total debt relief from the package is worth around $2 billion and will free up $100 million per year for the next three years for expenditures on health care and other social services. The Doba oil project, for which construction recently began following World Bank approval in June 2000, is expected to aid Cameroon's economy; total government revenues from the 28-year project are expected to range from $500 million to $900 million.

Cameroon plays an important role in promoting increased economic integration among the Central African Economic and Monetary Community's (CEMAC) six member states. Cameroon hosted CEMAC's first trade fair and is a potential site (along with Gabon) for the establishment of a CEMAC joint stock exchange.

Oil

Cameroon is sub-Saharan Africa's fifth largest oil producer with crude oil production of just over 76,600 barrels per day (bbl/d) in 2001, down from 84,000 bbl/d in 2000. Preliminary figures for the first half of 2002 show an average but steadily declining production of almost 71,200 bbl/d. Cameroon also faces long-term declining crude oil production as mature fields are exhausted and new discoveries diminish. The Edinburgh-based consulting firm Wood Mackenzie has predicted that Cameroon's crude oil output will fall to between 50,000 bbl/d and 60,000 bbl/d by 2005. However, the government is confident in the development potential of the Logone Birni and Douala basins, two areas that have been largely unexplored.

Bakassi Peninsula

Both Cameroon and Nigeria have claimed the Bakassi Peninsula, a 1,000-square-kilometer (600-square-mile) area located in the Gulf of Guinea that is believed to contain significant reserves of oil. Several oil discoveries have been made on the peninsula and its adjoining waters, but at present operations in the disputed area have been suspended. In February 1994, Cameroon submitted the dispute to the International Court of Justice (ICJ) for settlement, and Nigeria later followed with its own suit. In March 1998, the ICJ began formal hearings on the case concerning the Land and Maritime Boundary between Cameroon and Nigeria (Cameroon v. Nigeria), and in June 1999, Equatorial Guinea applied for permission to enter the proceedings. The ICJ authorized Equatorial Guinea to intervene in the case in October 1999 on the basis of the country's arguments to protect its legal rights in the Gulf of Guinea as the maritime boundary between Nigeria and Cameroon is determined. In its Order, the Court fixed April 4, 2001 as the time-limit for the filing of a written statement by Equatorial Guinea and July 4, 2001 as the time-limit for the filing of written observations by Cameroon and by Nigeria on that statement. Despite the continuation of the dispute, Nigeria's Cross River State, contiguous to Cameroon in the Bakassi peninsula area, and the government of Cameroon agreed in April 2000 to open the disputed border and construct roads linking the respective border towns in order to facilitate international trade and commerce. In September 2002, both nations agreed to abide by the ICJ ruling when it is issued, and recognized the need to demilitarize the peninsula.

Recent Upstream Developments

Bidding on Cameroon's third licensing round was closed on June 30, 2000, after a three-month extension. Due to lower-than-expected interest, the government of Cameroon announced that it will abandon the bidding process and instead adopt a permanent open-door policy for any company interested in developing oil in the country. After years of heavy exploration in Cameroon, international oil companies' lack of interest in the recent round of bidding is attributed by industry observers to the perception that major new finds of oil in Cameroon are highly unlikely.

Grynberg Petroleum Co. reached an agreement in July 2001 with the Societe Nationale des Hydrocarbures (SNH) to explore a 25,000-square-mile block in the Douala/Kribi-Campo basin. Recoverable hydrocarbon reserves are estimated at 80 billion cubic feet (Bcf) and 16 million barrels of condensate.

An exploration contract for the Ntem area was signed in March 2001 between Fusion Oil and SNH. No bids were submitted for the onshore blocks in the Logone Birni Basin. Fusion Oil holds 100 per cent equity in the Ntem contract, which has three exploration terms of two years. During the first two-year exploration period, Fusion Oil acquired a 2D seismic survey of the entire block and now has the final data for evaluation. This survey fulfills Fusion's obligations for the first exploration term.

Cameroon has revised its laws concerning oil and gas exploration to attract investors. Changes included a fixed division of profits under joint venture projects (40 per cent to the contractor), as well as incentives and tax breaks for the exploration and development of marginal and deepwater offshore fields. The incentives were sufficient to stimulate the development of Kribi F, operated by Perenco, and several satellites around Rio del Rey, including Bao South which began producing oil in October 2000. Bao South has an estimated proven capacity of 10 million barrels, with an output of 5,000 bbl/d. Production from Bao South should help to partially offset Cameroon's declining crude oil output.

Chad-Cameroon Pipeline

In November 1996, the consortium, then consisting of Exxon (now ExxonMobil; 40 per cent and operator), Shell (40 per cent) and TotalFina Elf (20 per cent), signed a memorandum of understanding (MOU) with the government of Chad. The MOU provides the terms of the development of the Doba basin fields, and the construction of a 1,050- kilometer (650-mile) export pipeline through Cameroon to offshore export facilities located near Kribi. The total cost of field development and construction of the pipeline and export facilities is estimated to be $3.5 billion. In April 2000, Chevron (now ChevronTexaco) and Malaysia's Petronas joined the consortium, replacing TotalFina Elf and Shell, who had divested their combined 60 per cent share in the consortium. Petronas acquired a 35 per cent interest, with Chevron taking the remaining 25 per cent. Exxon retained its controlling 40 per cent share in the consortium.

The project consists of two primary components, development of the oil fields and construction of the pipeline and export facilities. The Doba basin's three fields (Bolobo, Kome and Miandoun) will produce 900 million–1 billion barrels of low sulfur oil over the life of the project. The consortium plans to drill 300 wells. Production is forecast to continue for 25-30 years, with peak production projected at 225,000 bbl/d to 250,000 bbl/d. Crude oil from the three fields will be gathered, treated and blended at a facility located on the Kome field. Construction of the project began in October 2000. Production is set to begin in early 2004, with the cost of the field development estimated to be $1.5 billion. Originally production was set to begin in 2003, but contract problems and two strikes by workers have caused serious delays.

Construction of the pipeline and export facilities is expected to last two years and to cost $2.2 billion. The Chadian portion of the pipeline, approximately 170-kilometers (105 miles) long, and the first (located at Kome) of the three pumping stations will be built and owned by the Tchad Oil Transport Company (TOTCO). TOTCO, which was established in July 1998, is composed of members of the consortium and the government of Chad. The Cameroon Oil Transport Company (COTCO), will build and own the 880-kilometer (545-mile) Cameroonian section of the pipeline, including the remaining pumping stations and the export facilities. The export facilities will consist of an onshore pressure-reducing station near Kribi, a 12-kilometer (7-mile) sub-sea pipeline and an offshore floating storage and off-loading facility (FSO), the Marine Export Terminal. COTCO consists of the consortium members, the Cameroonian government, and the government of Chad. Deviations in the pipeline route may extend the total distance up to 663 miles.

The foreign consortium's share in bankrolling the project will amount to $2.2 billion (59.2 per cent of total costs), while commercial banks and export credit agencies will contribute $600 million (16.1 per cent) and capital markets will finance $400 million (10.7 per cent). The lynchpin of the project financing, however, is the World Bank loan of $93 million–$53.4 million for Cameroon and $39.5 million for Chad–agreed to in June 2000. In addition, the International Finance Corporation will loan the three companies $100 million and make another $300 million available through commercial banks. Finally, the European Investment Bank may also contribute $40 million to the project. A week after the World Bank's board "overwhelmingly" approved the project, the U.S. Export-Import Bank approved a $300 million loan guarantee to finance exports to build the pipeline.

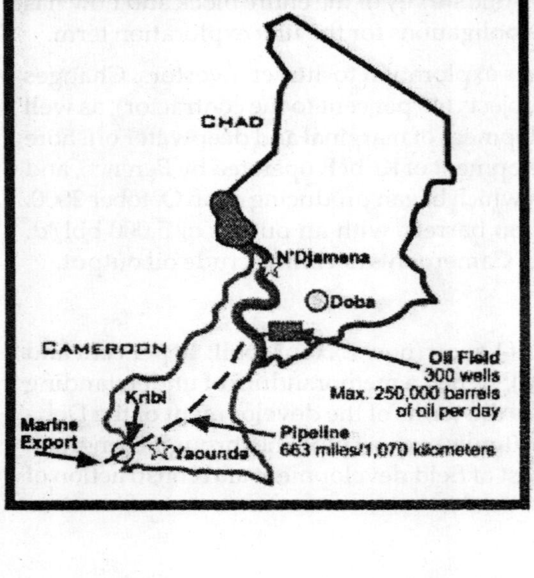

According to projections released by the World Bank, the Doba project is highly profitable. Based on a benchmark price of $15.25 per barrel of oil, total receipts for the Doba project are expected to reach $12 billion over a 28 year period. Net revenue would amount to $9.2 billion before debt servicing. Cameroon is expected to earn $0.46 on every barrel of Chad oil transited once the project is completed which will result in receipts between $500 million and $900 million over the life of the project. The World Bank decided in September 2002 to continue its endorsement of the pipeline, despite a report by internal sources that cites the project's environmental damage and failure to meet important World Bank criteria.

The Chad-Cameroon oil project has been the source of intense opposition from human rights and environmental groups. In July 1998, nearly ninety local, national and international non-governmental organizations (NGOs) submitted a petition to the World Bank to shelve plans for the pipeline project and in September 1999 the NGOs called for a 2-year moratorium on the project. Similar petitions and lobbying efforts continued through 2000. Many of the human rights and environmental concerns have focused on Cameroon, where the pipeline is set to traverse areas of dense jungle inhabited by the Bagyeli ethnic group, or Pygmies as they are popularly known. Consequently, the consortium has made numerous adjustments to the route of the pipeline, which will run entirely underground. To date, all local ethnic groups except the Bagyeli pygmies have been compensated for land they lost to the pipeline project. Compensation for this group has proven difficult since the Bagyelis are nomadic and generally do not own land. Their compensation will be based on intangibles, such as loss of access to certain plants and hunting areas.

Downstream

Cameroon's downstream oil sector is a significant part of the country's economy. Petroleum consumption was an estimated 22,000 bbl/d in 2001. TotalFina Elf, ExxonMobil, Shell, and ChevronTexaco market petroleum products in Cameroon. ChevronTexaco, which purchased Agip's retail assets, is the largest retailer in Cameroon, with 120 petrol stations. A single firm, the Cameroon Petroleum Depot Company (SCDP), handles petroleum product distribution. The government holds the majority in SCDP (51 per cent), and the retail marketers hold the remaining shares. SCDP launched a study in June 2000 to evaluate the possible terms of its privatization.

Cameroon's Société Nationale de Raffinage (SONARA) refinery, which is located in the port city of Limbe, is the source for the country's petroleum products. The government holds a 66 per cent interest in SONARA while TotalFina Elf, ExxonMobil, Shell, and ChevronTexaco hold the remaining shares. The SONARA refinery has utilized its spare capacity to refine products for export to other African nations, Europe, Latin America, and the United States. Bouygues Offshore acquired a contract in October 2001 to debottleneck the operation units at the SONARA refinery in Limbe. The project is scheduled to last 19 months, during which time the company will increase crude oil distillation capacity on light crude and obtain a new distillation fraction. SONARA previously invested $15 million to upgrade its port facilities. The new port allows tankers with capacity as large as 90,000 tons access to the refinery. As a condition of multilateral debt relief, SNH, which previously reported directly to the President of Cameroon, now includes its oil receipts in the national budget.

Electricity and Natural Gas

Along with the Democratic Republic of Congo (DRC), Cameroon has the greatest potential for hydroelectric power in Africa. Over 110 possible sites have been identified, with a combined potential capacity of 500,000 megawatts (MW). If Cameroon's electricity potential is developed, the country could become a net electricity exporter.

Cameroon has installed electric generation capacity of 817 MW, of which 88 per cent is hydroelectric and 12 per cent is thermal. The country's two main hydro stations, Edea and Song-Loulou, are located on the Sananga River. The much smaller Lagdo station is located in the north near

Garoua. Cameroon employs roughly 30 diesel power stations around the country as back-up facilities, particularly when droughts affect water levels behind the hydro station dams. The largest diesel power stations are located in Garoua (20 MW), Douala (15.4 MW), and Yaounde (10.8 MW).

Generation and distribution of electricity is handled by Société Nationale d'Electricité de Cameroun (SONEL), the partially state-owned utility. In July 2001, the U.S.-based firm AES acquired 56 per cent of SONEL's shares at a purchase price of $70 million. Cameroon's government retain the remaining 44 per cent to sell off to private investers. Under the terms of the contract, AES will be required to increase the number of electricity subscribers in Cameroon to at least one million. There are currently just 400,000 subscribers. In March 2002, AES-Sirocco, the parent company of AES-SONEL, was reported to be on the verge of bankruptcy, and AES-SONEL has been accused of diverting funds to its parent. Frequent power outages have occurred since AES took over, and may force Cameroon's government to scale back growth projections. AES was the only company out of five pre-selected firms to submit a financial offer to the government of Cameroon, which had stated that it had anticipated a purchase price of $80-$90 million for SONEL. This privatization scheme was one requirement of the IMF-sponsored structural adjustment plan of 2000.

Cameroon has estimated natural gas reserves of 3.9 trillion cubic feet (Tcf). Associated and non-associated gas reserves have been discovered in the Rio Del Rey, Douala and Kribi-Campo basins, but they have yet to be utilized. In June 2002, nine companies were invited to consider developing the Sanaga Sud gas field, which was originally discovered in 1979 by Ocelot Energy, now a subsidiary of PanOcean Energy. Six bids were received, and the contract will be awarded by SNH in October 2002. As a result of previous exploration in the area, Cameroon's government believes that the winning company will not have to drill appraisal wells, but will be able to move straight to production.

Through its wholly owned local subsidiary EurOil, Edinburgh-based developer BowLeven currently holds concessions for blocks MLHP 5, 6 and 7. EurOil is committed to drill one exploration well before March 2003 and is currently in the process of drilling a well to test whether an apparent oil sand in the MLHP-7 block has enough reserves to support a 200-MW open cycle gas turbine power plant. According to company officials, estimated proven reserves in MLHP 7 amount to at least 140 (Bcf) of gas. The project will cost approximately $290 million.

Country Overview

Independence: January 1, 1960 (from France)

Population (2001E): 15.8 million

Location/Size: Western Africa, bordering the Bight of Biafra, between Chad (on the north and east); the Central African Republic (on the east); the Republic of Congo, Gabon and Equatorial Guinea (on the south); and Nigeria (on the west) / 475,440 square kilometers (183,520 square miles), slightly smaller than the combined size of South Carolina, Georgia, Florida, and Alabama

Major Cities: Yaounde (capital), Douala, Nkongsamba, Maroua, Garoua

Languages: French (official),English (official), 100-200 other indigenous Congo-Kordofanian and Afro-Asiatic languages

Major Ethnic Groups: Hausa, Fulbe, Bamileke, Tiker, Bamoun, Fang, Ewondo, Boulou, Eton, Bassa, Bakoko, Douala

Religion: Traditional beliefs 40 per cent, Christian 40 per cent, Muslim 20 per cent

Defense (1998E): Army (11,500), Air Force (300), Navy (1,300), paramilitary forces (9,000)

Economic Overview

Currency: Communaute Financiere Africaine (CFA) franc

Market Exchange Rate (9/18/02): US$1 = 671.34 CFA francs

Real Gross Domestic Product (GDP) (2000E): $9.52 billion *(2001E):* $9.98 billion

Real GDP Growth Rate (2001E): 5.2 per cent *(2002E):* 4.9 per cent

Inflation Rate (2001E): 6.0 per cent *(2002E):* 2.8 per cent

Major Trading Partners: France, Spain, Italy, Netherlands

Merchandise Exports (2001E): $2.18 billion *(2002E):* $2.25 billion

Merchandise Imports (2001E): $2.13 billion *(2002E):* $2.32 billion

Merchandise Trade Balance (2001E): $50 million

Major Export Products: Crude oil and petroleum products, timber, cocoa, coffee

Major Import Products: Food, machinery, industrial goods, consumer goods

Current Account Balance (2001E): $64 million

Total External Debt (2001E): $9.5 billion

Energy Overview

Proven Oil Reserves (1/1/02E): 400 million barrels

Oil Production (2001E): 76,667 barrels per day (bbl/d), all of which was crude oil *(2002E):* 71,167 bbl/d of crude oil

Oil Consumption (2001E): 22,000 bbl/d *(2002E):* 21,000 bbl/d

Crude Refining Capacity (1/1/01E): 42,000 bbl/d

Net Oil Exports (2002E): 50,167 bbl/d

Natural Gas Reserves (1/1/02E): 3.9 trillion cubic feet (Tcf)

Natural Gas Production (2001E): None

Natural Gas Consumption (2001E): None

Electric Generation Capacity (1/1/00): 819 megawatts, of which 97.4 per cent is hydroelectric, 2.6 per cent thermal

Electricity Generation (2000E): 3.6 billion kilowatthours (bkwh)

Environmental Overview

Total Energy Consumption (2000E): 0.08 quadrillion Btu* (<0.1 per cent of world total energy consumption)

Energy-Related Carbon Emissions (2000E): 1.8 million metric tons of carbon (<0.1 per cent of world carbon emissions)

Per Capita Energy Consumption (2000E): 5.5 million Btu (vs. U.S. value of 351.0 million Btu)

Per Capita Carbon Emissions (2000E): 0.1 metric tons of carbon (vs. U.S. value of 5.6 metric tons of carbon)

Energy Intensity (1999E): 6,391 Btu/ $1990 (vs U.S. value of 12,638 Btu/ $1990)**

Carbon Intensity (1999E): 0.14 metric tons of carbon/thousand $1990 (vs U.S. value of 0.19 metric tons/thousand $1990)

Fuel Share of Energy Consumption (2000E): Natural Gas (0.0 per cent), Oil (55.1 per cent), Coal (0.0 per cent)

Fuel Share of Carbon Emissions (2000E): Oil (43 per cent), Natural Gas (57 per cent), Coal (0.0 per cent)

Renewable Energy Consumption (2000E): 0.0367 quadrillion Btu*

Status in Climate Change Negotiations: Non-Annex I country under the United Nations Framework Convention on Climate Change (ratified October 19th, 1994). Not a signatory to the Kyoto Protocol.

Major Environmental Issues: Water-borne diseases are prevalent; deforestation; overgrazing; desertification; poaching; overfishing.

Major International Environmental Agreements: A party to Conventions on Biodiversity, Climate Change, Desertification, Endangered Species, Law of the Sea, Ozone Layer Protection, Tropical Timber 83 and Tropical Timber 94. Has signed, but not ratified, Nuclear Test Ban.

The total energy consumption statistic includes petroleum, dry natural gas, coal, net hydro, nuclear, geothermal, solar, wind and waste electric power. The renewable energy consumption statistic is based on International Energy Agency (IEA) data and includes hydropower, solar, wind, tide, geothermal, solid biomass and animal products, biomass gas and liquids, industrial and municipal wastes.

Oil and Gas Industries

Organization: Société Nationale des Hydrocarbures (SNH)

Major Oil Fields: Asoma, Bavo, Betika, Boa, Ebome, Ekoundou, Kole, Kombo, Makoko, Rio del Rey

Refineries (capacity): Société Nationale de Raffinage (SONARA) (42,000 bbl/d)

Oil Terminals: Kole, Kribi (planned)

Pipelines: Cameroon Oil Transport Company (COTCO) pipeline 880 kilometers (545 miles) (planned)

Foreign Oil Company Involvement: ChevronTexaco, CMS Nomeco, TotalFina Elf, ExxonMobil, Globex Petroleum, Ocelot, Perenco, Petronas, Royal Dutch/Shell (Pecten), Trophy Petroleum

17

Canada

Canada is a net exporter of oil, natural gas, coal, uranium, and hydropower. It is one of the most important sources of U.S. energy imports.

General Background

Canada's real gross domestic product (GDP) for 2002 expanded approximately 3.3 per cent, indicating that the country has recovered from an economic slowdown in 2001, when GDP grew by only 1.5 per cent. Canada's economic slowdown in 2001 and subsequent recovery in 2002 were both largely impacted by fluctuations of the U.S. economy, which represents Canada's largest trading partner. Continued weak U.S. growth, in combination with strong appreciation of the Canadian dollar, SARS outbreak in Toronto and the ban on exports of Canadian beef due to a case of "mad cow" disease, has driven down first quarter 2003 economic growth forecasts for Canada. In 2002, approximately 85 per cent of Canadian exports went to the United States.

Energy

Canada was the fifth-largest energy producer in the world in 2001, behind the United States, Russia, China, and Saudi Arabia. Over the past two decades, Canada has become a significant net energy exporter. In 2001, about 31 per cent of Canadian energy production was exported, with the United States its main customer. In 2002, the United States imported more oil (including crude oil and petroleum products) from Canada than from any other country. The United States also imported about 3.8 trillion cubic feet (Tcf) of Canadian natural gas in 2002, with 94 per cent of total U.S. natural gas imports coming from Canada. In 2001, about 39 per cent of Canada's primary energy production was natural gas, followed by oil (25 per cent), hydropower (20 per cent), coal (11 per cent), and nuclear power (5 per cent). Alberta is Canada's largest producer of energy. Besides being a major producer, Canada also is a significant energy consumer.

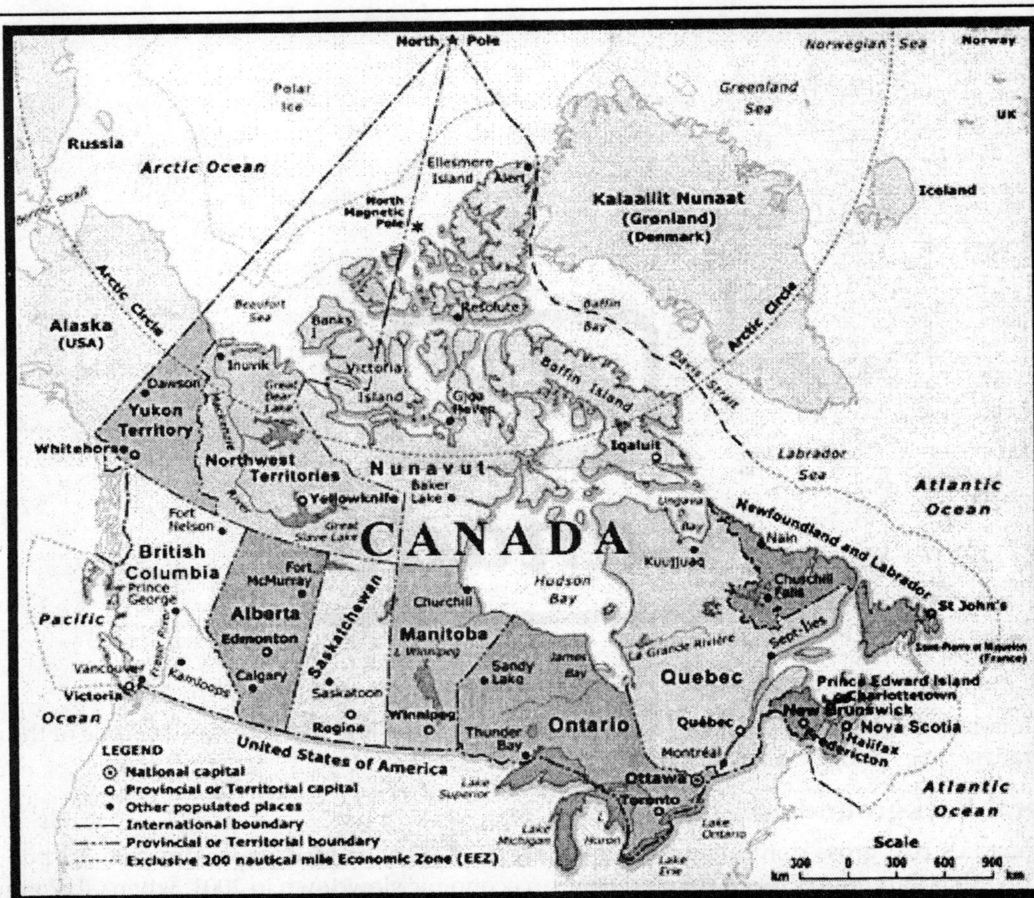

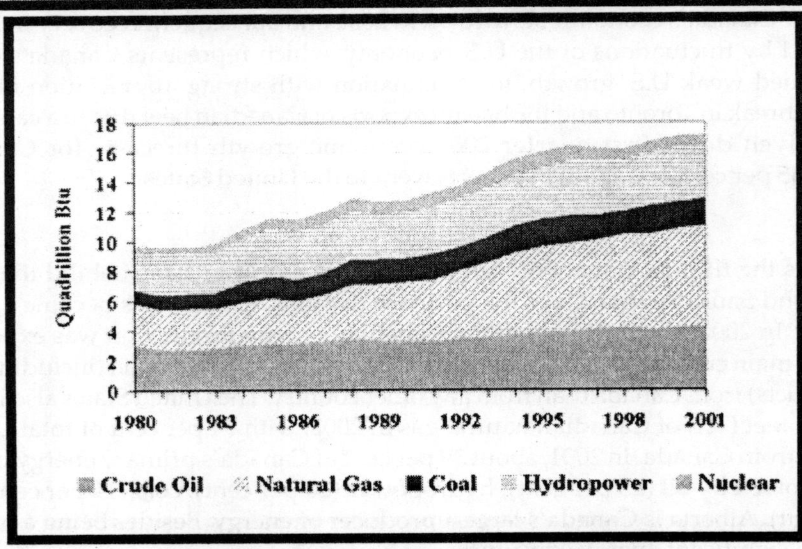

CANADIAN ENERGY PRODUCTION, 1980-2001

Oil

According to *Oil & Gas Journal*, Canada's total crude oil reserves stood at 180 billion barrels as of January 2003. The massive increase (reserves stood at 4.8 billion barrels in 2002) reflects the *Journal's* inclusion of Alberta's oil sands, which stood at 174.8 billion barrels as of January 2003, according to Alberta Energy and Utilities Board (EUB). Canadian Association of Petroleum Producers (CAPP) reported that conventional crude oil and condensate stood at an estimated 5.2 billion barrels. The change in reserves has pushed Canada into second place behind Saudi Arabia, which holds the most proven crude oil reserves in the world. However, some analysts have questioned the new assessment and whether it is accurate.

Oil production averaged 2.9 million barrels per day (bbl/d) during 2002, with estimated consumption of 2.0 million bbl/d. The province of Alberta, located in western Canada, is the country's leading oil producing region. While Alberta's conventional oil reserves are declining (the province now contains an estimated 31 per cent of the country's total conventional oil reserves), the province also contains huge oil sands deposits. Meanwhile, projects and potential projects in other provinces are shifting the oil industry's focus to include the eastern and northern parts of the country.

Canadian Oil Exports

Canada is a major source of U.S. oil imports. In 2002, the United States imported 1.97 million bbl/d of oil from Canada (1.45 million bbl/d of which was crude oil). This makes Canada the top petroleum supplier to the United States and the third-largest supplier of crude oil imports (behind Saudi Arabia and Mexico, and ahead of Venezuela). Canada has been the top supplier to the United States of refined petrôleum products, including gasoline, jet fuel, distillate, etc., since 1996.

A majority of U.S.-bound Canadian oil exports go to the Midwest (PADD II). In 2002, Canada exported 1.08 million bbl/d to PADD II, which accounted for approximately 66 per cent of the region's total oil imports. Canadian oil exports also dominate in the Rocky Mountain region, (PADD IV)

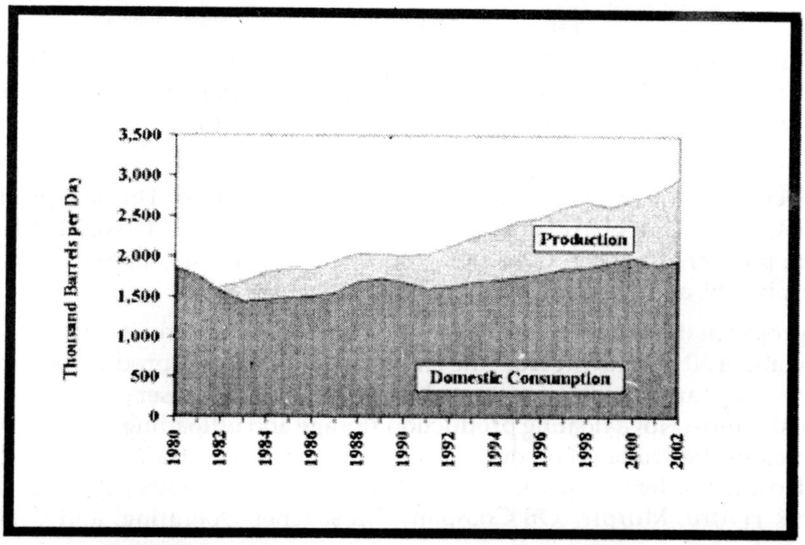

CANADIAN OIL PRODUCTION AND CONSUMPTION, 1980-2002

averaging 249,000 bbl/d in 2002, which represented 100 per cent of the region's total oil imports. Canada also exports its oil to the U.S. East Coast (PADD I)–480,000 bbl/d in 2002–as well as, to a lesser extent, to the Gulf Coast region (PADD III) and the West Coast (PADD V).

Sector Organization

Canada's oil sector has seen significant mergers and acquisitions in recent years. In 2001, U.S. firms purchased over $35 billion in Canadian oil and gas assets, including Houston-based Conoco's purchase of Gulf Canada for $8.9 billion and Devon Energy's (U.S.) acquisition of Canada's Anderson Exploration for $7.1 billion.

Canadian firms also have been busy reorganizing the country's oil patch. In April 2002, two of Canada's largest companies, Alberta Energy and PanCanadian, merged to create EnCana Corp., the world's largest independent oil and natural gas producer (by market value).

However, a number of companies, particularly U.S.- based, are beginning to seek buyers for assets that they acquired during the period leading up to 2001. Marathon Oil, Vintage Petroleum, Hunt Oil, ConocoPhillips, ChevronTexaco, Canadian Natural, and Murpy Oil all have put up both Canadian oil and natural gas properties for sale. Analysts speculate that companies are putting their Canadian assets up for sale because of declining exploration prospects.

Exploration and Production

The Western Canadian Sedimentary Basin (WCSB), underlying Alberta, Saskatchewan, and part of the Northwest Territories, has been the main source of Canadian oil production for the past 50 years. Alberta, however, has been the premier energy-producing region in all of Canada. In 2002, an estimated 50 per cent of conventional oil production came from Alberta. However, conventional oil production has been declining in the province, down 8.5 per cent in comparison to 2001 and 36 per cent since 1996. At the same time, oil production in the East, particularly offshore Newfoundland, has increased significantly in the last few years. Newfoundland production increased by 92 per cent in 2002 after the Terra Nova field came onstream.

Atlantic Coast

Exploration and production activity on Canada's east coast is focused on the Jeanne d'Arc Basin of the Grand Banks, offshore *Newfoundland*. The climate demands technologically advanced offshore oil platforms, which are able to withstand extremely cold temperatures and high winds, adding to production costs.

The first project in the area, *the Hibernia field*, came on stream in 1997. The field produces between 135,000-150,000 bbl/d of light (high API gravity), sweet (low sulfur) crude. ExxonMobil is the operator, with joint venture partners Chevron Canada Resources, PetroCanada, Canada Hibernia Holding Corporation, Murphy Oil, and Norsk Hydro.

The second project in the Jeanne d'Arc Basin, *Terra Nova*, began production in January 2002. The Terra Nova field is about 30 miles away from the Hibernia field and has a production capacity of about 115,000 bbl/d over six years. Crude oil extracted from Terra Nova's reservoir on the ocean floor is routed upward to the Terra Nova floating production storage and offloading vessel (FPSO), a storage vessel which processes the crude oil on deck. The petroleum is then transferred to shuttle tankers which take the oil to market. Terra Nova is owned by PetroCanada (operator), ExxonMobil, Husky Oil Operations, Norsk Hydro, Murphy Oil Company, Mosbacher Operating, and Chevron Canada Resources.

There are two more projects in the Jeanne d'Arc basin: *White Rose* and *Hebron/Ben-Nevis*. The White Rose field, which is operated by Husky Oil Operations, is expected to begin production in 2005 or early 2006. The field could reach an estimated 90,000 bbl/d at peak production. In February 2003, Chevron Canada Resources, ExxonMobil, Norsk Hydro and Petro-Canada announced that they would not continue exploration in the Hebron/Ben-Nevis fields for financial reasons and for significant challenges in developing scattered deposits.

Petro-Canada and partners Encana and Norsk Hydro prospected for oil in the *Flemish Pass Basin*. One well, the Mizzen L-11, did produce oil but not enough to make it commercially viable. As a result, the partners decided to abandon the second prospect, Tuckamore B-27, in early June 2003. Vintage Petroleum Canada plans to conduct seismic surveys onshore in *Cape Breton* at River Denys, Georges River, and in the north end of Boularderie Island. There are also potential reserves off the coast of Labrador.

Pacific Coast

On the west coast, the British Columbia government is pushing to lift a 30-year-old ban on exploration in the Pacific Ocean in order to begin production by 2010. The area near Queen Charlotte Basin is thought to hold as much as 10 billion barrels of oil. Queen Charlotte, Tofina, Winona and Georgia basins are expected to hold an estimated 43.4 trillion cubic feet (Tcf) of natural gas.

Synthetic Crude Oil

Oil sands contain deposits of bitumen, a heavy black viscous oil. To process bitumen, it first must be extracted from the ground. There are two extraction methods. Shallow oil sands deposits, which can be excavated from the surface, and deeper "in situ" (in situ means "in place", indicating that the bitumen is separated underground) deposits which require other recovery methods, such as cyclic stem stimulation and steam-assisted gravity drainage. These methods inject steam to help separate the bitumen from sand and clay and bring it to the surface. According to the Alberta Ministry of Energy, roughly two tons of oil sands must be dug up, moved and processed to produce one barrel of oil.

The extracted bitumen is then separated from sand and water, which it surrounds. Before it is sent to a refinery to be upgraded into a high-quality oil called "synthetic crude", the bitumen must be treated. Bitumen is so heavy and thick that it will not flow unless heated or diluted with lighter hydrocarbons. At the refinery, the bitumen will be treated again and upgraded into synthetic crude. The upgrading process also generates other products, such as petroleum coke which is either used to power the refinery or sold.

The *Athabasca Oil Sands deposit*, in northern Alberta, is one of the two largest oil sands deposits in the world. There are also oil sands deposits on Melville Island, in the Canadian Arctic and two smaller deposits in northern Alberta, Peace River and Cold Lake. In 2002, production of bitumen averaged 829,000 bbl/d, according to Alberta Energy and Utilities Board (EUB). As of December 2002, established recoverable reserves stood at 174 billion barrels, with ultimate recoverable reserves at an estimated 315 billion barrels, according to EUB.

Producers

Syncrude (a joint venture of eight companies with Canadian Oil Sands Investments Inc. having the largest stake) is Canada's largest producer of crude from oil sands. In 2002, the company produced an estimated 250,000 bbl/d. Syncrude is in the process of expanding its operations. The program includes a second production train at the producing Aurora mine and major expansions in upgrading capacity.

Suncor, the first company to begin processing Alberta oil sands in 1967, completed in 2001 its Project Millennium, which increased production capacity to 225,000 bbl/d. The company's next project, Voyageur, aims to increase production to between 500,000-550,000 bbl/d in 2010 to 2012. A third project, the Firebag In-situ Oil Sands Project, will initially add approximately 35,000 bbl/d. Production is expected to begin in 2004.

The Athabasca Oil Sands Project is a joint venture of Shell Canada Limited, Chevron Canada Limited (a wholly owned subsidiary of ChevronTexaco Corp.) and Western Oil Sands Inc. The project has two components: the Muskeg River Mine and the Scotford Upgrader. Albian Sands Energy Inc. is the operator of the Muskeg River Mine, which is expected to produce 155,000 bbl/d of diluted bitumen. There are plans to increase production to 525,000 bbl/d of bitumen.

Other Developments

EnCana has two projects: Christina Lake Thermal Project and Foster Creek. The first project is expected to produce 70,000 bbl/d and the second 20,000 bbl/d, with a potential of 100,000 by 2007. In May 2003, ConocoPhillips, along with partners Total and Devon Energy, received regulatory approval to start construction on an oil sands plant in northern Alberta. The group will decide later on this year whether to proceed with project. The plant would produce 25,000 bbl/d by 2006. A number companies are currently planning oil sand development projects in Alberta. A number of companies are in the process of developing oil sands projects, such as BlackRock Ventures, Deer Creek Energy, Husky Energy and Imperial oil.

Difficulties

A combination of high development costs, growing environmental concerns, and high natural gas prices have resulted in some companies shelving projects. The extraction of bitumen, particularly

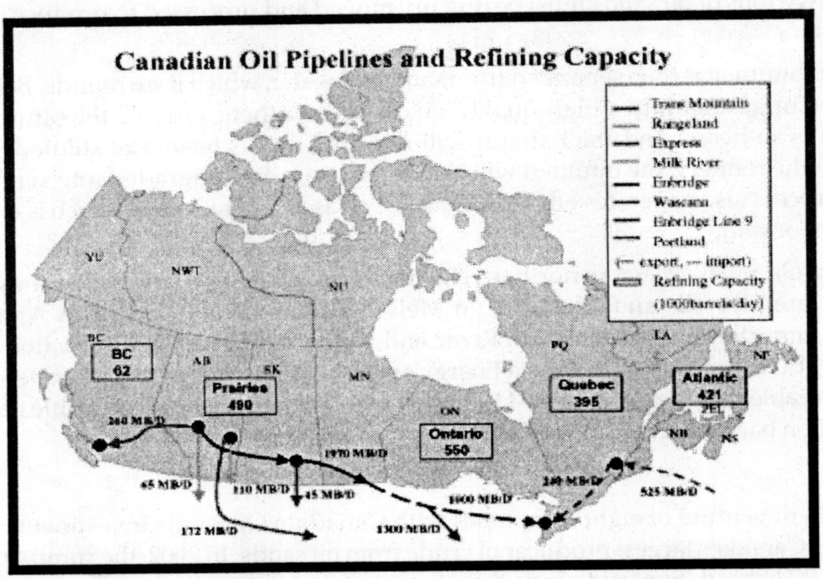

underground (in-situ), requires significant amounts of natural gas and water. Natural gas is used to generate steam which is injected into the ground to melt the mud away from the bitumen. Natural gas is also used to generate electricity to power the massive shovels used for pit-mining. Some analysts predict that if all proposed oil sand projects are realized, companies will require up to 2 billion cubic feet per day of natural gas. Future Kyoto requirements–Canada ratified the Protocol in December 2002–will also raise costs, as companies would most likely have to invest in emission reducing technologies or acquire carbon credits to offset emissions resulting from production. These potential costs have already resulted in three companie –TrueNorth Energy, Canadian Natural Resources and PetroCanada–delaying projects.

Pipelines

Although most Canadian oil is produced in western Canada (mainly Alberta), oil is consumed primarily in central and eastern Canada. As a result, Canada exports mostly crude oil from Alberta and imports crude oil and petroleum products on the east coast.

An extensive pipeline system transports western oil to eastern Canadian and U.S. markets. There are two major pipeline networks. The first is *Enbridge Pipelines Inc.* (formerly Interprovincial Pipe Line-IPL), an 8,700-mile network of piping and terminals, delivering oil from Edmonton, Alberta, east to Montreal, Quebec and eastern Canada, as well as to the U.S. Great Lakes region.

Enbridge is one of the largest crude oil and petroleum liquids pipeline systems in the Western Hemisphere, and the company is currently expanding its U.S. export capacity through its three-phase *Terrace Expansion program*. The first two phases of the program, which links Kerrobert, Saskatchewan to Clearbrook, Minnesota, were completed in January 2002, adding approximately 210,000 bbl/d capacity to the Enbridge network. Construction of phase three of the Terrace Expansion began in 2002 and will extend the line further into Minnesota, adding 140,000 bbl/d of capacity. Completion of phase three is expected by mid-2003.

The other major pipeline system is the *Trans Mountain Pipe Line (TMPL) system*, which delivers oil mainly from Alberta west to refineries and terminals in the Vancouver, British Columbia area, as well as to the Puget Sound area of Washington State.

Oil Sand Pipelines

Development of Alberta's massive oil sands deposits has required new pipelines to transport diluted bitumen from the mine to downstream processors and eventually to market terminals. In the near term, Canadian pipeline companies have focused on taking the Athabasca oil sands southward, to processing facilities in the Edmonton area. The first of these pipelines, the 344-mile *Athabasca pipeline* (Enbridge), was completed in April of 1999 and connects Suncor's oil sands operations to the Enbridge network. The pipeline has potential capacity of 525,000 bbl/d.

Construction of another similar pipeline, the *Corridor Pipeline* (TransMountain) connects the Muskeg River Mine (Shell Canada, Chevron Canada, Western Oil Sands) to Shell's Scotford Refinery, located in the Edmonton area, near market terminals. Oil began to flow through the Corridor Pipeline in May 2003.

In January 2002, BC Gas announced that it intends to build a new pipeline in conjunction with TransMountain, called the *Bison Pipeline*, to transport diluted bitumen from mines and refineries near Fort McMurray to pipelines and processing plants in the Edmonton area. The Bison Pipeline would cover about 320 miles and cost about $625 million. Initially, the pipeline is expected to transport

100,000 bb/d, with ultimate capacity reaching 450,000. The company will, however, decide in late 2003 whether it will proceed with the project.

Natural Gas

According to Oil and Gas Journal, Canada holds about 60.1 trillion cubic feet (Tcf) of proven natural gas reserves, as of January 2003. Canada produced about 6.6 Tcf of natural gas in 2001, making it the world's third largest natural gas producer (after the United States and Russia) and second largest natural gas exporter (after Russia). Canada's natural gas exports go almost exclusively to the United States. Canadian natural gas consumption, 2.9 Tcf in 2001, is projected to grow significantly in coming decades, largely for use in electricity generation and in production of oil sands.

Exploration and Production

Like the oil industry, Canada's natural gas industry is based primarily in the province of Alberta, which in recent years has accounted for nearly 70 per cent of the country's natural gas output. Some analysts, however, forecast little growth potential for natural gas in Alberta, as well as for the entire Western Canada Sedimentary Basin (WCSB) -includes most of Alberta, parts of British Columbia, Saskatchewan, Manitoba and the Northwest Territories, causing many industry players to shift their focus to newer projects in British Columbia, Atlantic Canada and the Arctic, as well as to new sources, such as imported liquefied natural gas (LNG) and coalbed methane. Many of these developments, however, are not on pace to offset declines. Furthermore, companies, such as ConocoPhillips, Marathon, Vintage Petroleum, and Murphy, have begun selling natural gas properties, citing that many wells have high depletion rates and operational costs, making them less economically viable.

Alberta

According to the National Energy Board's (NEB) short-term outlook for WCSB, natural gas production has decreased despite higher drilling activity. In 1996, companies drilled 4,800 wells

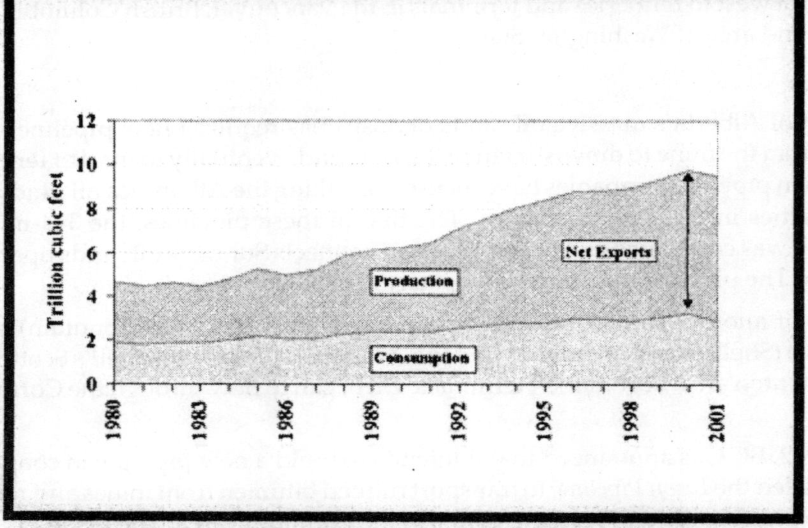

CANADIAN NATURAL GAS PRODUCTION AND CONSUMPTION, 1980-2001

compared to 12,400 in 2001. In 2002, natural gas production in Alberta alone decreased by 3.8 per cent, to 4.8 Tcf. According to Alberta Energy and Utilities Board (EUB), production is expected to increase in Alberta in 2004 but then will decrease by 2 per cent per year until 2012. As companies develop oil sand projects, natural gas requirements for bitumen recovery and upgrading to synthetic crude oil are expected to increase significantly by 2012, reports EUB. This would inevitably make less natural gas available to outside markets, including the United States.

In addition, the Gas Resources Preservation Act requires Alberta to set aside large volumes of natural gas for the province's use before exports are permitted. Companies seeking to export natural gas from Alberta are required to seek the approval of EUB. Exports of natural gas are not permitted unless there is a surplus and they do not jeopardize the needs of Alberta's core consumers for the next 15 years. In June 2003, EUB announced that it was considering a plan to shut-in 900 natural gas wells, or 90 Bcf per year by August 1, 2003 in order to maintain subsurface pressure to extract 100 billion barrels of crude bitumen in the Wabiskaw-McMurray area. EUB explained that when natural gas is extracted from pools among the oil sands, pressure is reduced and can result in hindering thermal recovery of bitumen. After the extraction of bitumen is completed, EUB indicated that natural gas pools would be returned to production after operations. Many companies oppose the proposed shut-in.

British Columbia

The *Ladyfern field* was touted as Canada's largest natural gas find in 15 years, after Murphy Oil drilled a well in 2002 that tested out 100 million cubic feet per day (MMcf/d). The find resulted in a number of companies (Murphy, Apache, Canadian Natural Resources, and Encana) buying property and subsequently over-drilling. The field's production, which peaked at about 700 million cubic feet per day (MMcf/d) in 2002, has declined to about 300 MMcf/d. Many analysts expect the field to be drained by 2004. On a positive note, Canadian Superior Energy reported in April 2003 that had made two new natural gas discoveries at the *East Ladyfern* field, located 22 miles southeast from the Ladyfern main field.

In May 2003, the British Columbia government announced a package of royalty credits in an effort to attract new investment in region. The government plans to double natural gas production by 2008 from the current 900 Bcf per year. As mentioned before, the British Columbia government hopes to lift the moratorium on *offshore* drilling, giving access to an estimated 43.4 Tcf of natural gas. British Columbia also has three onshore basins–Whitehorse Trough, Bowser, and Nechako–with combined estimated reserves of 23.2 Tcf.

Atlantic Coast

Similar to oil exploration in this region, companies have been reevaluating their offshore projects. Shell recently downgraded the reserve base of the *Sable Offshore Energy Project (SOEP)* by 11 per cent, to 700 Bcf. Shell had originally estimated Sable reserves to hold 1.1 Tcf, but in January 2002, that estimate was revised downwards to 800 Bcf. The project, which began production in late December 1999, is the second phase of development. The consortium, consisting of Mobil, Shell, Imperial, Nova Scotia Resources, and Moshbacher, plan to develop three more fields between 2003 and 2007, with Alma field coming on stream this year. Natural gas is transported through an underwater pipeline to the consortium's onshore processing facilities in Goldboro, Nova Scotia, and distributed to the Canadian provinces of Nova Scotia and New Brunswick and to the U.S.

EnCana announced in February 2003 that the company has asked federal and provincial regulators for a "time out" in the regulatory approval process in order to reassess the economics of the

Deep Panuke project, located off the Scotian shelf. The Deep Panuke field holds an estimated 0.9 Tcf of natural gas. Similar to the Sable Island project, natural gas from Deep Panuke will be routed to onshore facilities in Goldboro, and then connected to mainland distribution lines.

Despite these setbacks, Canadian Superior Energy, in conjunction with the El Paso Corporation, announced in June 2003 that it will proceed with three offshore Nova Scotia projects: Marquis, Mariner and Mayflower. Many analysts are becoming skeptical of the potential in the Atlantic region and without a big discovery, it will become more difficult to attract investment.

Arctic

Natural gas production in the MacKenzie Delta could begin in 2008 after the Aboriginal Pipeline Group became part owners in the Arctic pipeline project. The project is being built around three proven, but undeveloped, natural gas fields in the Delta, Imperial's Taglu field, (estimated 3 Tcf), Conoco's Parsons Lake (estimated 1.8 Tcf) and Shell's Niglintgak (estimated 1 Tcf). Other explorers include Petro-Canada, Anadarko, Devon, Apache and Burlington.

Liquefied Natural Gas

Irving Oil announced in June 2003 that it will proceed with an environmental assessment and other regulatory work for a proposed LNG facility at Canaport, near Saint John, New Brunswick. Access Northeast Energy also announced that it is interested in constructing a LNG regasification facility in Nova Scotia on the Strait of Canso. Northeast Energy would then utilize the Maritimes and Northeast pipeline to transport to Canadian and U.S. markets.

Coalbed Methane

Coalbed Methane (CBM) production is still in its infancy, with the first wells being drilled in 2002. EnCana, in a joint venture with MGV Energy, is the only group currently producing CMB. Total CBM reserve estimates range from 145 Tcf to 3,000 Tcf.

Pipelines

Canada has a number of natural gas pipelines connecting to the United States. The 1,875 mile-Alliance Pipeline links western Canada (Fort St. John, British Columbia) to the Chicago region. The pipeline has a transportation capacity of 1.3 billion cubic feet per day (Bcf/d). The Alliance pipeline entered into commercial service on December 1, 2000.

Atlantic Coast

The *Maritimes and Northeast Pipeline (M&NE)* came onstream in January 2000. M&NE delivers natural gas from Canada's Sable Island area to Nova Scotia, New Brunswick and New England. The pipeline interconnects with existing U.S. pipeline grid in Dracut, Massachusetts. There are plans to expand the pipeline's capacity, as well as link it to Prince Edward Island, Cape Breton, and northern New Brunswick; however, the lack of a recent major natural gas discovery has slowed expansion of the pipeline. The pipeline's current capacity is about 600 Mmcf/d.

In October 2001, El Paso Corporation announced plans to build an undersea natural gas pipeline off the coast of Nova Scotia–called the *Blue Atlantic Transmission System*. El Paso, however, decided in April 2003 to postpone development of the project, as major oil and natural gas companies operating off the coast of Nova Scotia had announced plans to forgo drilling activities for this year. The Blue Atlantic group plans to resume development when drilling activities continue. The proposed 750-

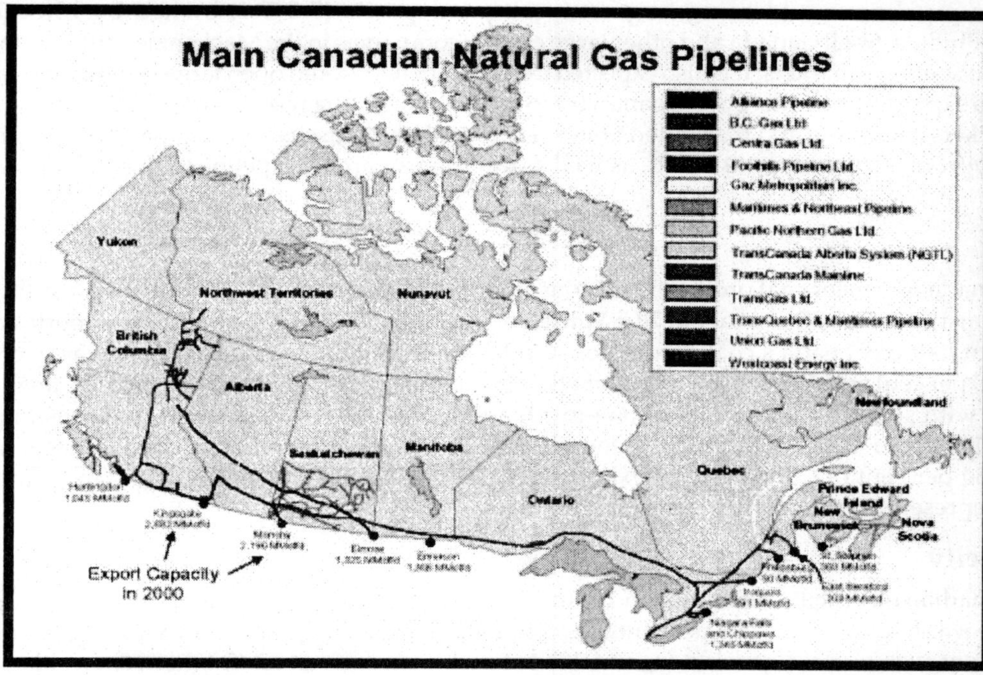

mile pipeline would have a transportation capacity of 1 Bcf/d, connecting first to Nova Scotia and then continuing underwater to New York and New Jersey.

In September 2002, the *Millennium Pipeline*, which would carry Canadian natural gas from Lake Erie to New York City, won U.S. regulatory approval from the Federal Energy Regulatory Commission (FERC). Though the pipeline won FERC endorsement, the New York Department of Environmental Conservation found that Millennium's proposed crossing of the Hudson River at Haverstraw Bay was not consistent with the Coastal Zone Management Act (CZMA). Accordingly, it denied a state permit. In January 2003, FERC granted the sponsors of the Millennium project more time to provide an environmental mitigation plan for the planned pipeline. The Canadian portion of the project is also waiting for approval from a Joint Review Panel established by the Federal Minister of the Environment and the National Energy Board. The panel will not make a decision until receiving all regulatory approvals in the U.S.

Arctic

In June 2003, TransCanada Pipe Lines agreed to lend the Aboriginal Pipeline Group (APG) funds to pay for its share in the preliminary feasibility study of the *Mackenzie Delta Pipeline*. The project had been delayed as APG had difficulty in raising the required funds for their share. The deal also gives APG one-third ownership in the 850-mile pipeline. In return for the loan, TransCanada, Canada's biggest pipeline firm, could end up with the natural gas it needs to ensure its main pipeline stays full as production declines in Alberta. The company also gets a small stake in the line along with the right to boost its ownership over time. With the agreement signed, the group of natural gas producers led by Imperial Oil filed a preliminary information package with regulator.

The pipeline would bring huge natural gas deposits (5.8 Tcf) owned by Imperial Oil, ConocoPhillips, Shell Canada and other large energy companies in the Mackenzie Delta northwest of Inuvik to southern markets. It is also expected that the pipeline would open large parts of the Canadian Arctic to exploration as energy companies try to replace dwindling reserves and feed growing natural gas markets in Canada and the United States. There is already discussion of expanding the pipelines capacity from 1.2 Bcf/d to 1.9 Bcf/d, as well as building a parallel pipeline to transport natural gas liquids.

Coal

Canada is a major coal producer and consumer, with estimated 2001 output of 77.7 million short tons (Mmst), consumption of about 72.6 Mmst, and reserves of 7.2 billion short tons. Alberta accounts for about half of Canada's coal production, while British Columbia and Saskatchewan account for about 30 per cent and 15 per cent, respectively. Bituminous coal makes up about half of Canada's coal output, with sub-bituminous (about one-third) and lignite the rest. Canadian coal consumption is primarily (90 per cent) for electricity generation, with the remainder mainly used for steel-making. About 80 per cent of Canada's coal exports are for metallurgical purposes, with Japan and South Korea representing the country's largest customers.

Electricity

Canadian electricity generation in 2001 totaled 566.3 billion kilowatt hours (Bkwh), of which 56 per cent was hydropower, 28.3 per cent was conventional thermal power (oil, natural gas, and coal), 13 per cent was nuclear generation, and 1.3 per cent was derived from other renewable sources. Canada was the largest producer of hydropower in the world in 2001. Trends in coming years are expected to favor thermal power generation, mainly from natural gas. Canadian nuclear output increased by 6 per cent to 72.9 bkwh in 2001, in comparison to 2000. Output, however, is much lower than its peak of 102.4 Bkwh in 1994. Ontario contains the bulk of Canadian nuclear capacity.

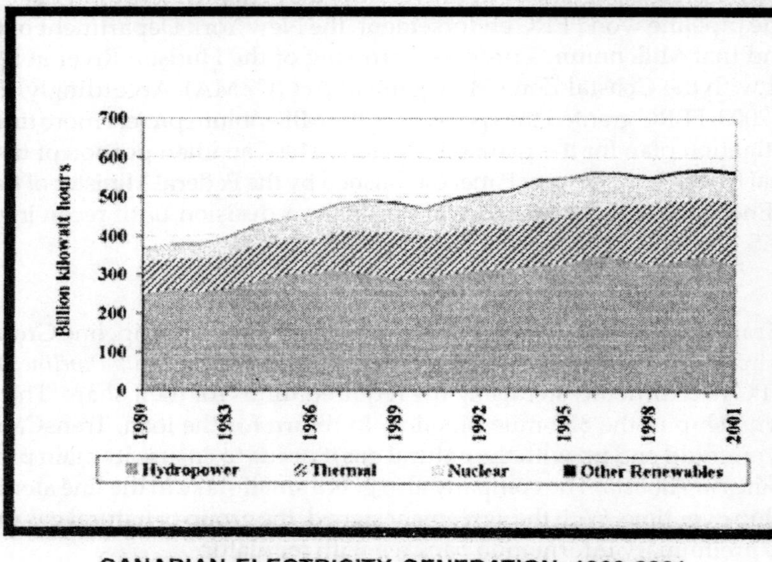

CANADIAN ELECTRICITY GENERATION, 1980-2001

Exports

Canadian electricity gross exports to the United States averaged 39.6 Bkwh between 1992 and 2001. Most of the exports originate from the hydro-rich provinces, such as Québec, Manitoba, and British Colombia. These provinces have also accounted for a majority of exports in recent years. The National Energy Board (NEB) reported that exports have begun to decline over the last few years as domestic demand increased and generation capacity remained the same.

Imports

There exists considerable reciprocity between the Canadian and U.S. power markets, as the United States also exports electricity to Canada (18 Bkwh in 2001). In 2002, imports from the U.S. increased in Ontario due to nuclear power plant closures. British Columbia and Québec have also increased imports, taking advantage of improved access to electricity market in the U.S. According to NEB, net exports have decreased significantly since 1996, as imports from the U.S. continue to increase. A number of the Provinces have expressed interest in constructing more electricity lines to the U.S., as well as in participating in U.S. Regional Transmission Organizations (RTO).

Restructuring

Under Canada's constitution, electricity regulatory authority falls under the jurisdiction of the provinces. In most provinces, a few dominant utilities provide the bulk of generation, transmission, and distribution. Although some of these utilities are privately owned, the provinces own most of them. There is also limited independent power producer (IPP) generation, mostly for sales to the larger utilities.

The process of restructuring has varied across the Canadian provinces. For example, in Nova Scotia, Newfoundland and Labrador, and Yukon Territory, Northwest Territories & Nunavut, public utility boards regulate electricity prices. Alberta and Ontario, which together account for about half of Canada's electricity market, have made efforts to deregulate their power sectors. Full deregulation began in Alberta in January 2001, and in Ontario in May 2002. In both provinces, electricity prices spiked shortly following initial deregulation, causing provincial governments to freeze the electricity prices charged to consumers, and to pay the difference to electric companies out of the government budget.

Environment

Canada's energy abundance has encouraged the development of a highly fuel-intensive economy based on natural resource extraction and processing. This heavy reliance on energy-intensive industries has led to serious environmental concerns, primarily regarding air pollution and climate change.

In 2001, Canada consumed 13 quadrillion Btu of energy and emitted 156 million metric tons of carbon. The industrial sector was the primary emitter of carbon dioxide and within the sector, six energy intensive industries accounted for over 80 per cent of these emissions. Per capita energy consumption ranks fourth among OECD countries, and per capita carbon emissions rank third.

Canada is proving to be a leader in addressing environmental concerns. Renewable energy sources, such as wind, are beginning to gain more attention as Canada works toward meeting international obligations to reduce greenhouse gas emissions.

Country Overview

Independence: July 1, 1867 (from UK)

Population (2002E): 31.3 million

Location/Size: Northern North America/3.85 million sq. miles (slightly larger than the United States)

Administrative divisions: 10 provinces and 3 territories*; Alberta, British Columbia, Manitoba, New Brunswick, Newfoundland, Northwest Territories*, Nova Scotia, Nunavut*, Ontario, Prince Edward Island, Quebec, Saskatchewan, Yukon Territory*

Major Cities: Toronto, Montreal, Vancouver, Ottawa (capital), Edmonton, Calgary, Winnipeg, Quebec

Languages: English (official), French (official)

Ethnic Groups: British Isles origin (28 per cent), French origin (36 per cent), other European (15 per cent), Amerindian (2 per cent), other, mostly Asian, African, Arab (6 per cent), mixed background (26 per cent)

Religions: Roman Catholic (45 per cent), Protestant (41 per cent)

Defense (8/98): Army (20,900), Navy (9,000), Air Force (14,000), Other (15,700)

Economic Overview

Exchange Rate (7/7/03): $1 U.S. = $1.34 Canadian dollars

Gross Domestic Product (GDP, 2002E, $U.S.): $735 billion

Real GDP Growth Rate (2002E): 3.3 per cent *(2003F):* 2.3 per cent

Inflation Rate (consumer prices, 2002E): 2.2 per cent

Unemployment Rate (2002E): 7.6 per cent

Merchandise Exports (2002E, $U.S.): $264 billion

Merchandise Imports (2002E, $U.S.): $227 billion

Merchandise Trade Balance (2002E, $U.S.) $37 billion

Current Account Balance (2002E, $U.S.): $15 billion

Major Export Products: Motor vehicles and parts, newsprint, wood pulp, timber, crude petroleum, machinery, natural gas, aluminum, telecommunications equipment, electricity

Major Import Products: Machinery and equipment, crude oil, chemicals, motor vehicles and parts, durable consumer goods, electricity

Major Trading Partners: United States, European Union

Energy Overview

Conventional Crude Oil Reserves (2002E): 5.2 billion barrels, according to Oil and Gas Journal

Oil Sands Reserves (2002E): 174.8 billion barrels, according to Oil and Gas Journal

Oil Production (2002E): 2.9 million bbl/d, of which 2.0 million bbl/d was crude oil

Oil Consumption (2002E): 2.0 million bbl/d

Net Oil Exports (2002E): 0.9 million bbl/d

Oil Exports to the United States (2002E): 1.94 million bbl/d, 1.43million bbl/d of which was crude oil

Oil Imports from the United States (January-March 2003): 140,000 bbl/d

Total Gross Oil Imports (2002E): 1.09 million bbl/d

Natural Gas Reserves (1/1/03E): 60.1 trillion cubic feet (Tcf)

Natural Gas Production (2001E): 6.6 Tcf

Natural Gas Consumption (2001E): 2.9 Tcf

Net Natural Gas Exports (2001E): 3.7 Tcf

Coal Reserves (2001E): 7.2 billion short tons

Coal Production (2001E): 77.7 million short tons (Mmst)

Coal Consumption (2001E): 72.6 Mmst

Electric Generation Capacity (2001E): 111 million kilowatts

Electricity Generation (2001E): 566.3 billion kilowatt hours (56 per cent hydro, 28.3 per cent thermal, 13 per cent nuclear, 1.3 per cent geothermal and other)

Electricity Consumption (2001E): 504.4 billion kilowatt hours

U.S. Gross Imports from Canada (2001E): 38 billion kilowatt hours

Environmental Overview

Total Energy Consumption (2001): 12.51 quadrillion Btu* (3.1 per cent of world total energy consumption)

Energy-Related Carbon Emissions (2001): 156.2 million metric tons of carbon (2.5 per cent of world carbon emissions)

Per Capita Energy Consumption (2001): 403 million Btu (vs U.S. value of 341.08 million Btu)

Per Capita Carbon Emissions (2001): 5.1 metric tons of carbon (vs U.S. value of 5.5 metric tons of carbon)

Energy Intensity (2001): 17,425 Btu/ $1995 (vs U.S. value of 10,736 Btu/ $1995)**

Carbon Intensity (2001): 0.22 metric tons of carbon/thousand $1995 (vs U.S. value of 0.17 metric tons/thousand $1995)

Fuel Share of Energy Consumption (2001): Oil (30.3 per cent), Natural Gas (23.6 per cent), Coal (13.5 per cent), Hydro (27.3 per cent) Nuclear (6.6 per cent)

Fuel Share of Carbon Emissions (2001): Oil (45.1 per cent), Natural Gas (27.9 per cent), Coal (27.03 per cent)

Status in Climate Change Negotiations: Annex I country under the United Nations Framework Convention on Climate Change (ratified December 4th, 1992). Under the negotiated Kyoto Protocol (signed on April 29, 1998 and ratified on December 17, 2002), Canada has agreed to reduce greenhouse gases 6 per cent below 1990 levels by the 2008-2012 commitment period.

Major Environmental Issues: Air pollution and resulting acid rain severely affecting lakes and damaging forests; metal smelting, coal-burning utilities, and vehicle emissions impacting on agricultural and forest productivity; ocean waters becoming contaminated due to agricultural, industrial, mining, and forestry activities

Major International Environmental Agreements: A party to Conventions on Air Pollution, Air Pollution-Nitrogen Oxides, Air Pollution-Sulphur 85, Air Pollution-Sulphur 94, Antarctic Treaty, Biodiversity, Climate Change, Desertification, Endangered Species, Environmental Modification, Hazardous Wastes, Marine Dumping, Nuclear Test Ban, Ozone Layer Protection, Ship Pollution, Tropical Timber 83, Tropical Timber 94, Wetlands and Whaling. Has signed, but not ratified, Air Pollution-Volatile Organic Compounds, Antarctic-Environmental Protocol, Law of the Sea and Marine Life Conservation

The total energy consumption statistic includes petroleum, dry natural gas, coal, net hydro, nuclear, geothermal, solar, wind, wood and waste electric power. The renewable energy consumption statistic is based on International Energy Agency (IEA) data and includes hydropower, solar, wind, tide, geothermal, solid biomass and animal products, biomass gas and liquids, industrial and municipal wastes. Sectoral shares of energy consumption and carbon emissions are also based on IEA data.

Oil and Gas Industries

Organization: private sector (major companies: ExxonMobil's Imperial Oil, Royal Dutch/Shell's Shell Canada, Petro-Canada, Suncor, EnCana).

Major Oil and Gas Producing Provinces: Alberta; British Columbia; Saskatchewan; Nova Scotia; New Foundland

Major Oil Pipelines: Trans Mountain; Enbridge

Oil Refining Capacity (January 2003): Ontario (545,200 bbl/d); Alberta (436,900 bbl/d); Quebec (449,900 bbl/d); New Brunswick (250,000 bbl/d); British Columbia (62,250 bbl/d); Newfoundland (105,000 bbl/d); Nova Scotia (82,200 bbl/d); Saskatchewan (52,000 bbl/d)

Major Gas Pipeline Companies: Enbridge, TransCanada PipeLines Ltd.

18

Caribbean Fact Sheet

The Caribbean Sea region plays an important role as a petroleum processing and transhipment area, with several major refineries and independent storage facilities. Furthermore, Trinidad and Tobago is becoming an increasingly significant supplier of liquefied natural gas (LNG) to regional markets and to the United States.

General Background

The islands of the Caribbean basin, with a total population of approximately 37 million in 2002, are predominantly net energy importers, with the exception of hydrocarbon rich Trinidad and Tobago. Agriculture and natural resource extraction activities continue to constitute the basis of the islands' economies, though tourism and service sectors are growing. In the larger economies, manufacturing is also important, such as oil and natural gas production in Trinidad; pharmaceuticals and cement in Puerto Rico; and textiles in the Dominican Republic and Jamaica. Economic growth, however, varies across the region. While many of the islands' economies contracted in 2002, Trinidad and Tobago and the Dominican Republic's economies continued to grow, with real gross domestic product (GDP) growth rates of 2.6 per cent and 4.1 per cent, respectively.

Over the past decade, the Caribbean states have made efforts to integrate their economies. The major regional organization is the Caribbean Community and Common Market (CARICOM), whose members include the South American states of Guyana and Suriname and the Central American State of Belize, and the Caribbean islands of Antigua and Barbuda, The Bahamas, Barbados, Dominica, Grenada, Haiti, Jamaica, Montserrat, St.Kitts and Nevis, St. Lucia, St. Vincent and the Grenadines and Trinidad and Tobago. All other islands, with the exception of Cuba, have either associate or observer status. The Caribbean Community has three objectives: (a) economic cooperation through the Caribbean Single Market Economy; (b) coordination of foreign policy among the independent Member States; and (c) common services and cooperation in functional matters such as health, education and culture, communications, and industrial relations. CARICOM countries have steadily reduced tariffs among members. Besides CARICOM, the other main organizations of the region include the Association of Caribbean States, the Eastern Caribbean Currency Union and the associated Eastern Caribbean Central Bank.

Oil and Gas Industries

In 2001, the islands of the Caribbean region consumed a combined total of 2.2 quadrillion Btu of energy. Oil is the dominant fuel, accounting for about 92 per cent of total 2001 energy consumption. The Caribbean relies on imported oil for most of its energy needs. Barbados, the Dominican Republic, Haiti, and Jamaica are party to the San Jose Pact, under which Mexico and Venezuela supply crude oil and refined products under favorable terms. Natural gas and hydropower are used in countries that have these domestic resources. Natural gas is used most extensively in Trinidad and Tobago, where natural gas-intensive industries, such as steel, fertilizer, and petrochemicals are important to the country's economy. Puerto Rico and the Dominican Republic import liquefied natural gas (LNG) from Trinidad and Tobago for power generation.

Production

Only three Caribbean countries have oil and natural gas reserves: Barbados, Cuba, and Trinidad and Tobago. Of these, Trinidad and Tobago is the only significant exporter.

Table 18.1

	*Proven Reserves as of 1/1/03**		*Production*	
	Crude Oil (Million Barrels)	*Natural Gas (Billion Cubic Feet)*	*Oil (Crude, Liquids, Refinery Gain) (1,000 Barrels per Day, 2002)*	*Natural Gas (Billion Cubic Feet, 2001)*
Barbados	2.5	5	1.2	1
Cuba	750.0	2,500	49.3	17.7
Trinidad & Tobago	716.0	23,450	141.5	490
Total	**1,468.5**	**25,955**	**192**	**508.7**

Trinidad and Tobago

Trinidad and Tobago is the Caribbean's largest producer of oil and gas. In 2002, oil production averaged about 141,500 barrels per day (bbl/d), of which 121,833 bbl/d was crude oil. Crude oil reserves, at an estimated 716 million barrels, are expected to last only another decade unless new reserves are found. BP is the nation's largest oil and gas producer. Petrotrin, the state oil company, is the second largest oil producer.

In contrast, the country's natural gas reserves are expected to increase significantly in coming years. Proven natural gas reserves currently stand at 23.5 trillion cubic feet (Tcf). In 2001, the country produced 536 billion cubic feet (Bcf) of natural gas.

Trinidad and Tobago has become one of the major natural gas development centers in the world. Natural gas is expected to surpass oil as the main revenue earner for the country in the near future. According to the Trinidad and Tobago Ministry of Energy and Energy Industries, about 36 per cent of its natural gas was exported in 2002 as liquefied natural gas (LNG), while the rest was used domestically, particularly for ammonia and methane manufacturing. Trinidad now has nine ammonia complexes, six methanol units, a urea plant, and an iron and steel complex. In early 2004, the largest methanol plant in the world, Atlas, is expected to be completed. Atlas will be surpassed by an even bigger methanol project, M5000 facility, which is due to enter production in 2004. Trinidad is the world's leading exporter of both ammonia and methanol.

Exploration and Production

Both natural gas and oil exploration activities in Trinidad and Tobago have continued at a fast pace over the last three years. During 2000-2001, British Petroleum Trinidad (BPTT) alone added more than 6 Tcf to its own reserves. EOG Resources (U.S.) announced in May 2002 a new discovery in its SECC block, located off Trinidad's eastern shore. The new discovery is expected to add 250-350 Bcf of natural gas to the field's existing reserves. Repsol-YPF and BPTT discovered in September 2002 nearly 1 Tcf in the Iron Horse field, located east of Trinidad and Tobago.

Recent Developments

The Trinidad and Tobago government granted approval in March 2003 to BHP Billiton, TotalFinaElf, and Talisman Energy to proceed with the $726 million first phase development of the Greater Angostura project, located off the northeastern coast of Trinidad. The companies will be developing Block 2(C). Billiton estimates that the block contains up to 160 million barrels of oil and 1.75 Tcf of natural gas. It is expected that this field could produce 75,000 to 100,000 bbl/d, significantly increasing the country's crude oil production.

In April 2002, BG International, Billiton, Talisman, and TotalFinaElf signed a Production Sharing Contract (PSC) with the government for the adjacent Block 3(a). The group plans to drill two wells in the Block later this year.

Deepwater Exploration

In 1998, Shell, BP, Arco and ExxonMobil received licenses to drill off the eastern coast of Trinidad in depths ranging from 2,460 to 4,921 feet. The four deepwater Blocks–25(a), 25(b), 26 and 27–have yet to produce oil after drilling seven wells. In March 2003, Shell returned Block 25(a) to the Trinidad government, after the well proved to be noncommercial. The Trinidad government is not giving up hope of finding commercial deposits of oil. In June 2002, the government, in conjunction with 12 foreign firms, undertook a new seismic study of previously unexplored regions with depths ranging from 5,576 to 9,480 feet.

New Licensing Round

The Trinidad and Tobago government plans to launch a nine-block licensing round, which is expected to run from May 23 to September 30, 2003. The first two Blocks, 23(a) and 24(b), are in ultra-deep water. They are located near the area where the government recently conducted a seismic study (mentioned above). Blocks 2(ab), 3(b) and 4(a) are located in shallow waters to the east of Trinidad.

Blocks 1(a) and 1(b) are west of Trinidad in the Gulf of Paria. The finally two Blocks, 22 and 24, also located near 23(a) and 24(b), but are not as deep.

Joint Field Development with Venezuela

Trinidad and Tobago has been in negotiations with Venezuela to develop and possibly combine the two countries' gas reserves in the Deltana region, located of the Paria Peninsula. These negotiations, however, have been placed on hold.

Pipelines

The Trinidad and Tobago government plans to build an undersea natural gas pipeline throughout the eastern Caribbean states. The proposed 600-mile pipeline would run north from Trinidad, connecting to other islands en route to the French island of Guadeloupe. The pipeline would serve the islands Martinique, Guadeloupe, Barbados, Grenada, St. Vincent, Grenadines, St. Lucia and Dominica. There has also been discussion of potentially extending the pipeline to Florida.

LNG

Trinidad and Tobago is the largest LNG exporter to the United States. The Atlantic LNG Company of Trinidad and Tobago currently operates three LNG trains on the islands. The Atlantic plant is considered the largest single-train plant ever built, and the first LNG producer in the Latin America and Caribbean region.

The plant recently underwent a $1 billion expansion to triple its capacity from 3 million to 9 million tons per year by adding a second and third train. Train 2 began operation in August 2002, while Train 3 came on line in April 2003. 62.5 per cent of the total output of Trains 2/3 is committed to the Spanish market. The remaining 37.5 per cent is sold in the U.S. market, mainly in southeast, through the Elba Island Terminal and to Lake Charles. The Dominican Republic also imports LNG from Trinidad.

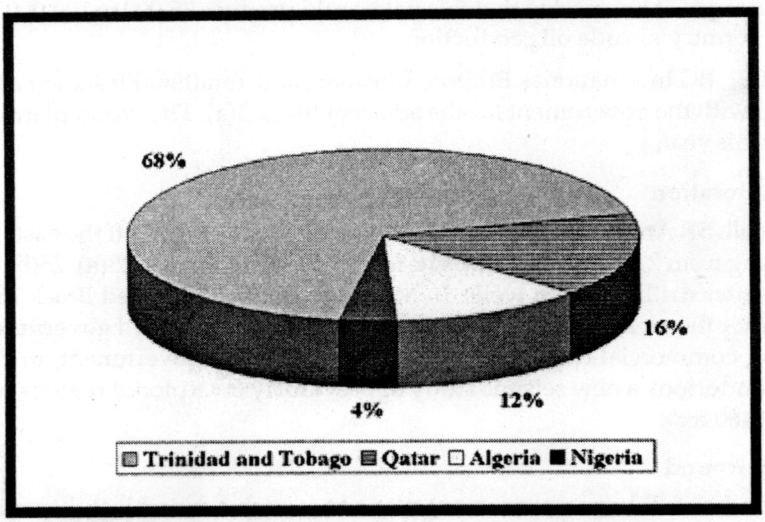

PERCENTAGE OF LNG DELIVERY TO THE US, 2002

BP is building one of the world's largest offshore gas processing units, the Kapok platform, as part of a $600 million project to supply gas to these additional trains. BPTT is also constructing a 30-mile long pipeline, the "Bombax". The pipeline will feed natural gas from Kapok to Atlantic LNG's Trains 2 and 3, and to the Atlas Methanol plant. The pipeline will be one of the world's largest, with a transport capacity 2 billion cubic feet per day (Bcf/d).

Partners of Train 1 are BP, BG, Repsol, Tractebel, and National Gas Company of Trinidad and Tobago (NGC). Partners of Train 2 and 3 are BP, BG and Repsol.

Additional Trains

Negotiations for the construction of a fourth train, which, according to design plans, would raise Trinidad and Tobago's LNG production capacity to 13.8 million tons per year, are ongoing. The construction of the fourth train was originally scheduled to begin by the end of 2002 and come online at the end of 2005. The government has also indicated that it would be interested in constructing a fifth and sixth train.

Cuba

Cuban oil production has more than doubled since 1991. In 2002, Cuban oil production averaged 49,300 bbl/d while oil consumption averaged 157,000 bbl/d. Most domestic crude oil production consists of a heavy, sulfur-laden oil, which is converted into usable fuel. Refineries process imported crude oil, mainly from Venezuela and Mexico. Cuba has a financial arrangement with Venezuela, allowing for the sale of oil under preferential conditions. Cuba imports 53,000 bbl/d of crude oil and refined products from Venezuela, representing approximately a third of the island's oil demand. The contract, which was signed in October 2000, is set to expire in 2005. Natural gas production has also increased significantly in Cuba since the early 1990s. As of January 2003, proven natural gas reserves stood at 2.5 Tcf.

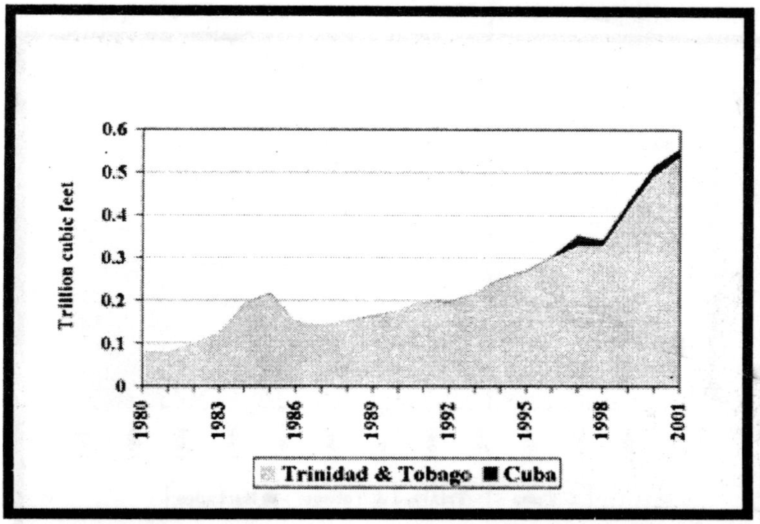

CARIBBEAN NATURAL GAS PRODUCTION, 1980-2001

Exploration and Production

The Cuban government plans has set a goal to generate all of its electricity demand from domestic crude oil resources, making the island self-sufficient in energy. Increases in oil and natural gas production already fulfill more than 50 per cent of domestic energy demand, according to the Cuban government. The government has forecasted that the combined oil and natural gas production will increase by 17 per cent in 2003.

Cuba's territorial waters in the Gulf of Mexico are expected to hold the most promising oil prospects. In early 2000, Cuba offered 59 offshore blocks in its deepwater economic exclusive zone (EEZ) for oil exploration by international oil companies. The Cuban government has awarded Repsol-YPF (Spain) and Sherrit International (Canada) blocks in the EEZ. The United States maintains an economic embargo against Cuba, and oil companies from other countries may be subject to U.S. sanctions under the Helms-Burton Act of 1996 if they conduct significant transactions in Cuba.

New Developments

Sherrit International announced in March 2003 that it plans to invest $110 million in its oil and gas operations in Cuba. Petrobras of Brazil expressed interest in restarting oil exploration in Cuba after the Cuban government offered Petrobras a deep-water exploration block in Gulf of Mexico. Two years ago, Petrobras abandoned exploration efforts in Cuba following disappointing results. Pebercan (Canada) doubled crude oil production at its Block 7 joint-venture (Sherrit) in 2002. Pebercan expects to increase its production as it brings more wells online.

Barbados

Oil production for 2002 totaled 1,200 bbl/d. Oil production has declined slightly since 2001 despite efforts of Barbados National Oil Company (BNOC) to expand oil production. BNOC contracts with a consortium led by Triassic. As Barbados has no refining capacity, its oil is refined in Trinidad, and then returned for domestic consumption.

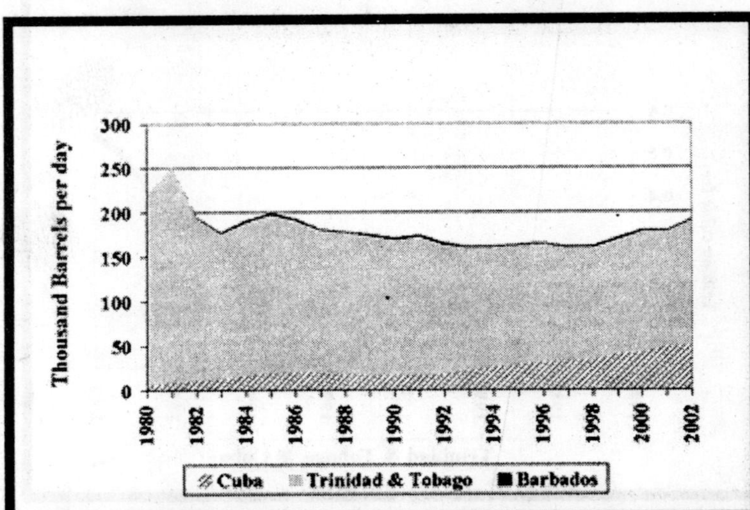

CARIBBEAN OIL PRODUCTION, 1980-2002

Other Regional Developments

Jamaica announced in April 2003 that it plans to restart exploration efforts off its southern coast. Jamaica expects to receive support from Ecuador and Venezuela.

U.S. companies AES Ocean LNG and El Paso and Belgium's Tractebel have plans to build regasification terminals in the *Bahamas*, from where it would be piped to Florida.

If approved, AES would construct a regasification terminal at Ocean Cay in the Bahamas, along with a 54.3-mile pipeline to Florida. The pipeline would have a capacity of 842 million cubic feet per day (Mmcf/d). AES has already received provisional permission from both the U.S. Federal Energy Regulatory Commission (FERC) and the Bahamian government. Environmental impact assessments will determine whether the project will be able to proceed.

Tractebel, which also received approval from FERC but is still waiting approval from Bahamian authorities, would build a LNG regasification terminal in Freeport, on Grand Bahama Island, as well as a 54-mile pipeline (Calypso), with a capacity of 832 Mmcf/d.

El Paso has already received provisional approval from the Bahamian government but not yet from FERC. El Paso also plans to build its terminal on Grand Bahama Island. The pipeline (Seafarer) would have a capacity of 1 Bcf/d. The Bahamian government indicated, however, that it would only accept two of the three proposed projects.

In November 2002, the Bahamian government granted the oil company Kerr McGee a license to drill oil in Bahamian waters.

Refining

Crude oil refining capacity in the Caribbean exceeds 1.7 million bbl/d. Smaller refineries are geared towards meeting local demand, while the larger refineries in Aruba, the Netherlands Antilles, Trinidad and Tobago, and the U.S. Virgin Islands serve both local and export markets. The Hovensa refinery of St. Croix, owned by Hess and the Venezuelan state oil company, PdVSA, is among the largest in the Western Hemisphere. PdVSA also operates the Curacao Isla refinery (the Netherlands Antilles), which the company is in process of expanding. PdVSA is investigating the possibility of building two new refineries in the Dominican Republic, as well as reactivating the Cienfuegos refinery in Cuba.

In August 2002, the Trinidad and Tobago government granted Soreco Inc. approval for the construction of a 224,000-bbl/d refinery. The refinery, the Sabaneta Refinery Project, is expected to be completed in 2005. However, it remains unclear whether the project will move forward due to difficulties of securing $2 billion in loans for the project. U.S. energy company El Paso plans to sell its refinery on Aruba, Coastal Aruba Refining Company, in 2003.

Storage

The Caribbean region has a number of independent petroleum storage facilities, with the capacity to store approximately 100 million barrels of crude oil and petroleum products. In addition to long-term storage arrangements, these facilities offer logistical options for petroleum shipments. Islands with storage facilities include: Bahamas, Trinidad, Puerto Rico, Saint Lucia, Aruba, and St. Eustatius, uracao, and Bonaire of the Netherlands Antilles.

Table 18.2: Crude Oil Refining Capacity (January 1, 2003)

	Company/Location	*Capacity (barrels/day)*
*Aruba (NETH)**	Coastal Aruba Refining Co./San Nicolas	280,000
Cuba	Cienfuegos	76,000
	Ermonos Dias/Santiago	101,500
	Niko Lopes/Habana	121,800
	Serhio Soto/Cabaiguan	2,100
	Subtotal, Cuba	301,400
Dominican Republic	Falconbridge Dominicana/Bonao	16,000
	Refineria Dominicana de Petroleo/Haina	32,300
	Subtotal, Dominican Republic	48,300
Jamaica	Petrojam/Kingston	34,200
Martinique (FR)	Societe Anonyme de la Raffinerie des Antilles/Fort-de-France	17,000
Netherlands Antilles (NETH)	Refineria Isla Curazao/Emmastad	320,000
*Puerto Rico (US)**	Caribbean Petroleum Refining/Bayamon	48,000
	Shell Chemical/Yabucoa	45,000
	Subtotal, Puerto Rico	93,000
Trinidad & Tobago	Petroleum Co. of Trinidad & Tobago/Pointe-a-Pierre	160,000
U.S. Virgin Islands	Hovensa/St. Croix	525,000
Total	**14 Plants**	**1,778,900**

Exports to the United States

In 2002, the continental United States imported about 390,000 bbl/d of petroleum from the Caribbean, of which about 83 per cent were petroleum products. The U.S. Virgin Islands was the largest single regional exporter to the United States (about 236,000 bbl/d of petroleum products), followed by The Netherlands Antilles (about 74,000 bbl/d of petroleum products), and Trinidad and Tobago (80,000 bbl/d of total crude and petroleum products). Trinidad and Tobago is the only exporter of crude oil (68,000 bbl/d) from the region.

Electricity

Installed electric generating capacity in the Caribbean exceeds 17 gigawatts (GW). In general, the region needs additional capacity. Several countries (including the Dominican Republic, Haiti, and Cuba) experience power outages on a regular basis. Electricity demand in the region is expected to grow substantially in the coming decade.

The Dominican Republic

The Dominican Republic has been plagued by electricity blackouts for years. In 1999, the former President Leonel Fernandez privatized parts of the state-owned Dominican Electricity Corporation (CDE) in the hope of eliminating the country's power outage problems. However, privatization has yet to assure a steady supply of electricity to the population.

Table 18.3: Electricity in the Caribbean, 2001

	1/1/2001 Installed Capacity (million kW)	*2001E Net Generation (Billion kWh)*
Antigua and Barbuda	0.03	0.12
Aruba (NETH)	0.09	0.53
Bahamas, The	0.40	1.56
Barbados	0.17	0.78
Cayman Islands (UK)	0.09	0.38
Cuba	4.49	14.39
Dominica	0.02	0.07
Dominican Republic	3.08	9.19
Grenada	0.03	0.14
Guadeloupe (FR)	0.42	1.16
Haiti	0.24	0.58
Jamaica	1.40	6.27
Martinique (FR)	0.12	1.15
Montserrat (UK)	0.001	0.003
Netherlands Antilles (NETH)	0.21	1.06
Puerto Rico (US)	4.9	20.90
Saint Kitts and Nevis	0.02	0.10
Saint Lucia	0.02	0.12
Saint Vincent/Grenadines	0.01	0.09
Trinidad and Tobago	1.42	5.32
Virgin Islands, U.S.	0.32	1.03
Virgin Islands, British (UK)	0.01	0.04
Total	**17.49**	**64.98**

A dispute between the Dominican government and the private companies which acquired parts of CDE in 1999 reportedly is partly responsible for the blackouts. When CDE was split up, Union Fenosa (Edenorte and Edesur subsidiaries) and AES (Edeeste subsidiary) bought into the distribution network. In September 2002, the government and customers began falling behind on their payments, leaving the private companies without enough money to pay generators. The distributors periodically stopped supplying electricity, resulting in blackouts. Although current President Hipolito Mejia renegotiated contracts between the government and private power companies in an effort to solve the country's power supply problems in September 2002, the situation has not improved.

Despite the need for further structural reform, foreign firms continue to invest in the Dominican Republic's power infrastructure. AES Andres, a subsidiary of AES, has invested $400 million to construct an LNG import terminal, a regasification facility, a pipeline and a 300-MW combined cycle power plant, located outside of Santo Domingo. The LNG terminal received its first shipment in February 2003. However, since the 300-MW power plant will not be completed until summer 2003, the imported natural gas has been going to the Los Mina power plant in Santo Domingo, which AES converted from fuel oil to natural gas.

In April 2003, Canada signed two $35 million agreements with the Dominican government to build hydroelectric and renewable energy projects.

Jamaica

In March 2001, the U.S.-based utility, Mirant Corporation, completed an 80 per cent acquisition of formerly government-owned Jamaica Public Service Company, the island's main power provider. Mirant has stated that it is making progress in reducing the number of blackouts currently affecting Jamaica due to lack of capacity. Including independent producers, installed capacity is 700 MW, and Mirant expects to add 385 MW to the system in 2003. Last year, Jamaica made the decision to begin replacing fuel oil with natural gas for its power plants and bauxite and alumina sectors. The Jamaican government plans to import LNG to meet natural gas demand.

Table 18.4: Primary Energy Consumption in the Caribbean, 2001

Country/Territory	*Total (quadrillion Btu)*	*Petroleum*	*Natural Gas*	*Coal*	*Other*
Antigua and Barbuda	0.007	100%	---	---	---
Aruba (NETH)	0.013	100%	---	---	---
Bahamas, The	0.049	100%	---	---	---
Barbados	0.024	96%	4%	---	---
Cayman Islands (UK)	0.005	100%	---	---	---
Cuba	0.390	92%	6%	0.2%	1.8%
Dominica	0.0016	75%	---	---	25%
Dominican Republic	0.273	95%	---	2%	3%
Grenada	0.002	100%	---	---	---
Guadeloupe (FR)	0.027	100%	---	---	---
Haiti	0.025	90%	---	---	10%
Jamaica	0.147	98%	---	0.7%	1.3%
Martinique (FR)	0.028	100%	---	---	---
Montserrat (UK)	0.001	100%	---	---	---
Netherlands Antilles (NETH)	0.158	100%	---	---	---
Puerto Rico (US)*	0.428	93%	5%	1%	.4%
Saint Kitts and Nevis	0.001	100%	---	---	---
Saint Lucia	0.005	100%	---	---	---
Saint Vincent/Grenadines	0.003	90%	---	---	10%
Trinidad and Tobago	0.475	10%	90%	---	---
Virgin Islands, U.S.	0.138	95%	---	5%	---
Virgin Islands, British (UK)	0.001	100%	---	---	---
Total/Average*	**2.202**	**92%**	**4.8%**	**0.4%**	**2.3%**

*Note: Because data and percentages were rounded, not all will total 100 per cent.

Puerto Rico

In January 2000, Puerto Rico's primary electric power producer and grid operator, publicly-owned Puerto Rico Electrical Power Authority (PREPA), initiated a $2.4 billion program to increase the island's electric generation capacity. Economic growth in the late 1990s averaged 3 per cent, resulting in concerns that electricity demands associated with economic growth would exceed generation capacity. In response, PREPA began signing contracts with independent power producers (IPPs) to increase the island's generation capacity from about 4.9 GW to 5.2 GW by 2003. A new 454-MW power station, built by U.S.-based AES, is the first coal-fired plant in Puerto Rico. The plant has the additional benefit of being able to sell its steam to Phillips Petroleum's Puerto Rico subsidiary. The EcoElectrica and AES facilities are part of a general plan to reduce Puerto Rico's dependence on oil for electricity generation. In August 2000, the first shipment of LNG from Trinidad and Tobago arrived at Punta Guayanilla, near Ponce, where there is an LNG receiving terminal used to supply gas to the plant.

Renewable Energy

Only Jamaica and Cuba had significant amounts of power generated from non-hydro renewable (geothermal, solar, wind, wood and waste) electric sources in 2001. Cuba produced approximately 0.8 billion kWh of renewable power while Jamaica generated 0.1 billion kWh. The Dominican Republic was the largest producer of hydroelectricity in the Caribbean, with 0.7 billion kWh. This was greater than hydropower produced by Cuba, Haiti, Jamaica, and Puerto Rico combined.

19

Caspian Sea Region

The Caspian Sea region, including the Sea and the states surrounding it, is important to world energy markets because of its potential to become a major oil and natural gas exporter over the next decade. However, this potential has been complicated by several factors, including a lack of adequate export infrastructure, disagreement over new export routes, and border disputes between the littoral states.

General Background

The Caspian Sea has become a landmark for the former Soviet South's untapped oil and natural gas potential. Herein, the Caspian Sea Region is defined to include the Sea's littoral states of Azerbaijan, Kazakhstan, and Turkmenistan, as well as parts of Russia and Iran and also Uzbekistan –which although not a littoral state, is the region's largest natural gas producer. The oil and gas potential of these states is sizeable. Proven oil and natural gas reserves have been estimated at up to 3 per cent and 4 per cent of the world total, respectively, and analysts expect that by 2010 regional oil production will meet or exceed 2002 capacity in Venezuela, South America's largest oil producer.

At the moment, however, the countries of the Caspian Sea region are relatively minor world oil and gas producers, struggling with difficult economic and political transitions. Following the break-up of the Soviet Union, the countries' economies languished as regional trade collapsed. And while, the region has enjoyed overall economic growth since the latter half of the 1990s, gross domestic product (GDP), remains below 1992 levels in Azerbaijan, Kazakhstan, and Turkmenistan (when measured in constant $1995 dollars). Uzbekistan, however, has shown some growth. Moreover, in the region's two biggest oil producers, Azerbaijan and Kazakhstan, 49 per cent and 26 per cent of the populations, respectively, were estimated to be living below the poverty line in 2001. Improving these conditions depends, in large part, on the successful development of the region's oil and natural gas potential.

Although there is still no overarching agreement between the five Caspian littoral states on division of the Sea's resources, three states have come to a trilateral agreement on sub-surface boundaries and collective administration of the sea's waters. In May 2003, Russia, Azerbaijan, and Kazakhstan divided the northern 64 per cent of the Caspian Sea into three unequal parts along a median line principle, giving Kazakhstan 27 per cent, Russia 19 per cent, and Azerbaijan 18 per cent. Accordingly, development of the northern Caspian Sea's hydrocarbon potential, where most of the region's oil reserves and largest international projects are found, will likely move forward despite the lack of a comprehensive regional consensus. Meanwhile, offshore development in Turkmenistan and Iran, which were present at these negotiations but refused to sign on, could fall even further behind.

Oil

Estimates of the Caspian Sea Region's proved crude oil reserves vary widely by source. For this reason, we have estimated proven oil reserves as a range between 17 and 33 billion barrels, which is comparable to OPEC member Qatar on the low end, and the United States on the high end. In 2002, regional oil production reached roughly 1.6 million barrels per day, comparable to annual production from South America's second largest oil producer, Brazil. By 2010, the countries of the Caspian Sea Region are forecast to produce between 3 and 4.7 million barrels per day, which exceeds annual production from South America's largest oil producer, Venezuela.

Growing oil production since independence (an increase of roughly 70 per cent since 1992) has come primarily from the north Caspian states of Kazakhstan and Azerbaijan. Development of the region's oil resources has been led by two major projects: Kazakhstan's *Tengiz*, and Azerbaijan's *Azeri, Chirag, and deepwater Gunashli (ACG)*. Combined, these two projects produced about 410,000 barrels per day in 2002, one-fourth of the regional total, and are expected by the operating companies to produce 1.7 million barrels per day for world oil markets by 2010. Development of these vanguard projects, which are each roughly ten years old, has given rise to the influx of new investment and infrastructure development that constitutes the "second Caspian oil rush," the first having occured in the late 1800s. Following theses discoveries, major new finds were announced in Azerbaijan at Shah Deniz in 1999 ("potential recoverable resources" of roughly 14 Tcf of natural gas), and in Kazakhstan at Kashagan in 2000.

By comparison, other countries in the Caspian Sea region have made little progress towards developing their hydrocarbon resources since independence. Turkmenistan's and Uzbekistan's leading oil projects are significantly smaller and markedly less developed than those in Azerbaijan and Kazakhstan. This is because proven oil reserves in Turkmenistan and Uzbekistan are considerably smaller than those in their neighboring states, and the political regimes in Ashgabat and Tashkent have been considered less favorably by foreign investors. As a result, while multinational oil companies have initiated numerous large-scale projects in Azerbaijan and Kazakhstan, Turkmenistan and Uzbekistan have achieved only smaller-scale deals.

Oil and gas development in Russia's and Iran's sectors of the Sea has been similarly stunted, although exploration efforts are reportedly underway. Russian oil company, LUKoil, began exploration of the north Caspian in 1995 and is reportedly working on a drilling program for 2004-2010, with initial production of natural gas expected by as early as 2005. In July 2003, two Russian oil and gas firms, LUKoil and Gazprom, established a joint venture with Kazakhstan's state oil company, KazMunaiGaz, to develop the Tsentralnaya hydrocarbon structure, located on the border of the Russian and Kazakhstani offshore sectors. According to LUKoil, the Tsentralnaya structure holds recoverable reserves of roughly 20 Tcf of natural gas, with drilling expected to begin in 2007. Russian natural gas monopoly, Gazprom, is also party to another project in the offshore Caspian Sea, Kurmangazy. The Kurmangazy field also sits astride the border between Russia and Kazakhstan's sectors of the sea, and accordingly is also being developed in conjunction with Kazmunaigaz, as well as Rosneft (Russia). Exploratory drilling at Kurmangazy is expected to begin by the end of 2003 or early 2004. Iran has made less progress towards developing its Caspian Sea resources. In June 2003, a representative from Iran's Caspian Oil Company told reporters that the company intends to begin developing its Caspian Sea shelf within two years.

Natural Gas

The Caspian Sea region's natural gas potential is, by some measures, more significant than its oil potential. Regional proven natural gas reserves are estimated at 232 trillion cubic feet (Tcf), comparable to those in Saudi Arabia. Natural gas production in 2001 was approximately 4.5 Tcf, comparable to the combined production of South America, Central America, and Mexico. But companies and governments have thus far shown greater interest in oil than in natural gas–owing, in part, to the greater capital expenditures necessary to start up new natural gas projects, as well as the region's lack of existing infrastructure. As a result, regional natural gas production has increased only modestly since independence. With only one major foreign investment focused primarily on gas, the region will still need considerable investment in upstream projects and export infrastructure before its full potential can be realized.

Since independence, regional natural gas production has been characterized by modest annual increases from Uzbekistan, and a dramatic collapse (then partial recovery) from Turkmenistan. This is because, after 1991, natural gas from the Caspian Sea region, mostly from Turkmenistan, became a competitor with Gazprom, the Russian state natural gas company. Because all of the pipelines connecting the region to world markets were owned by Gazprom and routed through Russia, Caspian natural gas was squeezed out of the hard currency market.

As a result, Turkmenistan's incentives for increasing its production of natural gas disappeared. The country's output dropped throughout the 1990s, plummeting from 2.02 Tcf in 1992 to just 466 billion cubic feet (Bcf) in 1998 when the country was locked in a pricing dispute with Russia over the export of Turkmen natural gas. In 1999, a Turkmen-Russian agreement took hold, and in 2000, production skyrocketed to 1.64 Tcf before reaching 1.7 Tcf in 2001. In April 2003 the country signed new agreements with Uzbekistan and Russia to increase exports to both countries substantially over the next twenty years.

In contrast, Uzbekistan has secured modest annual production growth by avoiding Russia's pipeline system and by concentrating on domestic consumption and neighboring natural gas markets. Uzbekistan is the third largest natural gas producer in the Commonwealth of Independent States, and one of the top ten natural gas-producing countries in the world. Since becoming independent, Uzbekistan has ramped up its natural gas production by nearly 50 per cent, from 1.51 Tcf in 1992 to 2.23 Tcf in 2001. Uzbekistan's state oil and gas concern, Uzbekneftegaz, is attempting to slow the natural depletion of its existing fields, and to develop new projects, but has failed thus far to secure a major collaboration with foreign investors. Uzbekneftegaz signed its first Production Sharing Agreement (PSA) in April 2001 with Britain's Trinity Energy (through a specially formed subsidiary known as UzPEK Ltd). The project entails the development of fields in Uzbekistan's central Ustyurt and Southwest Gissar regions, and the partners expect natural gas production of roughly 71 billion cubic feet by 2006.

The region's leading oil producers, Azerbaijan and Kazakhstan, are both net natural gas importers, purchasing gas primarily from Uzbekistan and Russia. However, both states plan to increase their own natural gas production significantly over the decade and become net natural gas exporters. Azerbaijan's major natural gas production increases in the future are expected to come from the development of the aforementioned Shah Deniz field. The Shah Deniz offshore natural gas and condensate field, is thought to be one of the world's largest natural gas field discoveries of the last 20 years, and contains "potential recoverable resources" of roughly 14 Tcf of natural gas according to the project's operator, British Petroleum (BP). The field is expected to begin producing natural gas for

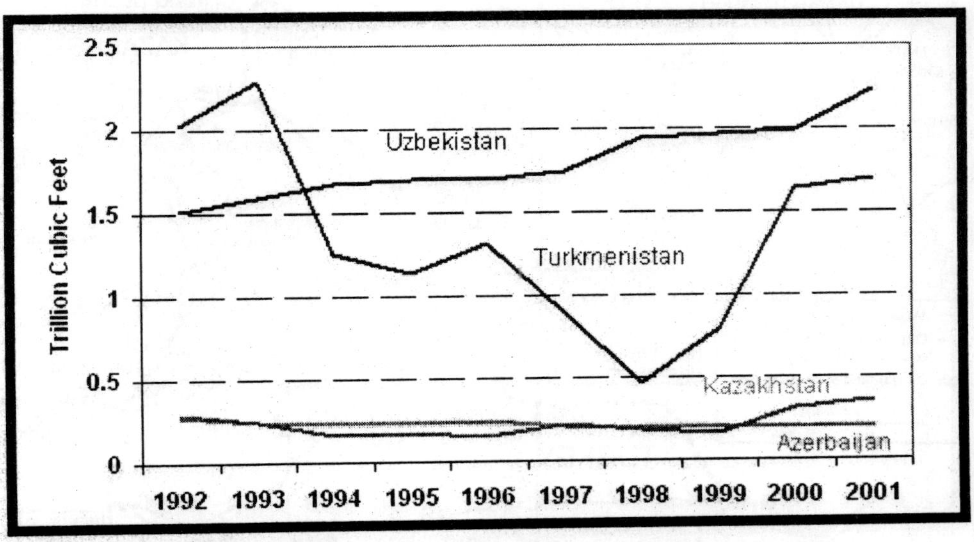

NATURAL GAS PRODUCTION IN THE CASPIAN SEA REGION

export in 2006 and will be capable of producing approximately 296 Bcf of natural gas per year. Kazakhstan's natural gas production increases are expected to come primarily from associated gas at Kazakhstan's three largest fields, Tengiz, Karachaganak, and Kashagan.

Export Issues

As increasing exploration and development in the Caspian Region leads to more production, the littoral states (and Uzbekistan) will have large new quantities of oil and natural gas available for export. Earning hard currency from these resources is essential to regional development plans, as well as to recouping the huge investments made by multi-national oil companies. However, for these purposes, the infrastructure left after the collapse of the Soviet Union is inadequate. Numerous new pipelines and pipeline expansions in each of the cardinal directions have been proposed, and some have been constructed.

West

Changing the region's energy flow to an East-West axis towards Europe, from the existing North-South axis towards Russia (which will effectively minimize dependence on Russia), is integral to the development goals of these newly independent states. The region's three biggest pipeline projects, the Caspian Pipeline Consortium Project (CPC), the Baku-Tiblisi-Ceyhan oil pipeline (BTC), and the South Caucasus natural gas pipeline (a.k.a.Baku-Tiblisi-Erzurum –BTE) are establishing the framework for this new axis.

The *Caspian Pipeline Consortium* project connects Kazakhstan's Caspian Sea area oil deposits with Russia's Black Sea port of Novorossiysk. Oil loaded at the port of Novorossiysk is then taken by tanker to world markets. Although the pipeline transverses Russia and was developed in conjunction

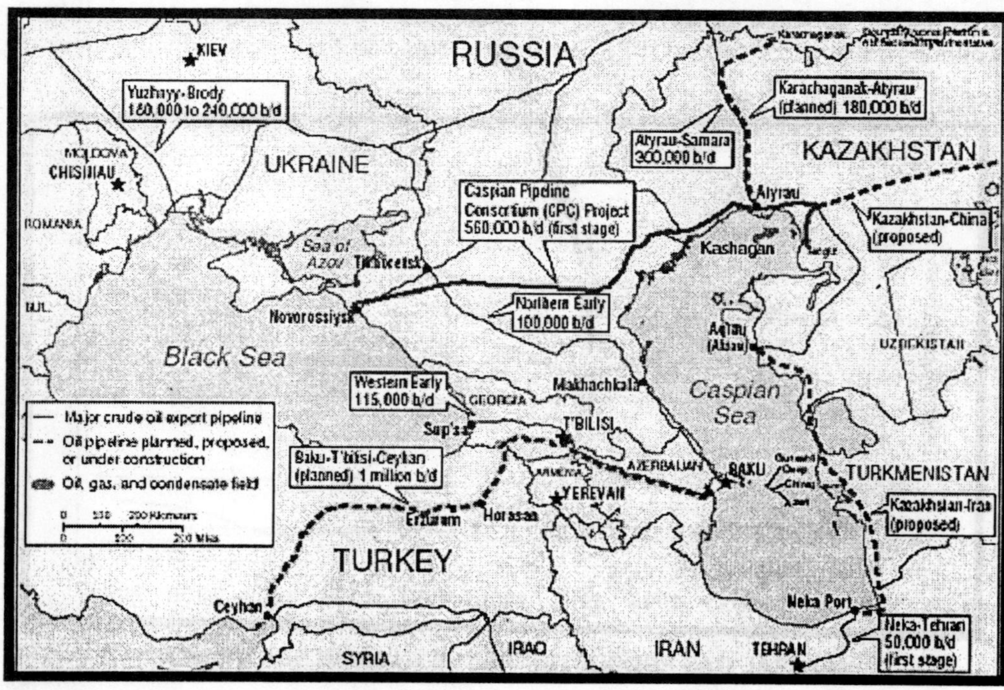

with the Russian government, development of the CPC has, for the first time, given the Caspian Sea region a viable alternative to the Russian dominated northern export routes (namely Atyrau-Samara, see below). The pipeline's first crude oil was loaded at Novorossiysk on October 15, 2001, and the pipeline was officially opened on November 27, 2001. Most of the oil routed through CPC comes from Kazakhstan's Tengiz field, however, several other Kazakhstani fields are developing spur connections to the pipeline. Accordingly, these new connections are expected to increase throughput from 260,000 bbl/d, in 2002, to between 350,000 and 415,000 bbl/d by the end of 2003. Current capacity of the CPC pipeline is 560,000 bbl/d, with plans to increase capacity to 1.34 million bbl/d.

However, additional Caspian oil exports through the CPC pipeline will increase congestion in Turkey's Bosporus Straits, which connect the Black Sea to the Mediterranean. Turkey has raised concerns about the ability of the Bosporus Straits, already a major chokepoint for oil tankers, to handle the additional tanker traffic, given that most of Russia's existing oil export pipelines also terminate at Novorossiysk. Turkey has stated its environmental concerns about a possible collision (and ensuing oil spill) in the Straits as a result of increased tanker traffic from the launch of the CPC's pipeline. As a result, there are a number of options under consideration for oil transiting the Black Sea to bypass the Bosporus Straits.

Another westward pipeline, *Baku-Tbilisi-Ceyhan*, will export Azeri (and quite possibly Kazakhstani) oil along a 1,040-mile route from Baku, Azerbaijan via Georgia to the Turkish Mediterranean port of Ceyhan, allowing oil to bypass the Bosporus Straits. Construction of the 1-million-bbl/d pipeline, which is estimated to cost $2.9 billion, is scheduled to be completed in 2004, with oil to begin flowing in 2005.

The BTC project has faced numerous challenges in its development. After allaying initial fears that the pipeline was technically infeasible given its considerable distance through rugged terrain, the projects developers now face criticism from local and international non-governmental organizations which consider the project to be environmentally hazardous, threatening to regional archeological treasures, and in violation of international human rights. Critics of the pipeline have taken theses grievances to the major international financial institutions involved in the project, the World Bank and the European Bank for Reconstruction and Development (EBRD), and have asked that they withdraw their support. According to statements from BTC Co., the pipeline's developer's, plans are still on schedule for completion in fourth quarter of 2004.

A third natural gas pipeline, known as the "South Caucasus Pipeline," a.k.a "*Baku-Tbilisi-Erzurum*", or "BTE", will run parallel to the Baku-Tbilisi-Ceyhan oil pipeline for most of its route before connecting to the Turkish infrastructure near the town of Erzurum. The South Caucasus pipeline is designed to carry natural gas from Azerbaijan's Shah Deniz field, and have a capacity of 258 Bcf per year. The project is scheduled to be completed in time for the Shah Deniz project's first contracted exports to Turkey in 2006.

East

However, there is some question as to whether westward to Europe is the right direction for Caspian oil and natural gas. Oil demand over the next 10 to 15 years in Europe is expected to grow by little more than 1 million bbl/d. Oil exports eastward, on the other hand, could serve Asian markets, where demand for oil is expected to grow by 10 million bbl/d over the next 10 to 15 years. In particular, Chinese oil consumption is projected to rise dramatically.

But, supplying this Asian demand would necessitate building some of the world's longest pipelines. Geographical considerations would force any pipelines to head north of the impassable mountains of Kyrgyzstan and Tajikistan across the vast, desolate Kazakh steppe, thereby adding even more length (and cost) to any eastward pipelines.

Scuth

An additional way for Caspian region exporters to supply Asian demand would be to pipe oil and natural gas south. This would mean sending oil and natural gas through either Afghanistan or Iran. The Afghanistan option, which Turkmenistan has been promoting, would entail building pipelines across war-ravaged Afghan territory to reach markets in Pakistan and possibly India. With the ouster of the Taliban in Afghanistan in December 2001, proposals to build a Trans-Afghan natural gas pipeline and the Central Asian Oil Pipeline have re-emerged. However, in July 2003, one of the Trans-Afghan natural gas pipeline's key supporters, the Asian Development Bank, called for additional feasibility studies, thus further delaying the project.

Development of a southern pipeline through Iran would be problematic under the Iran and Libya Sanctions Act, which imposes sanctions on non-U.S. companies investing in the Iranian oil and natural gas sectors. U.S. companies already are prohibited from conducting business with Iran under U.S law. In 1997, however, Turkmenistan and Iran completed the $190 million Korpezhe-Kurt Kui pipeline linking the two countries, thereby becoming the first (and so far, only) natural gas export pipeline from Central Asia to bypass Russia. Turkmenistan and Kazakhstan have initiated low-volume oil "swap" deals with Iran, delivering oil in tankers to refineries in Iran's northern regions in exchange for similar volumes of crude at Iranian ports in the Persian Gulf. Iran has recently undertaken efforts to upgrade its domestic distribution network to allow for swap capacity to increase from roughly 50,000 bbl/d to 150,000 bbl/d by the end of August 2003.

North or Northwest

For its part, Russia has proposed multiple pipeline routes that utilize its existing and proposed infrastructure. Shortly after independence, two new northwesterly pipelines were constructed known as the "Northern" and "Western" Early Oil Pipelines, which extend from Baku to Novorossiysk (Russia), and Baku to Supsa (Georgia), respectively, and have a combined capacity of roughly 215,000 bbl/d. Also, an existing northbound pipeline from Atyrau in Kazakhstan to Samara in Russia has been upgraded, but is expected to become relatively less significant as throughput at CPC increases. However, there are political and security questions as to whether the newly independent states of the former Soviet Union should rely on Russia (or any other country) as their sole export outlet, and Caspian region producers already have expressed their desire to diversify their export options.

Regional Conflicts

In almost any direction, Caspian region export pipelines may be subject to regional conflicts, an additional complicating factor in determining final routes. Despite the ouster of the Taliban government in December 2001, Afghanistan remains scarred and unstable after years of war. The Azerbaijan-Armenia war over the Armenian-populated Nagorno-Karabakh enclave in Azerbaijan has yet to be resolved. Separatist conflicts in Abkhazia and Ossetia in Georgia flared in the mid-1990's, and on August 9 2002, Russia's Defense Minister said that the Georgia's Pankisi Gorge had emerged as the world's second main "nest" of terrorism. Russia's war with Chechnya has devastated the region around Grozny in southern Russia. In addition, the Uzbek government has been cracking down on Islamic fundamentalism in Uzbekistan.

Environmental Issues

The spotlight on the Caspian region's oil and gas reserves also highlighted the appalling state of the environment in and around the Sea. Years of neglect have left the Sea and the surrounding region in a precarious position environmentally. Petrochemical and refining complexes on the Absheron peninsula in Azerbaijan are major sources of land-based pollution, and discharges and spills from oil and gas drilling–both onshore and in the sea itself–have had serious impacts on the environment. Untreated waste from the Volga River–into which half the population of Russia and most of its heavy industry drains its sewage–empties directly into the Caspian Sea, while pesticides and chemicals from agricultural run-off are threats to the Sea's flora and fauna. Thousands of seals that live in the Caspian Sea have died since 2000 due to pollution that weakened their immune systems, and overfishing, especially of the prized sturgeon, has caused a dramatic decline in fish stocks.

In addition to the existing problems, several other issues could compound the Caspian region's environmental difficulties. Oil and gas production in the Sea inevitably will result in the construction of pipelines and infrastructure to export these resources to consumers, raising the possibility of loss of habitats for marine life as well as the specter of accidental spills. The mysterious rise of the Caspian Sea could flood oil wells, rigs, and earth-walled reservoirs on the coastline, spilling into water tables and contaminating drinking water supplies. A lack of regional cooperation, highlighted by the still unresolved legal status of the Caspian Sea, as well as weak environmental laws and regulations and the inability to enforce them, already is affecting efforts to protect the Caspian's environment. Continued economic development, improved regional cooperation, and the implementation of modern technology will be required in order to improve the state of the environment in and around the Caspian Sea in coming years.

20

Caucasus Region

The Caucasus Region, comprising the newly independent states of Armenia, Azerbaijan, and Georgia, is important to world energy markets as a transit area for oil and natural gas exports from the Caspian Sea to Europe. Although the region has been beset by conflict, regional leaders hope that the development of several oil and natural gas export pipelines will bring peace and prosperity to the Caucasus.

General Background

The Caucasus region has been fought over by outside powers for centuries, but in 1991, Armenia, Azerbaijan, and Georgia received their independence from the Soviet Union. However, the Soviets bequeathed a number of problems on the three countries, including artificially drawn national borders and centrally-planned economies that were heavily dependent on Russia. Even before Azerbaijan and Armenia received independence, fighting broke out in 1988 between the then-Soviet republics over the disputed area of Nagorno-Karabakh, and separatist conflicts sparked in Georgia soon after independence. Although most of the regional conflicts that flared in different parts of the Caucasus throughout the 1990's are now dormant, none have been officially resolved.

Each of the Caucasus countries has experienced a severe economic downturn since the Soviet era. Conflicts in the region have discouraged foreign investment, and the lack of economic and political reform, along with the continuing hostilities between Azerbaijan and Armenia, have limited the region's economic growth. Although each of the countries is, to different degrees, trying to wean itself off dependency on Russia, Armenia, Azerbaijan, and Georgia remain economically linked to Russia.

Table 20.1: Economic and Demographic Indicators for the Caucasus Region

Country	*Gross Domestic Product (Nominal GDP), 2001E (Billions of U.S. $)*	*Real GDP Growth Rate, 2001 Estimate*	*Real GDP Growth Rate, 2002 Projection*	*Per Capita GDP, 2001E*	*Population 2001E (Millions)*
Armenia	$2.1	7.2%	5.4%	$542	3.8
Azerbaijan	$5.2	7.5%	7.0%	$646	8.1
Georgia	$3.1	4.0%	3.2%	$619	5.0
Total/weighted average	$10.4	6.4%	5.5%	$615	16.9

Despite their geographical proximity, Armenia, Azerbaijan, and Georgia all have different strategies to diversify their energy suppliers and reduce their dependence on Russia. While Azerbaijan looks to the development of its domestic oil and natural gas reserves for an economic boom, Georgia is merely trying to keep electricity on for its citizens. Thus, Georgia is counting on transit oil and natural gas transit from the Caspian Sea region to boost its growth outlook: tariffs from transit oil pipelines will feed the government's budget, while fuel received from natural gas pipelines will fire Georgian electric-generating plants. Armenia, which is fighting a blockade by Azerbaijan and Turkey, does not figure to benefit substantially from the east-west pipelines that are in development, is cultivating closer ties with Iran in order to diversify its energy sources.

With a number of regional energy projects in development, Georgian President Eduard Shevardnadze expressed his hope that additional cooperation in the Caucasus will lead to greater trust and increased prosperity for the three countries. Until economic and political reforms take root, however, and a lasting peace is implemented, economic growth in the Caucasus likely will continue to lag behind other regions undergoing the transition from communism to democracy.

Regional Energy Issues

The Caucasus region's energy importance stems from its location as a land bridge between the Caspian Sea and the Black Sea. Although the region itself does not have significant fossil fuel resources, except in the east of Azerbaijan, the Caucasus provides a link between the bountiful oil and natural gas reserves of the Caspian Sea region and the energy-hungry economies of Turkey and the European Union (EU).

The TRACECA Program (Transport System Europe-Caucasus-Asia, informally known as the Great Silk Road) was launched at a European Union (EU) conference in 1993, bringing together trade and transport ministers from the Central Asian and Caucasian republics to initiate a transport corridor on an east-west axis. As oil and natural gas production in Central Asia and the Caspian region increase, the Caucasus has the potential to become a major transit center for oil and natural gas supplies heading west to world markets.

Table 20.2: Energy Consumption and Carbon Dioxide Emissions in the Caucasus Region, 1999

Country	Total Energy Consumption (Quadrillion btu)	Petro-leum	Natural Gas	Coal	Nuclear	Hydro-Electric	Other Rene-wable Electric	Net Electricity Imports	Carbon Dioxide Emissions (Million metric tons of carbon)
Armenia	0.96	9.0%	48.4%	0.1%	25.6%	16.9%	0%	0%	0.8
Azerbaijan	0.55	56.5%	39.0%	0%	0%	4.2%	0%	0.4%	12.7
Georgia	0.16	31.9%	27.6%	0.1%	0%	42.4%	0%	-2.0%	1.6
Total/weighted average	1.67	26.8%	43.3%	0.7%	14.7%	15.2%	0%	-1.6%	

Table 20.3: Energy Supply Indicators, Caucasus Region

Country	Proven Crude Oil Reserves, 1/1/02 (Million Barrels)	Natural Gas Reserves, 1/1/02 (Trillion Cubic Feet)	Coal Reserves, 1/1/01 (Million Short Tons)	Petroleum Production, 2001 (Thousand Barrels Per Day)	Natural Gas Production, 2000 (Billion Cubic Feet)	Coal Production, 2000 (Million Short Tons)	Electric Generating Capacity, 2000 (Gigawatts)	Crude Oil Refining Capacity, 1/1/02E (Thousand Barrels Per Day)
Armenia	0	0	minimal	0	0	.003	2.7	0
Azerbaijan	1,178	4.4	0	311	200	0	4.8	442
Georgia	35	0.3	minimal	2	2.1	.022	4.5	110
Total	**1,213**	**4.7**	**minimal**	**313**	**202.1**	**.025**	**12.0**	**552**

Oil Transit

With the collapse of the Soviet Union and the increase in oil production from the Caspian Sea region, the Caucasus region has gained in importance from an energy perspective. Previously, the only way for Caspian Sea region oil exports to reach European consumers was via the Russian pipeline system. The United States has supported the principle of multiple export options for Caspian exporters, and a number of Caspian region oil export pipelines are planned or already have been built. Many of these pipelines are routed via the Caucasus, and the region has become the central component of a European Union strategy to direct Caspian oil to European consumers.

Baku-Ceyhan Pipeline

In November 1999, Azerbaijan, Georgia, and Turkey signed agreements affirming the Baku-Ceyhan route as the Main Export Pipeline (MEP) for Azeri oil exports. The planned 1-million-bbl/d capacity pipeline, which has received backing from the United States, will run 281 miles through Azerbaijan, 135 miles through Georgia, and 622 miles through Turkey, and is expected to cost between $2.8 billion and $2.9 billion.

Despite initial opposition to the pipeline, which several oil companies criticized as too costly and uneconomical with the planned volumes from Azerbaijan, construction on the pipeline appears set to begin in the summer of 2002, with a planned launch date at the end of 2004. Detailed engineering work for the pipeline began in June 2001 and is slated to finish in June 2002, identifying a 50-yard-wide corridor through each of the three countries to lay the pipeline. Tenders have been announced to lay the pipeline in Georgia and Azerbaijan (Botas will lay the Turkish section of the pipeline).

Baku-Supsa Pipeline

On March 8, 1996, Georgian President Eduard Shevardnadze and Azerbaijani President Heydar Aliyev signed a 30-year agreement to pump a portion of the Azerbaijan International Operating Company (AIOC)'s "early oil" via Georgia to its Black Sea port of Supsa. The Georgian International Oil Company, a subsidiary of the AIOC, made substantial upgrades to the existing 515-mile pipeline along this route and built the $565-million Supsa terminal on the Black Sea.

The so-called "western route", which became operational in April 1999, had an original design capacity of 100,000 bbl/d, but recent upgrades have raised capacity closer to 145,000 bbl/d. Officials from BP, the British-based operator of AIOC, said that the consortium exported approximately 130,000 bbl/d in 2001, with virtually all of its oil available for export being shipped to Supsa. The Baku-Supsa route, however, was designed to carry only the early oil from the AIOC's development of the Azeri-Chirag-Gunashli fields, and although there has been discussion of increasing the pipeline's capacity to 300,000 bbl/d or even 600,000 bbl/d, AIOC is planning to export its future production via the Baku-Ceyhan Main Export Pipeline, once it becomes operational.

Baku-Novorossiisk Pipeline

The 100,000-bbl/d-capacity Baku-Novorossiisk pipeline, also known as the "northern route", opened in 1997. The pipeline runs 868 miles from Baku via Chechnya to the Russian Black Sea port of Novorossiisk. Initial exports through the pipeline were limited to approximately 40,000 bbl/d, however, owing to pumping limitations, disputes over transit tariffs, and the conflict in Chechnya. Up to 70,000 bbl/d of oil was forced to bypass Chechnya by rail from Dagestan to Stavropol.

The ongoing conflict and instability in Chechnya prompted Russian pipeline operator Transneft to construct a 120,000-bbl/d Chechnya pipeline bypass (160,000 bbl/d including rail links). This bypass, which was completed in 2000, includes an 11-mile spur to Russia's Caspian Sea port of Makhachkala. The pipeline and spur could eventually transport up to 360,000 bbl/d of oil, enabling additional exports from Azerbaijan, Kazakhstan, and Turkmenistan to flow through the pipeline from Baku and Makhachkala.

In 2000, Azerbaijan's SOCAR committed itself to throughput of 46,000 bbl/d, but in the end only transported around 10,000 bbl/d, prompting Transneft to accuse Azerbaijan of not fulfilling its commitment to export oil along the bypass. The AIOC, which also was expected to export via Baku-Novorossiisk, has been reluctant to send its oil along this route because it is longer and more expensive than the Baku-Supsa route, and also because the northern route mixes AIOC crude with other crude oils while in transit to Novorossiisk, reducing its value.

SOCAR exported approximately 50,000 bbl/d via the Baku-Novorossiisk route in 2001, and plans to maintain that rate in 2002. Russia says the the capacity on Baku-Novorossiisk can be increased to 300,000 bbl/d, but SOCAR will not have sufficient volumes to fill the pipeline, even at its present capacity, in the next few years. Future Azeri oil production, mainly from the AIOC, is slated to be exported via the Baku-Ceyhan pipeline, although Transneft claims that exporting via the Baku-Novorossiisk pipeline would be cheaper.

Other Oil Transit

Prior to the opening of the CPC pipeline in the fall of 2001, ChevronTexaco had been delivering oil from the Tengiz field in Kazakhstan via the Caucasus. ChevronTexaco sent its oil across the Caspian by barge to the Dubendi terminal in Azerbaijan, where it was further transported via a pipeline to Ali-Bayramly (Azerbaijan), and then to Georgia's Black Sea port at Batumi in rail cars. In September 1999, Chevron (as it was then known) and Georgian company Geoengineering signed an agreement on the preparation of a feasibility study for the reconstruction of the Khashuri-Batumi, with an eye towards using the pipeline for transiting Tengiz crude. Together with an upgrade of the Batumi refinery, the project was estimated to cost $100 million.

With the launch of the CPC, however, ChevronTexaco decided in May 2001 to cancel the project to reconstruct the Khashuri-Batumi pipeline, saying that the pipeline was economically unfeasible, especially since most of the Tengizchevroil exports are now routed via the CPC. Nevertheless, Tengiz crude has been replaced at the Batumi port by high-quality Kumkol crude, supplied by Euro Asian Trading, and the lower-quality Buzachi blend, produced by Kazakhstan's Mangistaumunaigaz, both of which reach Batumi via a combination of barge, pipeline, and rail across the Caspian and the Caucasus. Turkmenistan also exports occasional cargoes of Cheleken and Okarem crude, which are mostly blended with the Kazakh oil either at the Batumi terminal or on barges, forming a "synthetic Urals" blend.

Rail cars loaded with oil from Ali Bayramly are beginning to overwhelm the Georgian Black Sea ports. Although throughput at the Batumi terminal was 120,000 bbl/d in February 2002, with 400 tank cars per day offloaded, Azerbaijan was sending more than 500 tank cars per day, leaving over 100 tank cars per day unloaded. The backlog in offloading tank cars at the Georgian ports has led to a shortage of tank cars in Azerbaijan. In May 2001, the EBRD agreed to finance the construction of a $20-million oil terminal at the Black Sea port of Poti. The Poti terminal will be able to handle up to 50,000 bbl/d, proving an alternative to the main port at Batumi.

In addition, Georgia and Turkey are working on plans to utilize a 172-mile railway line between Tbilisi and Kars, Turkey, to transport up to 200,000 bbl/d of crude oil from the planned Baku-Ceyhan pipeline to Turkish refineries. The railway plan, which could cost $400 million, will require refurbishing an existing line from Tbilisi to Akhalkalaki for $200 million, as well as extending the rail line 77miles to Kars.

Natural Gas Transit

The Caucasus region also is set to become a major transport corridor for natural gas. Most of this will come from increased Caspian Sea region natural gas production, especially from Azerbaijan. As Azerbaijan begins to exploit the Shah Deniz natural gas field, which has estimated reserves of 35.3 trillion cubic feet (Tcf), its natural gas production will rise dramatically, and the country's March 2001 agreement with Turkey will allow Azerbaijan to boost its natural gas exports by piping Shah Deniz supplies via Georgia to Turkey. In addition, Armenia and Iran are developing a natural gas pipeline to connect the two countries, with the pipeline possibly continuing further northwards to connect to Georgia and then to the Russian pipeline system.

Baku-Erzurum Pipeline

On March 12, 2001, Azerbaijan and Turkey reached a natural gas export deal whereby Azerbaijan will supply Turkey with 3.1 trillion cubic feet (Tcf) of natural gas from 2004 through 2018. To deliver the natural gas to Turkey, BP, the operator of the Shah Deniz field, announced on March 14, 2001, plans to build a 1.06-Tcf-per-year-capacity pipeline from Baku to Erzurum, Turkey, via Georgia. Originally, Azerbaijan had hoped to use part of the existing Gazi-Magomed-Gazakh pipeline to transport the natural gas across Azerbaijan, but technical inspection of the pipeline deemed that costly repairs would first be necessary, and the huge volume of natural gas that Azerbaijan agreed to export–exceeding the capacity of the Gazi pipeline–necessitated that a new pipeline be built.

After lengthy negotiations that threatened to negate Azerbaijan's export deal with Turkey, in September 2001, Georgia and Azerbaijan cleared a major hurdle for the pipeline by signing a transit agreement. The agreement calls for Georgia to receive 5 per cent of the natural gas in the pipeline, as well as preferential rights to purchase additional natural gas from the pipeline for 20 years, in exchange for transit rights. The Azeri parliament ratified the transit agreement in October 2001, followed by the Georgian parliament in December 2001. In January 2002, Georgian announced it will build two, 88.3-Bcf-capacity underground natural gas storage facilities in the east and southwest of the country as part of the pipeline project.

The Baku-Erzurum pipeline will stretch some 630 miles, including 290 miles in Azerbaijan and approximately 170 miles in both Georgia and Turkey. Detailed engineering for the pipeline, which will mirror the Baku-Ceyhan oil pipeline to Erzurum, is in progress, with tender proposals being prepared for construction of the Azeri and Georgian sections of the pipeline (Turkey's Botas will construct the Turkish section of the line). Currently, the pipeline project is estimated to cost $1 billion. Credits to be drawn from international financial institutions, including the World Bank, European Bank for Reconstruction and Development (EBRD), International Finance Corporation (IFC), and investors from the United States and Japan are expected to cover 70 per cent of the pipeline's construction costs, while shareholders in the development of the Shah Deniz field development will contribute the remaining 30 per cent.

Construction of the Baku-Erzurum pipeline is scheduled to begin in late 2002 and to be operational by the end of 2004. Initial capacity on the pipeline is slated to be 777 billion cubic feet (Bcf) per year, with capacity eventually rising to 1.06 Tcf per year. With natural gas production in the first stage of exploitation of the Shah Deniz field expected to be 282 Bcf per year, the Baku-Erzurum pipeline will have excess capacity to pipe additional Caspian Sea region natural gas exports, possibly from Turkmenistan if the Caspian littoral states agree on a legal regime for the Sea, allowing the proposed Trans-Caspian Gas Pipeline to be built. In addition, Turkey is in consultations with Greece to extend the Baku-Erzurum pipeline into the territory of the European Union.

Iran-Armenia Pipeline

Armenia and Iran are pushing ahead with a long-standing project to build an 84-mile natural gas pipeline linking the two countries. The pipeline project, dating back to intergovernmental agreements signed in 1992 and 1995, would allow Armenia to diversify and stabilize its natural gas supply sources, as well as expedite the closing of Armenia's Metsamor Nuclear Power Plant, which the European Union considers dangerous.

Implementation of the project has been delayed for years due to disagreements between the two sides over natural gas prices and the location of the pipeline. The original proposal was for the pipeline to be laid from northwestern Iran to the existing Armenian pipeline that terminates in Kadzharan, requiring the pipeline to transit the Azeri exclave of Nakhichevan. However, the still unresolved hostilities between Azerbaijan and Armenia over Nagorno-Karabakh made this plan impossible. In addition, Armenia balked at the high price of natural gas (from $84 to $90 per 1,000 cubic meters) that the Iranian side set for the pipeline.

Although the two sides have not been able to agree on a price for Iranian natural gas, in December 2001, Armenia and Iran signed a transit agreement to allow Armenia to import Turkmen natural gas via Iran. Turkmenistan is linked to the Iranian natural gas pipeline network through the Korpezehe-Kurt Kui pipeline, which opened in December 1997. The Iran-Armenia pipeline is now slated to cover 84 miles, running from Tabriz to Khadzaran and bypassing Nakhichevan. An initial feasibility study for the pipeline, with initial capacity of 35.3 Bcf and possibly up to 106 Bcf per year, has been completed.

The estimated cost of the Iran-Armenia pipeline is $120 million. The European Union has declared its readiness to assist in financing the pipeline's construction, but Gazprom and Itera, both of which previously expressed an interest in participating, announced that they would participate in the construction only through their partial ownership of Armrosgazprom, Armenia's natural gas distributor. France's Gaz de France previously announced its intention to invest in the pipeline, but has not committed to the project. Any significant investment in an Iranian oil project may be subject to the Iran and Libya Sanctions Act, which the U.S. Congress renewed in August 2001.

Once financing for the pipeline is secured, construction can commence. During the first stage of construction, which Armenia hopes to begin in 2002, Armenia will lay a 24-mile pipeline section from Kadzharan to the southern border at Megri for approximately $26 million. Ukraine, which currently imports significant amount of Turkmen natural gas via Russia, has suggested that the Iran-Armenia pipeline be linked to its proposed route from Turkmenistan through Iran and Armenia. In addition, once the Iran-Armenia pipeline is completed, it will link the Iranian and Russian natural gas transportation systems, allowing for possible Iranian natural gas exports to Europe through the Russian pipeline system.

Other Natural Gas Transit in the Caucasus

In November 2000, Georgia approved a project for a 37-mile pipeline to carry Russian natural gas to Turkey via the Georgian Black Sea coast. After a September 2001 meeting, Georgian officials announced that representatives from Conoco and Turkey's Acsoy Group were ready to invest in the pipeline, which would transport 35.3 Bcf per year of natural gas from Kobuleti, Georgia, to Hopa, Turkey.

Georgia also has held discussions with Gazprom on refurbishing the existing North Caucasus-Transcaucasian natural gas pipeline and extending it into a trans-Georgian pipeline to bring Russian natural gas to Armenia and Turkey. However, this idea has lost some support as Russia focuses on delivering its gas to Turkey via the "Blue Stream" natural gas pipeline under the Black Sea.

The Caucasus region may also serve as a transit region for liquefied natural gas (LNG). Conoco, together with the Acsoy Group, began a project to supply liquefied gas from Russia and the Caspian region to Turkey in 1999. However, the volume of natural gas supplies was small, and Conoco withdrew from the project. In September 2001, Batumi Oil Terminal Ltd., which owns the Batumi terminal, said that it would continue developing the project, where LNG will be barged from Kazakhstan and Turkmenistan across the Caspian to Baku, then transported using rail and sections of the natural gas pipelines in Georgia to Batumi. From there, the LNG will be sent to Turkey and other Mediterranean customers.

ARMENIA

Following a severe economic decline in the early 1990's, Armenia is continuing its slow recovery from the collapse of the Soviet Union and the effects of Armenia's six-year war with Azerbaijan over Nagorno-Karabakh. In 1994, Armenia launched an ambitious International Monetary Fund-sponsored economic program that has resulted in positive growth rates for the past eight years, despite the Azerbaijani-led economic blockade of Armenia.

Armenia's small- and medium-sized enterprises, most of which have already been privatized, have spurred continuing economic growth. The country's economy registered strong growth in 2001, with real gross domestic product (GDP) increasing 7.2 per cent, up from 6 per cent growth in 2000. Armenia's real GDP is forecast to grow by 5.4 per cent in 2002. Inflation rose only slightly in 2001, to 3 per cent, from 0.4 per cent in 2000, and Armenia's unemployment rate dropped from 11.7 per cent in 2000 to 10.5 per cent in 2001. At its current growth rate, by 2005, Armenia's absolute GDP will reach the same level as in 1991, the year that the Soviet Union and its central economic planning system collapsed.

Oil

Armenia has no oil production, known reserves, or refineries, making the country completely dependent on imports of refined petroleum products. In addition, because there are no oil pipelines into Armenia, all of the country's petroleum products must be imported by rail or by truck. Since the end of subsidized oil supplies after the collapse of the Soviet Union, Armenia's oil consumption has dwindled from 48,400 barrels per day (bbl/d) in 1992 to just 4,000 bbl/d in 2001, most of which comes from the Batumi refinery in western Georgia.

Although the planned "Main Export Pipeline" (MEP) for Caspian region oil is not slated to transit Armenian territory, Armenian officials occasionally have spoken of potential cost savings if the MEP were built through northern Armenia, since it would shorten the export route considerably. However, the lack of a peace agreement between Armenia and Azerbaijan over Nagorno-Karabakh makes this idea extremely unlikely. Azerbaijani officials have dismissed the idea altogether, noting that the route through Georgia already has been decided.

Natural Gas

With no proven natural gas reserves, Armenia is reliant on imports to meet its domestic natural gas demand. In 2000, Armenia consumed 49.8 Bcf of natural gas. Since the Nagorno-Karabakh conflict erupted, Azerbaijan has ceased natural gas shipments into Armenia from the Soviet-era natural gas pipeline linking the two countries, forcing Armenia to import all of its natural gas via the Georgian and Russian natural gas pipeline networks to the north. The private natural gas trading company Itera has been Armenia's main natural gas supplier since 1996. The proposed Iran-Armenia natural gas pipeline will allow Armenia to diversify its natural gas suppliers, with Turkmen natural gas to be piped via the Iranian pipeline network to Armenia.

Natural gas in Armenia is distributed by Armrosgazprom, a closed joint-stock company owned by the Armenian government (45 per cent), Russia's Gazprom (45 per cent), and Itera (10 per cent). Armrosgazprom planned to re-invest $6 million from its own funds in 2001 in an effort to rehabilitate Armenia's natural gas sector. In February 2001, Armenia and Russia reached an agreement on Armenia's $7-million debt for natural gas shipments dating back to 1999, which Itera supplied on Gazprom's behalf. Since Gazprom and Itera owed Armenia their contribution to the incorporation capital of Armrosgazprom, in July 2001, Itera agreed to write off Armenia's debt in exchange for the Armenian government transferring its natural gas pipelines to the joint enterprise towards Russia's share.

Armenia has continued to rack up natural gas debts to Itera, prompting the company to reduce supplies to Armenia in October 2001 to force payment. In late January 2002, Itera again threatened to reduce natural gas shipments to Armenia by two-thirds unless Armenia stayed current in its payment for supplies. In mid-February 2002, Itera announced that it had decided not to alter the level of natural gas supply to Armenia because the country honored a January 11, 2002, agreement to pay its debt for the natural gas acquired in 2001 and January 2002.

Coal

Armenia has no coal reserves or coal production. Armenia's coal consumption, most of which is used for home heating, totaled 3,307 short tons in 2000.

Natural Gas

Armenia's power sector has a total installed generating capacity of 2.7 gigawatts (GW). The country has two large thermal power plants–at Yerevan (550 megawatts, MW) and Hrazdan (1,110 MW)–as well as a smaller plant at Vanadzor (96 MW). Armenia also has significant hydroelectric power-generating ability and one nuclear power plant. In 2000, Armenia generated 5.7 billion kilowatt-hours (Bkwh) of electricity and consumed just 4.9 Bkwh, but since power is not provided to all regions of the country on a regular basis, the country's potential demand for electricity outpaces supply. In addition, the thermal plants have exceeded their projected operating lifespan, are in need of renovation, and are often low on fuel, since Armenia must import all of its natural gas.

Iran and Armenia already have linked their electricity grids, allowing for power sales in both directions driven by seasonal differences in demand between the two countries. In summer, Armenia exports its power to Iran and gets it back in winter. Armenia also supplies some of its surplus seasonal electricity to Georgia. Closer ties with Iran could give Armenia an additional source of electricity as Iran, Turkmenistan, and Armenia explore whether their power grids can be linked. Armenia could receive electricity from Turkmenistan via Iran's energy system at less than the price of power produced by its own power stations.

Privatization

In 1998, Armenia's parliament passed a law allowing for the sale of the country's electricity transmission and distribution networks, while keeping power generation under government control. In an effort to support Armenia's privatization efforts, on December 5, 2000, the European Bank for Reconstruction and Development (EBRD) signed an agreement with the Armenian government to take 20 per cent equity stakes in each of the country's four electricity distribution companies: Yerevan, Northern, Southern, and Central. The agreement, which lasts for five years, includes a clause giving the Armenian government the right to buy back the EBRD shares should the agreement be abrogated.

U.S.-based AES Silk Road, ABB Energy Ventures of Sweden, and Spanish Union Fenosa Acex had been among the companies initially interested in the networks when the privatization tender was announced. However, the privatization process stalled in March 2001, when Armenian authorities announced they had not received any bids for the 75 per cent stake in the first two distribution networks. In April 2001, the second stage of the tender also failed to attract any bidders. Armenia is now looking to implement additional needed reforms, including possibly unifying the distribution grids into one, before proceeding with another privatization tender for the distribution networks. In September 2001, Armenian Energy Minister Karen Galustian said that Russia's Unified Energy Systems, Gazprom, and Itera likely will participate in the eventual privatization.

Metsamor Nuclear Power Plant

Armenia has one nuclear power plant, the controversial Metsamor Nuclear Power Plant (NPP). The power plant, with two VVER-design reactors and a combined capacity of 815 MW, was shut down in March 1989 by the Soviet Union because of safety fears following the devastating earthquake that struck Armenia in December 1988.

However, faced with a deepening energy crisis due to the country's lack of fossil fuels and the economic blockade imposed by Azerbaijan and Turkey, on November 5, 1995, Armenia decided to resume operation at the 440-MW second unit. The plant, which was built in 1980 with a design life of 30 years, now supplies between 40 per cent and 45 per cent of the country's electricity.

Since the Metsamor NPP was inactive for six years, Armenian and Russian nuclear officials believe that the lone reactor functioning at the plant could operate through 2016. The European Union, however, is pressuring Armenia to shut the plant earlier, since the EU considers Metsamor to be a safety risk due to flaws in the plant's Soviet-designed reactors and the region's seismic activity. In July 2001, Unit 2 at Metsamor was halted for planned maintenance and was supposed to be back in operation by the end of August, but debts delayed the reactor's launch: Armenia owed Russia $17 million for nuclear fuel already supplied, with the cost of new fuel set at $14 million. Armenia finally re-started the reactor in November 2001. Experts have estimated that required safety upgrades at the plant will cost about $1 billion over the next 15 years.

Rather than increasing spending to maintain the Metsamor NPP, the Armenian government has pledged to decommission the plant by 2004, provided the country has sufficient alternative energy sources by that time. The EU is pledging Armenia 100 million euros ($91 million) to build alternative power-generating facilities to replace Metsamor. However, Armenian Energy Minister Karen Galustian said that the country will need up to $1 billion from foreign investors and donor countries to safeguard Armenia's energy security after closing Metsamor.

Hydroelectric Power

Hydroelectricity accounted for almost 25 per cent of Armenia's electric power generation in 2000. Armenia has several hydroelectric plants on the Hrazdan River, including the Sevan-Hrazdan hydroelectric plant, and has plans to develop several additional hydroelectric projects. Armenia is

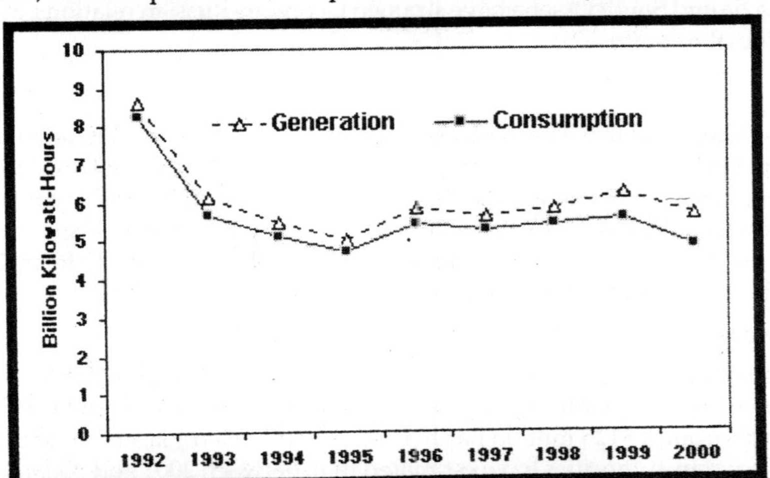

ARMENIA ELECTRICITY GENERATION AND CONSUMPTION, 1992-2000

undertaking a program to construct 38 small and three large hydroelectric power plants, with an overall capacity of 296 MW. The cost of this program will be $300 million, part of which will be financed by the World Bank and the EBRD.

Of the three large hydropower plants, two–Lori Berd and Shnokh–will be built in the Armenian northeast, one with a capacity of 60 MW and an annual output of 192 million kilowatt-hours, and the other with a capacity of 75 MW and an annual output of 300 million kilowatt-hours. The third proposed hydropower plant, at Megri on the Araks river on the Armenian-Iranian border, is slated to have a capacity of 78.9 MW and to generate 469 million kilowatt-hours of electricity a year. Armenia and Iran have set up a joint company to construct the Megri hydroelectric power station, which will cost from $60 million to $80 million and take over five years to build. However, Azerbaijan has objected to the proposed plant, arguing that its Nakhichevan exclave will have its water supplies severely decreased if the Megri hydroelectric power station is constructed.

AZERBAIJAN

GEORGIA

More than ten years since Georgia's independence from the Soviet Union, the country continues to suffer from political turmoil, civil strife, and a weak economy. Although President Eduard Shevardnadze restored order following the overthrow of Georgia's first democratically elected president, separatist struggles in Abkhazia (northwest Georgia) and South Ossetia (north central Georgia), along with entrenched corruption, discouraged foreign investment throughout the 1990's. As a result, Georgia's economy, which was already reeling from the loss of Soviet subsidies after independence, suffered through bouts of hyperinflation and severe economic contraction–by 1995, Georgia's GDP dropped to 20 per cent of 1990 levels.

Since the mid-1990s, however, Georgia has progressed slowly, bringing inflation under control, introducing a stable currency, and experiencing moderate economic growth. In 2001, Georgia's real GDP grew by 4 per cent, and the country's economy is projected to grow by 3.2 per cent in 2002. Georgia's agricultural sector rebounded in 2001 after a drought in the summer of 2000 had disastrous effects, both on food supplies and on the country's hydropower potential, Georgia's only sizable internal energy resource. Power generation continues to be a problem in Georgia, and the unresolved conflicts in Abkhazia and South Ossetia have strained Georgian-Russian relations, as well as hindered economic growth in the country.

Oil

Georgia's has limited oil reserves, with approximately 35 million barrels, and the country's small oil industry does not produce enough to meet domestic needs. In 2001, Georgia produced approximately 2,000 bbl/d of oil, far short of the 25,000 bbl/d of oil that that the country consumed. CanArgo-Georgia, also known as the Georgian-British Oil Company (GBOC), is Georgia's leading producer, with an average of 1,200 bbl/d of oil produced from the Ninotsminda field in 2001. Saknavtobi (Georgian Oil), the state oil company, produced just 120 bbl/d of oil in 2001.

Georgia is expected to produce around 2,000 bbl/d in 2002.

As Georgia continues its recovery from civil strife in the mid-1990s, oil consumption is on the rise, but so is investment in the country's oil sector. According to the Georgian National Oil and Gas Regulating Agency, around $125 million has been invested in Georgia's oil production sector in the past five years. Georgian authorities have estimated that, between 2001 and 2005, an additional $453 million will be invested in oil and natural gas exploration and production in Georgia by nine joint ventures. The country is pinning its hopes on a dramatic increase in domestic oil production in order to meet rising demand in the next decade.

Active exploration is underway in Georgia, both along the Black Sea coast and onshore. The joint ventures are conducting operations in a number of blocks, including Ninotsminda, Manavi, Rustavi Kartli, Samgori, Patardzeuli, Mtiani Kakheti, Mirzaani, Taribana, Patara Shiraki, Nazvrebi, Supsa, Chaladidi, and Shromisubani. Saknavtobi and Anadarko (U.S.) have worked out a package of production-sharing agreements on the Black Sea shelf, where exploration engineers have discovered roughly 580 million tons of oil (4.25 billion barrels), including about 200 million tons (1.47 billion barrels) offshore.

Georgia has negotiated production-sharing agreements and joint ventures with a number of companies, including an agreement with the German company GWDF International to develop the Chaladidi, Supsa, and Shromisubani fields in western Georgia, and with Frontera Eastern Georgia, a Georgian-American joint venture, to develop the Taribana field in Kakhetia. According to a program developed by Saknavtobi based on investment projects currently being carried out by oil joint ventures in Georgia, the country could produce up to 4 million tons of oil between 2001 and 2005 (an average of 16,000 bbl/d).

Downstream/Refining

Georgia has two refineries, a 106,000-bbl/d refinery at Batumi, and a smaller, 4,000-bbl/d refinery at Sartichala, the Georgian-American Oil Refinery (GAOR). The Batumi refinery, however, currently is undergoing a $250-million modernization and expansion by Japan's Marubeni Corporation and the JGC Corporation, forcing Georgia to import over 90 per cent of the petroleum products it consumes.

After sitting idle for much of 2001, the GAOR refinery, which was built in 1998 and is owned by Canadian CanArgo (51 per cent), Saknavtobi (28 per cent) and GBOC (21 per cent), ramped up operation in July 2001, processing about 8,000 tons of crude oil (an average of 1,928 bbl/d for the month) into gasoline, diesel and fuel oil. However, difficulties in selling the gasoline, as well as the plant's less than 50 per cent utilization rate, forced management to begin to shut down the facility again in August

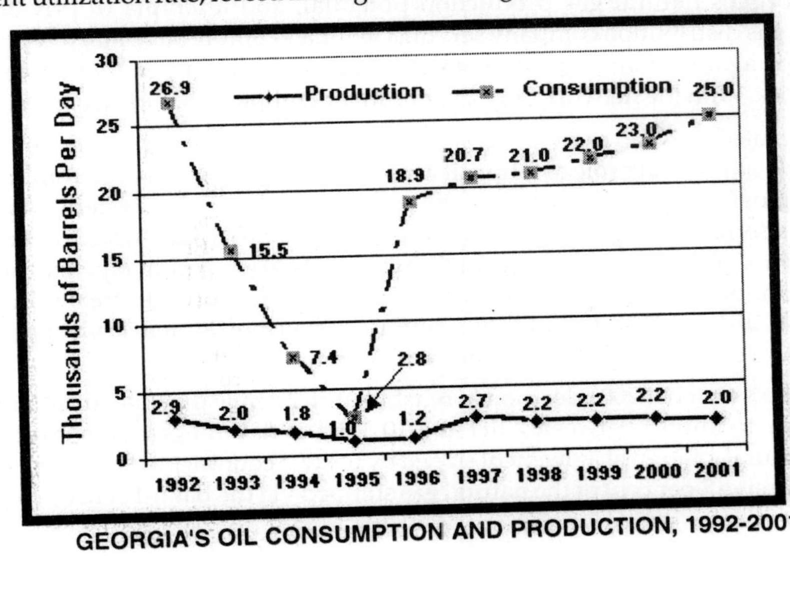

GEORGIA'S OIL CONSUMPTION AND PRODUCTION, 1992-2001

2001. In September 2001, CanArgo announced it was closing the GAOR refinery, saying that prices for crude oil and the saturation of the domestic market with cheap oil products from Azerbaijan and Russia had made operation of the small refinery unprofitable.

Rather than continuing operations at the GAOR refinery, CanArgo announced plans to build a larger refinery at the same site. The Canadian company announced it would build a $200-million refinery that would be able to process up to 1.5 million tons of crude oil per year (30,100 bbl/d). Additionally, Switzerland's National Petroleum Limited (NPL), which has been developing Georgia's biggest oil field with Sakvnavtobi since 1996, revealed plans in July 2001 to construct a 40,000-bbl/d refinery in Georgia in the next few years. Georgia also has awarded Frontera the right to construct a new refinery near Tbilisi as part of its production-sharing agreement.

Natural Gas

Georgia has approximately 300 Bcf in natural gas reserves, and during Soviet times the country did not produce any natural gas. Since Georgia became independent in 1991 and stopped receiving subsidized fuel, the country's natural gas consumption has plummeted, from 177 Bcf in 1992 to just 42.7 Bcf in 2000. Although the country's natural gas sector increased output from zero in 1997 to 2.1 Bcf in 2000, Georgia remains heavily dependent on foreign suppliers to meet its domestic demand.

In addition, Georgia's inability to pay its suppliers has limited the country's consumption, as both Russia and Turkmenistan at times have cut off natural gas supplies to Georgia due to payment arrears. Turkmenistan has left the Georgian natural gas market, claiming that Georgia still owes it for past supplies, leaving Russia as the country's sole supplier. Itera, the Gazprom-affiliated natural gas trader, has been supplying Georgia for the past year, but the company repeatedly has reduced supplies to Tbilisi in order to force Georgia to pay its bills. As of December 2001, Georgia owed Itera about $90 million for previous natural gas supplies, prompting Itera to require prepayment for natural gas shipments to Georgia after January 1, 2002.

Georgian leaders hope to decrease this reliance on natural gas imports in coming years by courting foreign investors to develop the country's natural gas deposits and by reforming the country's natural gas distribution system. However, to date, there has been little interest among international energy companies in Georgia's natural gas production potential, and attempts to private Tbilgaz, the municipal natural gas distribution company serving Georgia's capital, have failed repeatedly. In June 2001, Georgia once again offered an international tender for an 85 per cent stake in the Tbilisi distribution network. With the network only 25 per cent operational, no bids were received.

Energy companies have had more interest in Georgia's natural gas transmission sector, mainly due to the country's burgeoning role as a transit center for natural gas exports from Azerbaijan. The Georgian International Gas Corp., which runs the country's transmission system, has been working on a program to modernize Georgia's internal natural gas pipelines, which stretch over 6,000 miles, in order to pump Azeri natural gas via Georgia for the planned Baku-Erzurum pipeline. In October 2001, the Georgian International Gas Corp. and Russia's Gazprom joined forces to create Gruzrosgazprom, a joint venture that will develop and operate the natural gas transport system in Georgia. The Georgian International Gas Corp. owns 51 per cent of the joint venture.

With Azerbaijan's October 2001 decision to rehabilitate a 23-mile pipeline to the Georgian border, Georgia will be able to import between 7 Bcf and 10.5 Bcf of natural gas from Azerbaijan in 2002. Additionally, the natural gas transit agreement signed by the two countries in September 2001 stipulates that Georgia will receive 5 per cent of the natural gas shipped via the Baku-Erzurum pipeline when it comes online in 2004. Under the transit agreement, Georgia also will have the right to purchase 500 million cubic meters (17.7 Bcf) of natural gas per year for 20 years at a cost of $55 per 1,000 cubic meters (35,300 cubic feet).

Coal

Georgia's coal output, already low by world standards in 1991 when the country became independent, has decreased in the past decade. Between 1992 and 1997, the country's coal production declined from 220,462 short tons to 5,952 short tons–a drastic 97 per cent reduction–although production has rebounded slightly, to 22,046 short tons in 2000. Similarly, Georgia's coal consumption plummeted 95 per cent from 1992 to 1997, from 480,607 short tons to just 22,000 short tons, before climbing back to 37,479 short tons in 2000.

Electricity

Georgia's energy system includes about 60,000 miles of transmission lines, 53 hydroelectric power stations, and three thermal power plants, for an overall generating capacity of 4.5 GW. However, fuel shortages and aging power plants means that the sector is only able to operate at 40 per cent of capacity. Georgia consumed more electricity than it generated in 2000: 7.9 Bkwh consumed compared to 7.4 Bkwh generated. In addition, because of inefficient and deteriorating power lines, power outages are a daily occurrence in much of the country, and parts of Georgia do not receive any electricity at all.

Georgia's total electricity demands have been estimated at nearly twice the amount that is actually generated, and the Georgian Energy Ministry estimates that 40 per cent of all power that is generated is wasted due to equipment and maintenance problems in the transmission sector. To meet some of its power needs, Georgia imports electricity from Armenia, Azerbaijan, and Russia. Georgia has run up large electricity debts to each of the countries. In addition, because Georgia's natural gas supplies frequently are cut off due to the country's payment arrears, Georgia's own natural gas-fired power

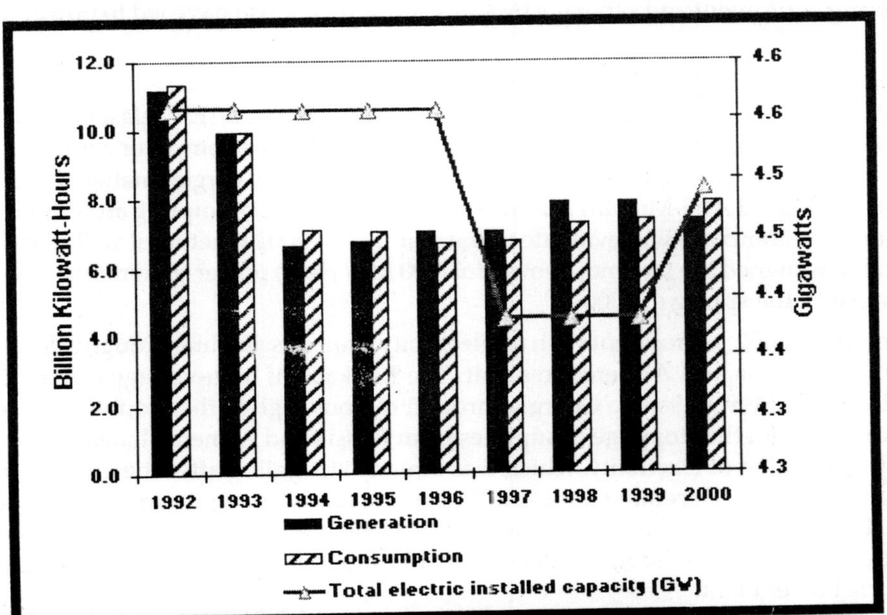

GEORGIA'S POWER SECTOR, 1992-2000

plants are often short of fuel. As a result, electricity supplies, even to Tbilisi, are erratic, leaving customers in the dark for 20 hours (or more) per day. Consumers, who use kerosene for heating, have responded to rate increases and inadequate power supplies by refusing to pay their electricity bills.

Privatization

The significant problems in Georgia's power sector have hampered the country's economic growth, making energy sector reform a government priority in 2002. In a January 2002 meeting with IMF representatives, Georgian Minister of State Avtandil Jorbenadze said that problems in the energy sector had been caused by rampant corruption and that the process of dealing with them should start with the establishment of a proper wholesale market, the speeding up of privatization of energy sector installations, and the drawing up of clear mechanisms for repayment of the energy sector's internal and external debts.

In an effort to resolve the problems in the power sector, Georgia is trying to reform and privatize Sakenergo, the state energy and power company. With support from the World Bank and the EBRD, most of Georgia's hydro and thermal generation units have been restructured as joint-stock companies, and the Georgian Ministry for the Management of State Property is proceeding with privatization of energy distribution companies. In January 1999, AES (U.S.) purchased a 75 per cent stake in the Telasi electricity distribution company, which serves the Tbilisi area, for $25.5 million. As part of the privatization agreement, AES made a made a further commitment to pay $10 million of Telasi's debt and $80 million in investment to provide consumers with 24-hour electricity.

Power Outages

Although Telasi's debts have stopped growing since AES took it over, AES has not been able to meet its pledge to provide 24-hour electricity because the company continues to receive erratic power supplies from power-generating companies. In addition to the inherent problems of Georgia's power sector, faltering equipment and sabotage by secessionist rebels also have led to power outages across Georgia.

On December 22, 2001, the Unit 10 at the Tbilisi Power Plant exploded, causing severe damage that will take at least a year to restore. AES-Mtkvari, a subsidiary of AES-Telasi that operates the 300-MW generating unit, said that the explosion occurred because the computer control system and the mechanical portion were incompatible. However, in January 2002, Georgia's national energy regulatory commission suspended AES-Mtkvari's license to operate the unit, prompting an AES Vice President to announce that the company will evaluate in the summer of 2002 whether it will remain in Georgia. AES-Mtkvari has owned the two most powerful (300 MW each) power-generating units at the Tbilisi Power Plant since the spring of 2000.

On January 2, 2002, an accident with an electricity transmission line brought electricity supplies to Georgia from Russia and Armenia to a halt. The Kavkasioni high-voltage electricity line, which supplies electricity from Russia to Georgia through regions high in the mountains, broke down for unknown reasons, but the stoppage of supplies from Russia and Armenia had an immediate impact on Georgia's provision of electricity, which was already struggling after the explosion at the Tbilisi Power Plant. After the December incident, Russia increased power supplies to Georgia via the Kavkasioni line. Georgian authorities did not rule out sabotage in the Kavkasioni incident, noting that in November 2001 secessionist rebels destroyed a bridge over the Pankisi Gorge near the Khador Hydroelectric Power Plant.

Hydropower Projects

Georgia is seeking foreign equity participation for both new capacity and rehabilitation hydropower projects. Hydropower accounts for 80 per cent of Georgia's electricity generation, and the country has a substantial amount of untapped hydroelectric potential that could be exploited. Georgia already has made plans to build two new hydroelectric plants on the Rioni River (the 250-MW Namakhvani and the 100-MW Zhoneti), and the country is also hoping to build the proposed 40-MW Minadze station on the Kura river.

In February 2001, Chinese and Georgian officials signed an agreement on the construction of the 24-MW Khador Hydroelectric Power Plant near the Georgian-Russian border in the eastern Kakheti region. In September 2001, the Georgian-Chinese *Energokorporatsia Vostoka* company opened the first phase of the Khador cascade of mini hydroelectric plants. The first mini hydroelectric plant has capacity of 2 MW, and the entire facility will be launched by the end of 2002. Chinese investment in the project totaled $27 million.

In January 2002, Georgian Energy Minister David Mirtskhulava said that China's Sichuan Machinery, which is constructing the Khador Hydroelectric Plant, will invest $10 million in a second hydroelectric station in Georgia. The 9.3-MW plant will be built on the Chelta River in the Kakheti region. According to Mirtskhulava, construction of another hydroelectric plant in Kakheti would end the serious power shortages in the region, which is one of Georgia's biggest agricultural regions.

In addition, in November 2000, Georgia announced a tender for the rehabilitation of the existing Inguri Hydropower Plant, the country's largest. The project, which will cost an estimated $62 million, will boost the station's capacity from its current capacity of 400-450 MW, to 1,300 MW. The EBRD will provide $39 million in the form of a long-term credit, while the EU and Japan will give grants totaling $10 million. Georgia will finance the remaining $13 million.

21

Central Asia: Kyrgyzstan Energy Sector

General Background

Despite some progress in implementing economic reforms, Kyrgyzstan remains one of the poorest countries of the former Soviet Union. Kyrgyzstan's economy contracted sharply in the early 1990s, and while foreign assistance played a substantial role in the country's economic turnaround in the mid-1990s, Kyrgyzstan suffered severe economic aftershocks from the August 1998 financial crisis in Russia. The Kyrgyz government has enacted a number of measures to combat the country's economic problems, including efforts to stabilize rampant inflation, boost stagnant industrial production, and stimulate growth.

In 2001, Kyrgyzstan's inflation rate was reduced to 5.3 per cent (down from 39.9 per cent in 1999) and its real gross domestic product (GDP) grew 6.6 per cent. Still, the country's nominal GDP in 2001 was just $1.5 billion, meaning that per capita annual income in the country of 5 million people is approximately $290. President Askar Akayev, formerly praised by the West for his market reforms, was re-elected to a third term in office in October 2000 with 74 per cent of the popular vote in an election in which the main opposition candidates were prevented from running.

Oil

With estimated petroleum reserves of only 40 million barrels, Kyrgyzstan is reliant on imports to meet its domestic supply needs. Kyrgyzstan has seven developed oil fields and two oil/gas fields, but due to the country's mountainous topography, extraction is difficult, and water encroachment means that recovery rates are low. In 2000, Kyrgyzstan produced an estimated 2,100 barrels per day (bbl/d) of oil. Although the country's oil consumption has declined sharply since 1992, when it consumed

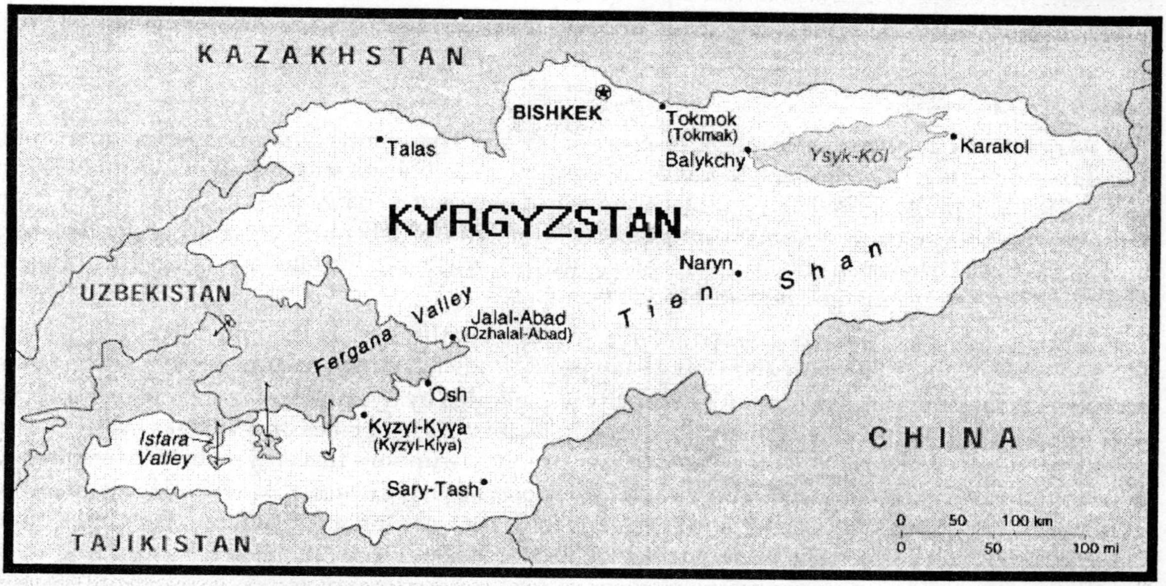

32,500 bbl/d, Kyrgyzstan's estimated oil consumption in 2000 of 12,000 bbl/d still required imported supplies to meet domestic demand.

Kyrgyzstan is looking to increase its oil production, and the government is undertaking a program of intensive oil extraction in order to meet the country's domestic petroleum needs. Oil reserves in the Fergana Valley are estimated at 733 million barrels, while 200-300 million tons (1.47-2.12 billion barrels) are thought to be deposited in the Chuy, Alay, Issyk-Kul, and At-Bashi depressions. Under the program to develop the country's oil sector, Kyrgyzstan is planning to produce 3,000 bbl/d by 2005.

In an effort to reach that target, Kyrgyzneftegaz, the state oil and natural gas company, is partnering with several foreign energy companies, as well as the Chinese government. A Kyrgyz-Austrian joint venture with Kyrgyzneftegaz and Action Hydrocarbons spent approximately $5 million on exploration work in 2001, and this may increase to $30 million in 2002. In addition, Chinese and Kyrgyz specialists are repairing more than 100 idle oil wells in Kyrgyzstan in 2002. Kyrgyzneftegaz also is planning to begin drilling exploration wells in the Dzhalalabad region in 2002, investing $30 million of its own money.

Downstream/Refining

Kyrgyzstan has one crude oil refinery, in Dzhalalabad, about 150 miles south of Bishkek. The refinery, which was built in 1997, is run by the Kyrgyz Petroleum Company, a joint venture between Kyrgyzneftegaz, the country's state-owned oil company, and Petrofac Resources International Ltd. (U.K.), which bought its share from Canadian-based Kyrgoil in June 2000. The 10,000-bbl/d-capacity refinery produces heavy fuel, diesel, and gasoline, but it has been hamstrung by difficulties in getting reliable supplies of crude oil from neighboring countries, especially Kazakhstan, amid the region's economic and political disorder.

A Kyrgyz-Kazakh joint venture, Bigmao Oil, is in the process of building a 400 bbl/d-capacity mini refinery for fuel oil in Kyrgyzstan. Abylaikham Group holds 50 per cent, Kyrgyzneftegaz holds

25 per cent, and private investors hold 25 per cent of the refinery, which will begin operating by end-2002.

Natural Gas

Kyrgyzstan has proven natural gas reserves of 200 Bcf. The country's natural gas sector is small, and domestic natural gas production has declined from 3.5 billion cubic feet (Bcf) per year in 1992 to only 0.5 Bcf in 2000. As a result, Kyrgyzstan is heavily dependent on natural gas imports, mainly from Uzbekistan, to meet its domestic consumption requirements (67.5 Bcf in 2000). Kyrgyzstan receives natural gas from Uzbekistan under agreements signed by Kyrgyzgaz, the state's natural gas distribution company, and Kyrgyzenergo, the state electric utility.

Since Uzbekistan began charging higher rates for its natural gas in the mid-1990s, Kyrgyzstan has fallen into payment arrears, and Uzbekistan periodically has cut off natural gas to Kyrgyzstan in response. While much of Kyrgyzstan's electricity is generated by hydropower in the warmer months of the year, natural gas is the primary fuel used in heating Kyrgyz cities and villages, as well as in electricity generation during winter. Thus, winter supply disruptions to Kyrgyzstan have resulted in blackouts and heating shortages. Kyrgyz and Uzbek officials have negotiated several barter deals to exchange Kyrgyz electricity, water, and/or goods for Uzbek natural gas, but these deals have often fallen through, causing tension between the neighboring states.

Coal

Kyrgyzstan's small coal industry includes 11 mines. From 1992 to 1999, the country's production and consumption of coal were on the decline, but both rebounded in 2000, with Kyrgyz coal production amounting to 0.7 million short tons (Mmst) while coal consumption totaled 1.7 Mmst. In addition, Kyrgyzkomur, the country's major coal producer, reportedly boosted its coal production by 12 per cent in 2001, with additional increases forecast for 2002. Under a government program passed in 1998 to develop the coal industry, Kyrgyzstan's coal production should be increased to 1.085 Mmst per year by 2005.

A 1999 report for Kyrgyzstan's Ministry of Foreign Trade and Industry suggested that the country could overcome its coal deficit by exploiting the Kara-Keche deposit in the Narynsk Region, one of Kyrgyzstan's 70 coal deposits. Although the Kara-Keche deposit would yield high-quality coal, high extraction costs and a lack of equipment have hindered development of the deposit. Analysts have estimated the cost of developing the Kara-Keche deposit at $52 million. The Kyrgyz government is seeking a strategic investor to develop the deposit, and is studying a comprehensive plan to invest in Kara-Keche as proposed by Tekhmashimpex.

Electricity

Kyrgyzstan's electric power industry is capable of meeting the country's domestic electricity needs while providing surplus electricity for export. Kyrgyzstan has two major electric power plants–a 1.2-gigawatt (GW) hydropower plant at Toktogul, and a 0.76-GW thermal plant at Bishkek, with plans for a major 6.8-GW hydropower station to be built by 2010. In 2000, Kyrgyzstan generated 14.7 billion kilowatt-hours (Bkwh) of electricity, up from 13.0 Bkwh in 1999, while the country consumed only 9.8 Bkwh in 2000.

Kyrgyzstan's abundant water resources give it significant hydroelectric potential. The energy potential of Kyrgyzstan's mountain rivers is estimated at 163 Bkwh per year, of which only about 10 per cent is currently exploited. Hydroelectric energy meets approximately 20 per cent of Kyrgyzstan's primary energy requirements and accounts for nearly 20 per cent of its total exports. With rapidly growing energy demand in neighboring countries, Kyrgyzstan's hydroelectric power potential is becoming more attractive to foreign investors. The long-delayed 450-megawatt (MW) Tash Kumyr Hydroelectric Plant was put into full operation in 2001, and Kyrgyzstan is working to secure financial resources to construct two power-generating units at the Kambar-Ata Hydroelectric Plant.

Although Kyrgyzstan has excess electricity generation, up to one-third of the power that the country generates is lost due to Kyrgyzstan's deteriorating power infrastructure. A lack of transmission-related equipment and inadequate pricing and cost recovery have contributed to problems in the power sector. The Kyrgyz government allowed Kyrgyzenergo to raise electricity tariffs in March 2002 in an effort to recoup generation costs, but already more than half of residents in the Kyrgyz capital are not able to pay because of previous rate increases.

Since the country's major hydroelectric power stations are located in the south, the north of the country typically depends on supplies of Kazakh electricity in the winter. After Kazakhstan withdrew from the Central Asian power grid in early 2002, northern Kyrgyz districts were left with insufficient electricity, prompting Kyrgyz government official to ask residents in the north to conserve electricity.

Besides the irregular natural gas supplies from Uzbekistan, Kazakhstan's decision to leave the regional power grid has given Kyrgyzstan additional incentive to shore up its power system. Kyrgyzstan already has embarked on a program to make the country self-sufficient in energy by 2005, seeking to increase its electric installed capacity and to modernize its distribution system. Kyrgyzstan has borrowed money from international development banks to build substations, the Alai-Batken, Kemin-Naryn, and Naryn-Torugart power lines, to rehabilitate/reconstruct heat and power grids and the Bishkek heat and power plant, and to buy equipment.

In 2001, Kyrgyzstan embarked on a restructuring of Kyrgyzenergo, splitting off the company's distribution networks and leaving the former monopoly as just an electricity generating company. Four joint-stock companies–Sever Elektro, Vostok Elektro, Osh Elektro, and Dzhalalabad Elektro–were created from Kyrgyzenergo in the different regions of the country. However, the new companies are still saddled by their own debts to Kyrgyzenergo and by consumers' failure to pay their electricity bills. Kyrgyzstan plans to privatize these regional electricity distribution companies as the next step in the reform process.

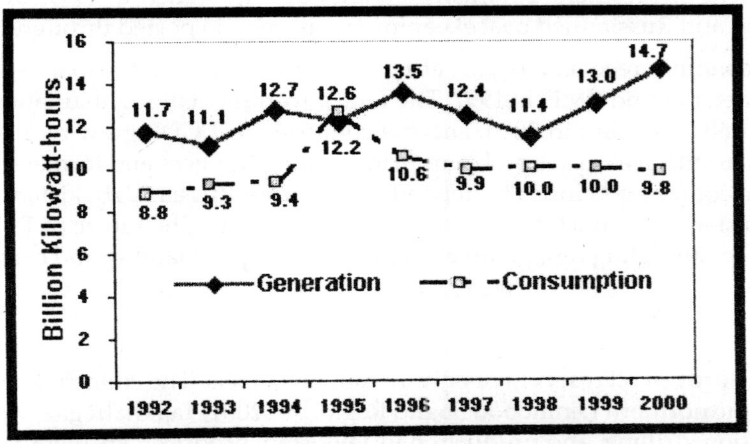

KYRGYZSTAN ELECTRICITY GENERATION AND CONSUMPTION, 1992-2000

22

Central Asia: Tajikistan Energy Sector

General Background

Tajikistan, which suffered through a civil war between Islamic conservatives and the secular government after becoming independent in 1991, has the lowest per capita gross domestic product (GDP) in the former Soviet Union. Although a peace agreement between the United Tajik Opposition and the government of President Emomali Rakhmonov was signed in 1997, implementation has progressed slowly, and Russian-led peacekeeping troops remain posted throughout the country.

A modest economic recovery began after Tajikistan concluded a loan agreement with the International Monetary Fund (IMF) in 1997. The Tajik government brought inflation down to 13.5 per cent in 2001 from 60.6 per cent in 2000, and the country's real GDP grew by 9.5 per cent in 2001. However, Tajikistan still faces major problems in integrating refugees and former combatants into the economy, and the country continues to depend on aid from Russia, Uzbekistan, and international humanitarian assistance for much of its basic subsistence needs. The future of Tajikistan's economy and the potential for attracting foreign investment depend upon stability and continued progress in the peace process.

Oil

Tajikistan has proven oil reserves of only 12 million barrels. The country's small oil industry is centered around the northern Leninobod Soghd Region. In 2001, Tajikneftegaz, which is responsible for all oil exploration, drilling, and production in Tajikistan, produced an average of just 350 barrels per day (bbl/d) of oil, continuing a downward trend that has seen the country's oil production drop

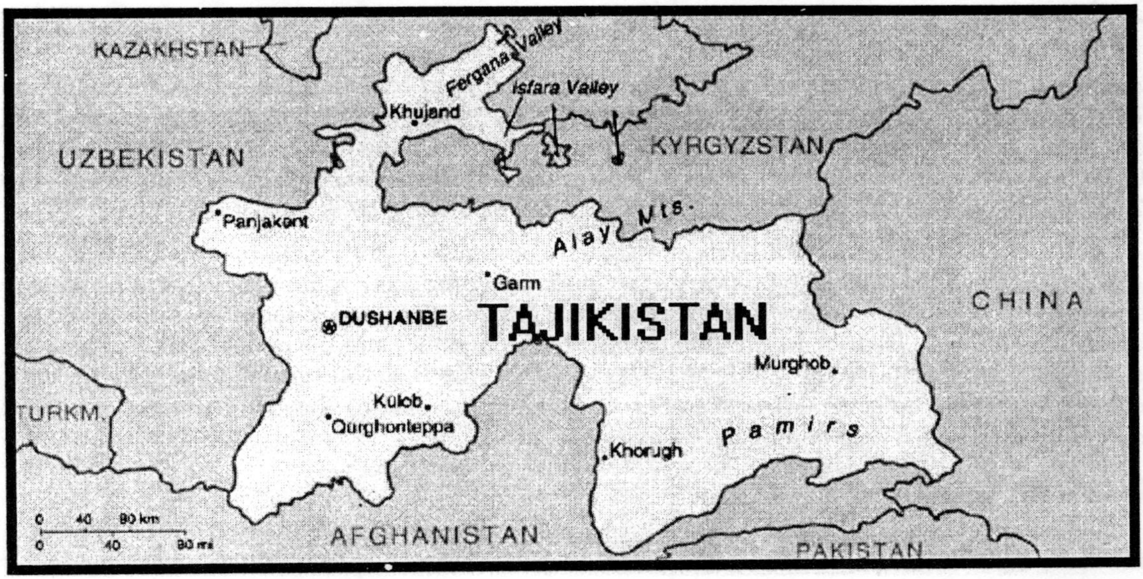

off from 1,311 bbl/d in 1992. Tajikistan's 1992-1997 civil war, coupled with economic contraction and a lack of investment to maintain the oil sector's infrastructure, has resulted in a 73 per cent decline in national oil production.

Tajikistan consumed approximately 29,000 bbl/d of petroleum products in 2001, of which nearly 100 per cent is imported. In July 2001, Tajikistan opened its first refinery, the small 400-bbl/d-capacity Konibodom refinery, which produces gasoline, diesel, kerosene, and fuel oil. However, the country still must import much of its oil as refined petroleum products. Uzbekistan supplies more than 70 per cent of Tajikistan's oil demand.

Natural Gas

With just 200 billion cubic feet (Bcf) in proven natural gas reserves, Tajikistan produces minimal amounts of natural gas domestically, leaving the country reliant on imports to meet domestic demand. In 2000, Tajikistan commissioned the Khoja Sartez natural gas field in the southern Khatlon Region, which, in combination with the increased utilization of the Qizil Tumshuq deposit in southern Khatlon Region's Kolkhozobod District, Tajikistan hopes will lead to increased domestic natural gas production. For 2000 as a whole, the country produced 1.4 Bcf of natural gas.

Tajikistan relies heavily on Turkmen and Uzbek natural gas to meet domestic demand, which stood at 44.1 Bcf in 2000. Due to distortions in the Tajik natural gas market, Tajikistan has continually run up debts to suppliers for natural gas already consumed. In addition, through April 2002, Tajikistan's population and industrial enterprises already had consumed about 80 per cent of the annual volume of natural gas envisaged under an intergovernmental agreement between Uzbekistan and Tajikistan. Tajikgaz blames the high natural gas consumption and nonpayment by individual consumers (only 18 per cent paid in 2001) on Soviet-era practices, when utilities were largely free. Tajikgaz has had to cut off nonpaying customers, as well as negotiate with suppliers for additional natural gas.

Coal

Tajikistan has 17 coal deposits, but the country's proven coal reserves are minimal. Since 1991, when Tajikistan produced approximately 430,000 short tons of coal, the country's coal production declined for seven consecutive years, to just over 11,000 short tons in 1998. Tajikistan's coal production rebounded to 22,000 short tons in 1999 and stayed at approximately that level in 2000. Preliminary 2001 figures reported by Tajikistan's Energy Ministry show that the country's coal production rose to 27,500 short tons, still below the country's consumption of 134,500 short tons of coal.

On April 16, 2001, President Rakhmonov signed a resolution on the establishment of Tojikangisht, a state coal enterprise uniting the country's two main coal producers, Leninobod Coal, a joint-stock company in the northern Soghd Region, and Fan-Yaghnob, a joint-stock company in the northern Soghd Region's Ayni District. In addition to these two companies and the mines of the same name, Tojikangisht also includes the Ziddi and Nazar-Ayloq coal fields in central Tajikistan. In 2001, Tojikanisht also set up several branches–Sayod, Hakimi, Miyonadu, Shurobod, and Saymiri–and embarked on opencast strip mining.

Electricity

Tajikistan had a total installed electricity-generating capacity of 4.4 gigawatts (GW) in 2000. Most of the country's electric power comes from seven large hydroelectric plants, which have a combined capacity of 4.05 GW. The Nureksk hydroelectric plant, which has nine units of 300 megawatts (MW) each, accounts for nearly 70 per cent of this power. Other hydroelectric plants include Golovnaya, Baipazan, Namadgud, Lenin, Pamir-1, and Qayroqqum. Tajikistan also has several thermal power plants, with combined capacity of approximately 350 MW. In 2000, Tajikistan generated 14.2 billion kilowatt-hours (Bkwh) of electricity and consumed 12.5 Bkwh.

A significant portion of Tajikistan's power sector infrastructure is in poor condition as a result of the civil war and the lack of proper maintenance, which has contributed to increased energy losses of nearly 15 per cent of generating capacity. Transformers are constantly breaking down due to overloads, and most power equipment has exhausted its service life; the Tajik government estimates that

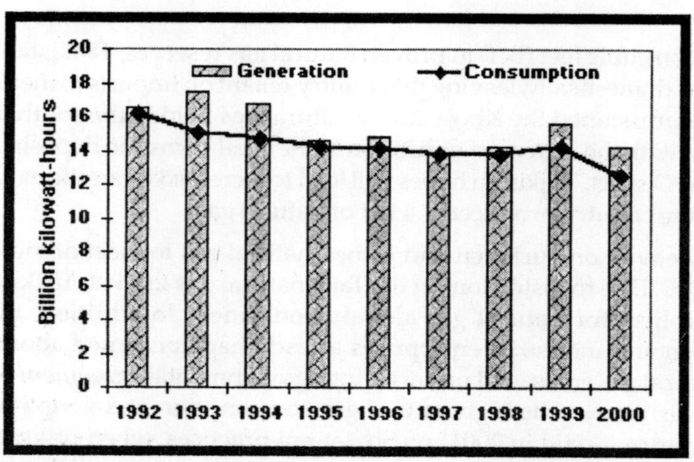

TAJIKISTAN ELECTRICITY GENERATION AND CONSUMPTION, 1992-2000

depreciation of energy equipment already has reached 75 per cent. In addition, hydroelectric plants have been operating at well below capacity due to severe weather and low water levels.

Barq-i Tojik is the state company that controls electricity generation, transmission, and distribution in Tajikistan. The country has two power grids–a unified energy grid in the southern part of the country and a grid in the northern Soghd Region that is powered by the Qayroqqum hydroelectric station on the Syr Dar'ya River. The Qayroqqum plant can cover only slightly more than 30 per cent of the northern region's energy needs, forcing the northern region to import power from Uzbekistan.

Tajikistan is hoping to modernize its power infrastructure by attracting foreign investment to the sector. The country is trying to increase its power-generating capacity and to reconstruct its energy grids. However, potential investors, which include international financial organizations and neighboring countries, are demanding that Tajikistan's power sector be privatized. Potential investors also want Tajikistan to change its rate policy: although the country's low electricity tariffs have increased bill collection to nearly 50 per cent of energy deliveries, the policy has also resulted in huge losses for Barq-i Tojik, since the tariffs do not cover production costs. Tajik industrial enterprises and residential customers owe Barq-i Tojik more than $100 million.

Already, Barq-i Tojik has embarked on a $62-million project to refurbish Tajikistan's electricity sector. The main components of the project are to rehabilitate the Nureksk hydroelectric power station and the Jangal and Novaya substations, to restore the power grid in the southern region of Tajikistan, to install electricity meters on inter-system transmission lines, and to improve the company's service. Tajikistan also will raise electricity tariffs by 25 per cent to 30 per cent in 2002, with another 60 per cent increase slated for 2003.

Completion of the Rogunsk and Sangtuda hydroelectric power stations are priorities for Tajikistan. The Rogunsk plant, which was begun during the Soviet period, has a design capacity of 3.6 GW, which will make it the 15th largest hydroelectric plant in the world. Construction of the 670-MW Sangtuda station, which also was begun before independence, has resumed with Russian and Iranian financing. Approximately $300 million still is needed to complete the Sangtuda power station. The Tajik government also is resuming a program to build 15 small hydroelectric plants, including Andarbak (250-MW capacity), Shkev (74 MW), Yemts (100 MW), Langar (60 MW), and Yamchun (150 MW).

23

Central Asia: Turkmenistan Energy Sector

General Background

Following several years of decline after Turkmenistan's independence from the Soviet Union in 1991, Turkmenistan's economy has rebounded in the past four years. Turkmenistan, whose economy relies heavily on oil and natural gas production, suffered a 25.9 per cent drop in its real gross domestic product (GDP) in 1997 when Russia closed off its natural gas pipeline network–Turkmenistan's sole natural gas export option at the time. Since the resolution of the dispute with Russia, Turkmenistan's natural gas exports have increased dramatically, spurring the country's economy to three straight years of double-digit real GDP growth, including an 18 per cent increase in 2001. Turkmenistan's economy is forecast to grow an additional 13 per cent in 2002.

Nevertheless, Turkmenistan's real GDP in 2001 was still only 70 per cent of its 1990 level, and economic and political reform have been stifled under the autocratic leadership of President Saparmurat Niyazov, a former communist who has ruled Turkmenistan since independence and was named president for life in 1999. The country's unemployment rate, although down to 14 per cent in 2001 from a high of 24.2 per cent in 1998, is still problematic, and foreign direct investment, over 90 per cent of which flows into the country's oil and natural gas sectors, has slowed over the past few years because of the restrictive conditions that Turkmenistan attaches to foreign investment. Privatization goals remain limited, and the country has not taken steps to diversify its economy to reduce its dependence on natural resource exports.

Oil

Turkmenistan has 546 million barrels in proven oil reserves, with possible reserves (mainly in the western part of the country and in undeveloped offshore areas in the Caspian Sea) of up to 1.7 billion barrels. The country's oil production, which steadily declined after independence, from 110,000 barrels per day (bbl/d) in 1992 to 81,000 bbl/d in 1995, has increased dramatically in the past six years, reaching 156,400 bbl/d in 1999 before leveling off in the past two years. In 2001, Turkmenistan produced 159,000 bbl/d of oil while consuming 52,000 bbl/d. Turkmenneft, the state oil company, produced approximately 90 per cent of this total, with the remainder coming from the state natural gas company, Turkmengaz, and several foreign oil companies operating under PSAs in Turkmenistan.

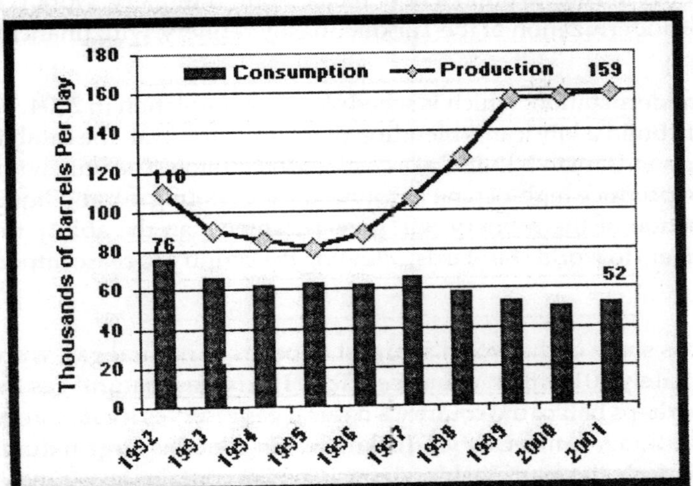

TURKMENISTAN OIL PRODUCTION AND CONSUMPTION, 1992-2001

In 2002, Turkmenistan is seeking to increase its oil output to 200,000 bbl/d, with additional production to come from newly developed wells in the western part of the country. Under a 10-year program dictated by President Niyazov, Turkmenistan aims to raise its oil production to nearly 1 million bbl/d by 2010. According to Turmenistan's Oil and Gas Industry and Natural Resource Minister, Kurbannazar Nazarov, Turkmenistan needs $25 billion in foreign investment to its oil and natural gas sectors between now and 2010. In an effort to create a better business climate to attract foreign investment, in June 1998 Turkmenistan restructured its oil and gas industries into several state-owned companies.

Although the country has attempted to ease restrictions on foreign investment, many layers of regulation remain in place. Turkmenistan maintains prohibitive rules that prevent companies using subsurface resources to export hydrocarbons. Since foreign investors do not have access to export pipelines (state-run Turkmenneft, Turkmengaz, and Turkmenneftegaz, the oil and natural gas marketing company, currently own all of the country's pipelines), they are forced to sell the oil and natural gas they produce in Turkmenistan through the state commodities exchange or send it to refineries. Oil and natural gas are sold in Turkmenistan at fixed prices that are well below world market levels.

As a result, several projects that could substantially increase Turkmenistan's oil production have stalled. Petronas (Malaysia), which is developing the Cheleken-1 oil and natural gas deposit under a PSA signed in 1996, suspended operations for more than a year, since the company determined it could not develop the field profitably under Turkmenistan's export restrictions. Swap arrangements, such as United Arab Emirate-based Dragon Oil's small-scale swap agreement with Iran, have proved modestly successful, but the Turkmen government has pledged to work on legislation that will expand the opportunities for foreign investors to export oil and natural gas, including liberalizing pipeline transport and easing the tax burden.

Downstream/Refining

Turkmenistan has two refineries, the 116,500-bbl/d refinery at Turkmenbashy and a 120,500-bbl/d refinery at Seidi. Both facilities are slated for modernization and expansion to meet the country's expected increases in oil production and demand, and Turkmen President Saparmurat Niyazov is planning to call a tender in 2002 to build a new 100,000-bbl/d refinery. Work is underway on a $1.4-billion upgrade and modernization of the Turkmenbashy refinery, with financing from German and Japanese sources.

As part of the modernization, which is scheduled for completion in 2004, France's Technip was awarded a contract to build a lubricants blending plant. In April 2001, the catalytic cracking unit was launched by Technip and Iranian NINISC at a cost of $300 million. The unit, with a capacity of 36,150 bbl/d, is designed to produce high-octane gasoline, diesel, heating oil, and liquefied petroleum gas. Complete reconstruction of the refinery will give Turkmenistan the ability to produce motor oil, lubricants, and polymers to world standards, allowing the country to cease importing lubricating oils.

Natural Gas

Turkmenistan has some of the world's largest deposits of natural gas, with proven natural gas reserves of approximately 101 trillion cubic feet (Tcf). The largest natural gas fields are in the Amu-Dar'ya basin, with perhaps half of the country's natural gas reserves located in the giant Dauletabad-Donmez field. In addition to Amu-Dar'ya, Turkmenistan contains large natural gas reserves in the Murgab basin, particularly the giant Yashlar deposit, which contains an estimated 27 Tcf. During the last 10 years, Turkmenistan also has discovered 17 new natural gas deposits in the Lebansky, Maryinsky, and Deashoguzsky regions of the country.

Turkmenistan was a substantial natural gas producer under the Soviet Union, but after the country became independent, Turkmen natural gas became a competitor with Russian natural gas. Since Turkmenistan's only natural gas export routes ran through Russia, Gazprom limited Turkmen natural gas exports, and as a result Turkmenistan's natural gas production sagged throughout the 1990's. Following the resolution of a pricing dispute with Russia in 1998 and the construction of an export pipeline to Iran, Turkmenistan's natural gas production began to climb steadily. In 2001, the country's natural gas production jumped to 1.64 Tcf against consumption of just 0.26 Tcf. Turkmengaz produced 85 per cent of this total, with Turkmenneft accounting for the remaining 15 per cent.

With its large natural gas reserves, Turkmenistan is counting on increased natural gas production and exports to fuel its economic recovery. In May 2001, Turkmengaz started exploration and prospecting work on a new field in Darganata, northeastern Turkmenistan. Commercial exploitation of the Gagarinskoye deposit in Zaunguz Karakum is scheduled to begin soon, while resumption of work in the Samantepe field on the right bank of Amu Dar'ya in eastern Turkmenistan is planned. Under a presidential program, Turkmengaz also is stepping up exploratory work in the Karakum and Kyzylkum deserts. Through the first two months of 2002, Turkmenistan already had produced 413 billion cubic feet (Bcf) of natural gas.

In order to reach its full natural gas production potential, however, Turkmenistan must solve the problem of getting its natural gas to consumers, as well as getting paid in hard currency. The country has been unable to capitalize on its natural gas resources because it lacks pipeline outlets to world markets. As a result, Turkmenistan is forced to sell its natural gas to ex-Soviet states that either cannot pay fully in cash or are tardy with payments for supplies already received; both Azerbaijan and Kazakhstan are indebted to Turkmenistan for natural gas supplies. In October 2000, Turkmenistan agreed to resume the export of natural gas supplies to Ukraine that had been suspended in May 1999 because of Ukraine's $281-million natural gas debt.

In a bid to secure a market for its natural gas, on May 14, 2001, Turkmenistan agreed with Ukraine on a major natural gas export deal. Under terms of the deal, Turkmenistan will provide Ukraine with 8.83 Tcf of natural gas between 2002 and 2006. Turkmenistan will sell Ukraine 1.41 Tcf of natural gas in 2002 and 1.77 Tcf in 2003, with remaining deliveries to be agreed later. Turkmen officials signed the deal on the condition that Ukraine makes timely payments for supplies. Ukrainian officials agreed to pay for the Turkmen natural gas 60 per cent in cash, with the remainder paid for through participation in 20 construction and industrial projects in Turkmenistan worth a total of $412 million.

Coal

Turkmenistan has no coal reserves, nor any coal production. Although the country consumed a minimal amount of coal during the Soviet era, in the aftermath of the collapse of the Soviet Union, Turkmenistan rapidly phased out its coal use, and the country's consumption fell from 551,000 short tons in 1992 to zero in 1998.

Electricity

With 3.9 gigawatts (GW) of installed capacity, 99 per cent of which is thermal, Turkmenistan has sufficient electricity-generating potential to power its own cities, unlike much of Central Asia. In 2000, Turkmenistan's power sector generated 9.3 billion kilowatt-hours (Bkwh) while Turkmen consumers used just 7.7 Bkwh, giving the country 1.6 Bkwh in surplus electricity. However, owing to the country's inefficient, Soviet-era power infrastructure that is in need of repair, power line losses wasted a significant portion of the electricity Turkmenistan generated in 2000, resulting in exports of only 0.9 Bkwh.

Most of the electricity that Turkmenistan exports is sent to southwestern Kazakhstan and northeastern Afghanistan, although Armenia, Turkmenistan, and Iran have discussed greater cooperation in the energy sphere. A power transmission line connecting Turkmenistan to northern Iran would allow Turkmen electricity exports to Iran and to Armenia, since Armenia and Iran's electricity grids are connected.

24

Central Asia: Uzbekistan Energy Sector

General Background

Since Uzbekistan gained its independence in December 1991, the government has sought to prop up its Soviet-style command economy with subsidies and tight controls on production and prices. Although this gradualist reform strategy has helped the country to avoid the dramatic economic contraction and drastic decline in living standards recorded in many other countries in the Commonwealth of Independent States, thus far it has failed to bring about much-needed structural changes. While Uzbekistan has now recorded six straight years of real gross domestic product (GDP) growth, the lack of significant macroeconomic and structural reforms, the country's rapid accumulation of external debt, as well as its declining level of foreign exchange reserves, makes this pattern unsustainable.

The government continues to have a dominating influence on the Uzbek economy. Uzbekistan tightened currency and export controls in its largely-closed economy following the Asian and Russian financial crises, further deterring foreign investors already shying away from the country because of a poor investment climate and Uzbekistan's non-convertible currency, the som. Analysts argue that continuing administrative and trade controls are inhibiting export growth and discouraging foreign direct investment. Foreign investment in Uzbekistan is significantly lower than in other energy-rich former Soviet republics, such as Azerbaijan and Kazakhstan.

Oil

Uzbekistan is estimated to contain 594 million barrels of proven oil reserves, with 171 discovered oil and natural gas fields in the country. The Bukhara-Khiva region contains over 60 per cent of

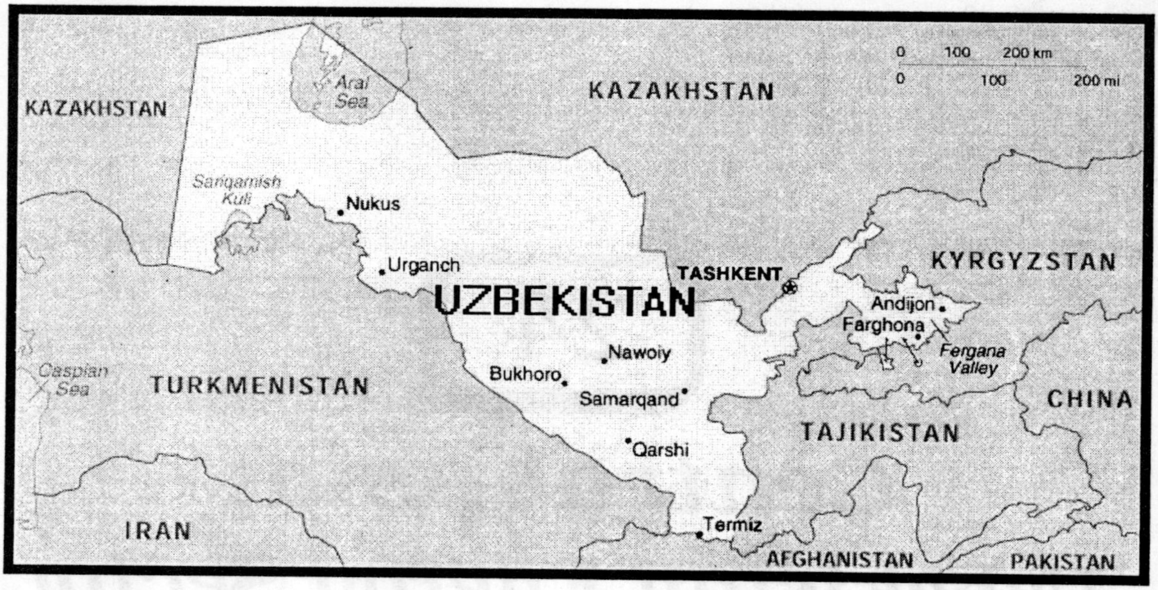

Uzbekistan's known oil fields, including the Kokdumalak field, which accounts for about 70 per cent of the country's oil production. In addition, the Fergana region contains another 20 per cent of the country's oilfields, and the Ustyurt plateau and the Aral Sea have been targeted for further exploration. Oil deposits in Kokdumalak, Shurtan, Olan, Urgin and south-Tandirchi (all in southwestern Uzbekistan) are being developed rapidly.

As a result, despite a drop in oil production in the past few years, Uzbekistan has more than doubled its petroleum output in the past decade. From 65,500 barrels per day (bbl/d) in 1992, Uzbekistan increased its oil production to 161,000 bbl/d in 1998. Combined with the country's decrease in oil consumption (from 190,400 bbl/d in 1992 to 130,000 bbl/d in 2000), in 1996 Uzbekistan became a net oil exporter. However, Uzbekistan's oil and gas condensate production has been declining in the past few years as existing fields are exhausted faster than new commercial reserves are discovered. Uzbekneftegaz, the state oil and natural gas company, expects liquid hydrocarbon production in the country to fall to 120,000 bbl/d in 2005.

In an effort to stem the decline in Uzbekistan's oil production, the Uzbek government is seeking foreign investment in the country's oil sector. Uzbekistan is offering a 49 per cent state in Uzbekneftegaz, the holding company that was created out of nine companies in 1998 to unite the country's entire oil and natural gas sector. Since independence, the Uzbek government has invested over $1.2 billion in modernizing Uzbekneftegaz, but the flow of money into the Uzbek upstream has been far slower than in other Central Asian nations due to Uzbekistan's strict currency controls.

Uzbekistan also is selling its 44 per cent stake of Uzneftegazdobycha (Uzbekneftegaz's oil and gas exploration arm), 44 per cent of Uztransgaz (oil and gas transport), 39 per cent of Uzneftepererabotka (oil refining), and 39 per cent of Uzburneftegaz (drilling company). This tender is part of an aggressive oil and natural gas investment bid launched by Uzbekistan on April 28, 2000, when President Karimov decreed that foreign companies involved in exploring and extracting oil and gas in Uzbekistan would receive tax exemptions and options to produce any oil or natural gas they discover within a set period of time.

The government is eager to attract $400 million through production-sharing agreements (PSAs) as well, with over 80 fields on offer. Of these, 78 of the fields are contained in 16 exploration blocks, and eight individual fields (with total remaining reserves of some 1.2 billion barrels of oil equivalent) have also been opened up for potential foreign participation. Those fields include four in the Southwest Gissar Basin (Dzharkuduk, Gumbulak, South Kizilbairak and South Tandircha) and four in the Amu Dar'ya region (North Shurtan, Shakarbulak, South Kemachi and Umid).

In addition, Uzbekistan is seeking investment to boost production at existing fields. Uzbekneftegaz already has teamed with oil services giant Baker Hughes in a joint venture to increase oil production at the country's North Urtabulak field to over 6,000 bbl/d. Baker Hughes, which will invest $8 million in the North Urtabulak project, also has the option to develop the Adamtash, South Kemachi, and Umid fields, with total investments of $120 million. UzPEC, a subsidiary of Britain's Trinity Energy, received licenses in 2001 to explore and develop oil and gas condensate fields in Southwest Gissar and Central Ustyurt. According to its PSA with Uzbekneftegaz, UzPEC will hold the licenses for 40 years and will be required to invest more than $400 million, including $200 million in the next five years.

Downstream/Refining

Uzbekistan has three refineries, at Fergana, Alty-Arik, and Bukhara, with a total refining capacity of 222,000 bbl/d. The Bukhara refinery, which was the first refinery built in the Commonwealth of Independent States since the breakup of the Soviet Union and cost in excess of $400 million, currently has a capacity of 50,000 bbl/d, although it is expected to expand to 100,000 bbl/d and refine both crude oil and gas condensate. Due to the country's decline in oil production in 2001, Uzbek refineries operated well below-capacity during the year. Uzbekistan's limited refined product exports move by rail and road to neighboring countries and to export ports on the Black Sea.

Along with joint ventures with foreign investors, Uzbekistan is looking to refinery modernization as a crucial component of the country's strategy to attain self-sufficiency in oil. In 1996, Texaco (now Chevron Texaco, U.S.) and Uzneftepererabotka formed the UZ-Texaco joint venture at the Fergana refinery to produce and market Texaco-branded engine, transmission, and hydraulic lubricants from local crude oil. In 2001, Mitsui (Japan) completed a $200-million reconstruction at the Fergana refinery to expand desulfurization capacity at the refinery.

Natural Gas

With estimated natural gas reserves of 66.2 trillion cubic feet (Tcf), Uzbekistan is the second largest natural gas producer in the Commonwealth of Independent States (after Russia) and one of the top ten natural gas-producing countries in the world. Uzbekistan produces natural gas from 52 fields in the country, with 12 major deposits–including Shurtan, Gazli, Pamuk, Khauzak–accounting for over 95 per cent of Uzbekistan's natural gas production. These deposits are concentrated in two general areas: the Amu Dar'ya Basin and in the Mubarek area of the southwest part of the country.

Since becoming independent, Uzbekistan has increased its natural gas production by over 30 per cent, from 1.51 Tcf in 1992 to 1.99 Tcf in 2000. According to preliminary 2001 data, Uzbek natural gas production increased to 2.03 Tcf for the year. However, Uzbekistan's natural gas fields were heavily exploited in the 1960's and 1970's by the Soviet Union, and as a result several older fields, such as Uchkyr and Yangikazgan, are beginning to decline in production. In order to offset those declines,

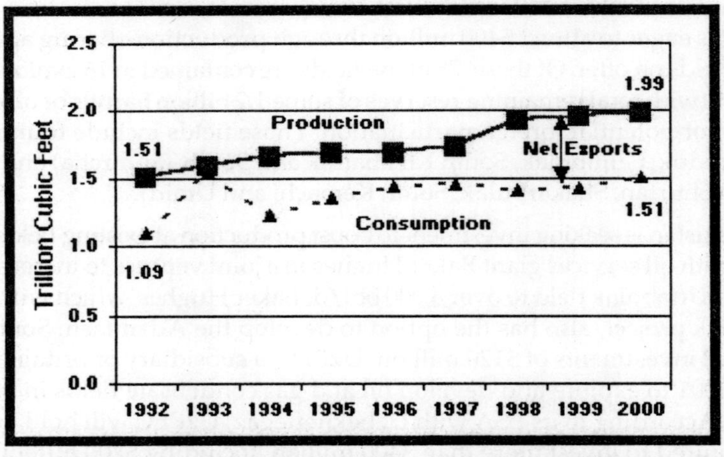

UZBEK NATURAL GAS PRODUCTION AND CONSUMPTION, 1992-2000

Uzbekistan is speeding up development at existing fields, such as Garbi and Shurtan, as well as developing new fields and exploring for new reserves. The Shurtan field, which began producing in 1980 and is the second biggest in the country after Gazli, accounted for approximately 36 per cent of Uzbekistan's total natural gas output in 2000.

Due to its high sulfur content, the majority of Uzbekistan's natural gas requires processing before it can be consumed. Much of Uzbekistan's natural gas is processed at the Mubarek processing plant, which has a capacity of over 1 Tcf/year. In December 2001, Uzbekneftegaz commissioned the Shurtan Gas-Chemical Complex, which includes installations to clean natural gas, a natural gas booster compressor station, and a plant with the capacity to produce 125,000 tons of polyethylene and 137,000 tons of liquefied natural gas per year. The complex, which is located by the Shurtan gas fields in the southwest part of the country in the Kashkadar'ya Region, was completed at a cost of $1 billion.

In addition to the Shurtan project, Uzbekneftegaz is undertaking several projects to ensure the country's natural gas sector will remain vibrant. The company's Kodzhaabad underground natural gas storage facility in Andizhan Region opened in 1999 at a cost of $72 million, allowing increased natural gas shipments to Uzbekistan's industrial heartland in the Fergana Valley. In January 2001, Trinity Energy (U.K.) committed to investing more than $400 million, over a 40-year period, in exploration and production of gas condensate deposits in the Plato Ustyurt region.

In March 2002, Russia's Itera and Lukoil signed a PSA with Uzbekneftegaz to form a joint-stock company to develop several new gas fields in Uzbekistan, including the giant Kandym field. Natural gas reserves at the fields covered by the PSA are estimated at 8.1 Tcf, including approximately 5.4 Tcf at the Kandym structure. Initial investments in the project are estimated at $377 million, with natural gas production rising from 159 billion cubid feet (Bcf) per year to between 280 Bcf and 350 Bcf per year at its peak. Itera and Lukoil each will hold 45 per cent shares in the company, with Uzbekneftegaz keeping a 10 per cent stake in the project.

Coal

Uzbekistan has estimated coal reserves of 4.4 billion short tons, the majority of which are located in just three deposits. Approximately 75 per cent of Uzbekistan's coal reserves are lignite and subbituminous brown coal. The Angren lignite coal field, which is in the Tashkent region and is the country's largest coal deposit, holds a proven 1.9 billion tons of commercially recoverable brown coal. In 2000, Ugol, the Uzbek national coal company, produced 3.2 million short tons (Mmst) of coal, 90 per cent of which came from the Angren mine. In the first nine months of 2001, Ugol produced 2.65 Mmst of coal, a 3.5 per cent increase over the same time period in 2000.

Uzbekistan's domestic coal consumption has declined by 50 per cent in the past decade–from 6.4 Mmst in 1992 to 3.2 Mmst in 2000–making Uzbekistan a net coal exporter, despite the country's drop in coal production from 5.1 Mmst in 1992. In order to increase coal production, the Uzbek government is implementing a program to update the country's coal sector by modernizing production facilities. In 2001, Krupp Fordertechnik GmbH (Germany) won a tender to refurbish the Angren coal mine, a project that will be implemented over 10 years in six stages. The project stipulates a transition from cyclical coal extraction technology to the flow-line method, which Uzbek officials hope will raise coal extraction to 5 Mmst/year and will cut production costs at Angren from $23/ton to $12/ton.

Ugol plans to upgrade mining operations at its other main deposits as well. The Shargun and Baisun deposits, both of which are located in the Surkhandarya region, are much smaller than the one at Angren. Additional investment at the Shargun deposit is expected to double or triple production of high-quality coal from current levels of over 200,000 short tons/year. Completion of a second mine at Baisun could quintuple the mine's production of over 100,000 short tons/year, and could ensure that Uzbekistan has a surplus of coal for export in the future.

Electricity

Uzbekistan has 37 power stations, with a combined installed generating capacity of 11.7 gigawatts (GW). Much of Uzbekistan's electric power is generated from natural gas-powered plants, with smaller amounts generated from coal and hydroelectric facilities. The largest natural gas-fired plants are the Syr Dar'ya (3,000 megawatts, MW) and Navoi (1,250 MW) plants, which together account for over one-third of the entire country's generating capacity. Several coal-powered facilities, including the 1,800-MW coal-fired Angren plant, are located near the Angren mine near Tashkent, while 25 small hydroelectric plants (the 620-MW Charvak station is the largest) supply almost 15 per cent of Uzbekistan's electricity.

Uzbekistan generated 44.1 billion kilowatt-hours (Bkwh) of electricity and consumed 41.9 Bkwh in 2000. Nevertheless, owing to significant line losses in the country's deteriorating power infastructure, much of the electricity that Uzbekistan generates never reaches customers. As a result, Uzbekistan is actually a net electricity importer. However, the Uzbek government has developed a plan to increase the country's electric-generating capacity by attracting foreign capital and loans to reconstruct and upgrade a number of Uzbek power plants.

In December 2001, Germany's Siemens completed reconstruction of the first of two power-generating units at the Syr Dar'ya Power Plant, with the second unit scheduled to be finished early in 2002. The modernization of two of the 10 units at the Syr Dar'ya Power Plant will increase the plant's power-generating capacity by 600 MW to 3,600 MW. Uzbekistan's plans also call for the modernization of Unit 1 at the Talimardjan Power Plant, as well as the construction of new units to increase the plant's installed capacity to 3,200 MW.

In March 2002, Uzbekenergo, the state power company, announced plans to call for an international tender later in 2002 to reconstruct the 1,860-MW Tashkent State Regional Power Plant. The $221-million project will include the construction of a new power-generating unit with a 370-MW steam gas turbine. The reconstruction will take 28 months, according to a feasibility study for the project prepared in 1999 by Japan's Mitsubishi Corporation and approved by the Uzbek government.

Uzbekistan also is attempting to attract foreign investment to revamp electric power systems and stations in Navoi, Mubarek, and other cities, as well as to modernize the electric power grid in Tashkent. ABB Lummus has begun a feasibility study of a $60-million project to rebuild the heat and power plant in Mubarek, increasing its capacity from 60 MW to 100 MW, and in January 2002, ABB signed a $17.4 million contract with Uzbekenergo on the construction of two electricity substations in Tashkent, as well as connections to the grid.

25

Chad

With its large undeveloped oil reserves, Chad has the potential to become a significant energy producer. Construction is currently underway on a 650-mile pipeline which will allow Chad to export an estimated 225,000 barrels per day of crude oil by the middle of the decade.

General Background

With an annual per capita income of $203, Chad is one of the poorest countries in the world and is ranked 155th out of 162 countries on the United Nations' 2001 human development index. Chad's economy is dominated by the agricultural sector. Cotton, the major cash crop, remains Chad's primary export. Due to increased revenues stemming from demand and investment associated with the Chad-Cameroon pipeline project, Chad's real GDP grew by 8.4 per cent in 2001, while inflation grew by more than 12 per cent. The pipeline project, for which construction began following World Bank approval in June 2000, may transform Chad's economy. In addition to roughly doubling the government's annual budget, the project is expected to double Chad's GDP by 2004. In May 2002, the International Monetary Fund (IMF) permitted Chad to draw $3 million in funds following a successful performance review under the Poverty Reduction and Growth Facility (PRGF) program. The original agreement was approved in January 2000 as a three-year, $49.9 million arrangement intended to support Chad's 1999/2002 economic program.

Chad is one of six members of the Central African Economic and Monetary Union (CAEMU). The sub-regional group, which is made up of Chad, Cameroon, the Central African Republic, the Republic of the Congo, Gabon, and Equatorial Guinea, was created in 1991. CAEMU's objectives include economic and political cooperation and eventual integration within the sub-region.

Chad is rich in natural and mineral resources. Currently, only deposits of natron (sodium carbonate) and kaolin are utilized commercially. Deposits of other minerals have been discovered in Chad, including gold, bauxite, tin, tungsten, titanium, and iron ore. Although Chad possesses

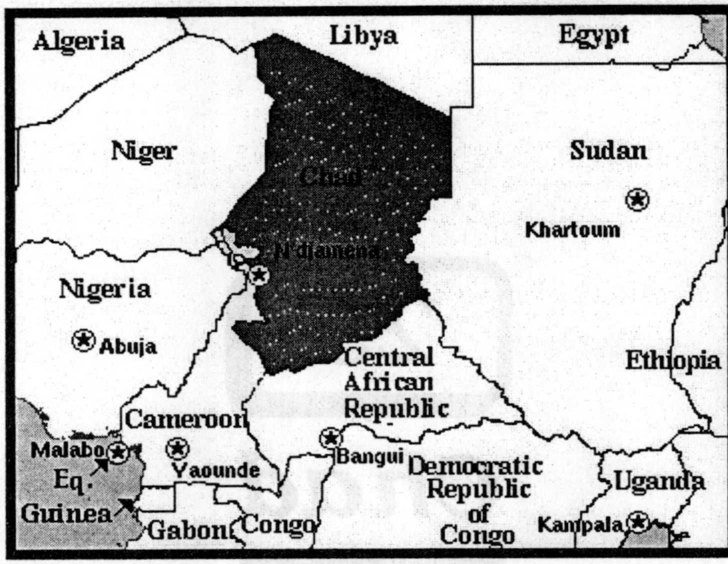

substantial proven oil reserves, it will be unable to produce or refine any of its own oil until the Doba oil project is completed, and is thus totally dependent on fuel imports, mainly from Nigeria and Cameroon.

A former province of French Equatorial Africa, Chad achieved its independence from France in 1960. Idriss Deby assumed state control in 1990. In April 1993, a National Conference, composed of Deby's political party and members of several opposition political groups, created a transitional government, with Deby remaining as head of state, adopted a transitional charter, and formulated a timetable for national elections. A new constitution was adopted by a national referendum in March 1996. The constitution calls for the establishment of a bicameral legislature (National Assembly) and the election of the president by direct universal suffrage. The president can serve a maximum of two five-year terms in office. Conflict began in 1998 when the Mouvement pour la Democratie et la Justice au Tchad (MDJT) began to protest Deby's government, though fighting was confined to the Tibesti region in the north. Deby was elected to his second term as president in 2001. In January 2002, the Chadian government and the MDJT signed a peace agreement calling for a cease-fire and giving amnesty to the rebel fighters.

Oil

Petroleum exploration in Chad began in the early 1970's, although exploration activities were suspended in 1979 due to the outbreak of civil war. A proposed plan to build a pipeline from the Lake Chad basin fields to a new refinery in the capital of N'Djamena was also put on hold. Major oil companies in Chad include ExxonMobil, TotalFina Elf, Shell, and ChevronTexaco.

In November 1996, a consortium consisting of Exxon (now ExxonMobil; 40 per cent and operator), Shell (40 per cent) and Elf Aquitaine (20 per cent), signed a memorandum of understanding (MOU) with the government of Chad. The MOU provided the terms of the development of the Doba basin fields, and the construction of a 1,050- kilometer (650-mile) export pipeline through Cameroon to offshore export facilities located near Kribi. The total cost of field development and construction of the pipeline and export facilities is estimated to be $3.5 billion. In April 2000, Chevron (now ChevronTexaco) and Malaysia's Petronas joined the consortium, replacing TotalFina Elf and Shell, who had divested their combined 60 per cent share in the consortium. Petronas acquired a 35 per cent interest, with Chevron taking the remaining 25 per cent. Exxon retained its controlling 40 per cent share in the consortium.

The Chad-Cameroon pipeline project consists of two primary components, development of the oil fields and construction of the pipeline and export facilities. The Doba basin's three fields (Bolobo, Kome and Miandoun) are expected to produce 900 million–1 billion barrels of low sulfur oil over the life of the project. The consortium plans to drill 300 wells. Production is forecast to continue for 25-30 years, with peak production projected at 225,000 to 250,000 barrels per day (bbl/d). Crude oil from the three fields will be gathered, treated and blended at a facility located on the Kome field. Construction of the project began in October 2000. Production is set to begin in early 2004, with cost of field development estimated at be $1.5 billion. Originally, production was set to begin in 2003, but contract problems and two strikes by workers have caused serious delays.

Construction of the pipeline and export facilities is expected to last just over two years and to cost $2.2 billion. The Chadian portion of the pipeline, approximately 170-kilometers (105 miles) long, and the first (located at Kome) of the three pumping stations are being built by the Tchad Oil Transport Company (TOTCO), which will also retain ownership. TOTCO, which was established in July 1998, is composed of members of the consortium and the government of Chad. The Cameroon Oil Transport Company (COTCO), will build and own the 880-kilometer (545-mile) Cameroonian section of the pipeline, including the remaining pumping stations and the export facilities. The export facilities will consist of an onshore pressure-reducing station near Kribi, a 12-kilometer (7-mile) sub-sea pipeline and an offshore floating storage and off-loading facility (FSO). COTCO consists of the consortium members, the Cameroonian government, and the government of Chad. Deviations in the pipeline route may extend the total distance up to 663 miles.

The foreign consortium's share in bankrolling the Chad-Cameroon project will amount to $2.2 billion (59.2 per cent of total costs), while commercial banks and export credit agencies will contribute $600 million (16.1 per cent) and capital markets will finance $400 million (10.7 per cent). The lynchpin of the project financing, however, is the World Bank loan of $93 million–$53.4 million for Cameroon and $39.5 million for Chad–agreed to in June 2000. In addition, the International Finance Corporation will lend the three companies $100 million and make another $300 million available through commercial banks. Finally, the European Investment Bank may also contribute $40 million to the project. A week after the World Bank's board "overwhelmingly" approved the project, the U.S. Export-Import Bank approved a $300 million loan guarantee to finance exports to build the pipeline.

According to projections released by the World Bank, the Doba project is highly profitable. Based on a benchmark price of $15.25 per barrel of oil, total receipts for the Doba project are expected to reach $12 billion over a 28 year period. Net revenue would amount to $9.2 billion before debt servicing. Chad would earn $2.5 billion over the life of the project with annual revenues of up to $200 million.

The Chad-Cameroon oil project has been the source of intense opposition from human rights and environmental groups. Many of the environmental concerns have focused on Cameroon, where the pipeline is set to traverse areas of dense jungle inhabited by the Bagyeli ethnic group, or Pygmies as they are popularly known. Consequently, the consortium has made numerous adjustments to the route of the pipeline, which will run entirely underground. To date, all local ethnic groups except the Bagyeli pygmies have been compensated for land they lost to the pipeline project. Compensation for this group has proven difficult since the Bagyelis are nomadic and do not generally own land. Instead, their compensation will be based on intangibles, such as loss of access to certain plants and hunting areas. Concerning Chad, human rights groups are fearful that revenue from oil exports will not go towards needed social programs but instead will be used to finance the ongoing civil war.

Chad's new laws on poverty reduction provide arrangements for the use of oil revenues generated by the pipeline. They state that 10 per cent of the royalties and revenues are to be held in trust for future generations, 80 per cent of the remaining funds are to be devoted to education, health and social services and rural development, and 5 per cent are to be earmarked for regional development in the oil-producing area. However, in November 2000, reports surfaced indicating that the Chadian government had spent $4 million of oil-related money on arms to fight the civil war. The money was part of a $25 million "entrance fee" paid by the Doba consortium. While technically not a violation of the terms of the World Bank loan, the arms purchase embarrassed World Bank officials who had defended the Chadian government against attacks from human rights groups. The government of Chad has said that the money spent on arms will be paid back from next year's state budget.

In response to many concerns over the project, the World Bank imposed several requirements as prerequisites to the loan's approval, including: a credible oil revenue management program; extragovernmental participation (including opposition parties and the private sector) in decisions concerning the use of oil revenues; government commitments to target revenue expenditures on education, health and infrastructure; resettlement of the handful of people displaced by the project; and consultation with residents and others affected by the project. An International Advisory Group, whose members were appointed by the World Bank in February 2001, will monitor the project. The oversight body is comprised of independent international experts who will be responsible for identifying potential problems in the use of public revenues, the adequacy of civil society participation, as well as environmental management and social impacts. The Group's findings will be made public following a review by senior World Bank officials. In September 2002, the World Bank opted to continue its support for the Chad-Cameroon oil project despite criticisms in the advisory group's report.

The Chad-Cameroon project has revived interest in development of the Sedigi basin fields, whose two wells have a combined capacity of 5,000 bbl/d. The development plan for Sedigi, which was included in the MOU signed by the consortium and Chad, called for development of the fields, and a pipeline linking them to a proposed refinery located in N'Djamena. However, the Sedigi project was uncoupled from the Doba project following the reorganization of the Doba consortium in April 2000. The following month, Concorp International Limited, headed by Sudanese businessman Mohamed Abdallah Jar al Nabi, took control of the Sedigi project. The refinery, with planned capacity of 2,000–3,000 bbl/d, would produce fuel to cover local transportation needs and supply a new 16-megawatt (MW) power station in Ndjamena.

In February 1999, Chad signed an exploration agreement with three firms, U.S.-based Trinity Gas, Carlton Energy, and Nigeria's Oriental Energy Resources, for the 108-million acre (430,000 sq. km) Block H. The group plans to spend $59 million on exploration activities on the block, which is roughly the size of Kansas and Oklahoma combined.

Downstream

Chad's downstream oil sector currently is totally dependent on petroleum product imports from neighboring Nigeria and Cameroon. Local companies (35 per cent) handle distribution and marketing of petroleum in Chad, followed by Shell (25 per cent), Exxon Mobil (20 per cent) and TotalFina Elf (20 per cent). Nearly all commercial energy requirements, including power generation, are provided by petroleum. If undertaken, the Sedigi project will be able to supply nearly all of Chad's petroleum requirements, and reduce its reliance on imported petroleum.

Electricity and Renewable Energy

Only 2 per cent of Chad's households have access to electricity. Chad currently has installed electric generation capacity of 29 megawatts (MW), all of which is petroleum-fired. Electricity production depends solely on the availability of imported petroleum products. A new power plant at Farcha is expected to be commissioned during 2003, and will help to alleviate power shortages like the one Chad experienced in 1998 due to disruption in the supply of petroleum from Nigeria and Cameroon. Generation and distribution of electricity in Chad is handled by the Societe Tchadienne D'eau et D'electricite (STEE), the state-owned utility. Under terms of the IMF-sponsored structural adjustment plan, STEE was slated for privatization. In September 2001, Vivendi-Dietsman began to implement a two year agreement to operate STEE, with the option to buy a majority share in the parastatal when the contract ends.

In June 1998, the World Bank approved a $5.3 million loan for the establishment of a Household Energy Project. The project is designed to provide affordable and sustainable supply of energy to Chadian households. Wood is the primary source of total energy in Chad, and the project hopes to establish a sustainable wood fuel and charcoal supply in nearly 100 villages located near N'Djamena.

Country Overview

Independence: August 11, 1960 (from France)

Population (2002E): 8.4 million

Location/Size: Central Africa, between Libya (on the north); Sudan (on the east); Central African Republic and Cameroon (on the south); and Nigeria, and Niger (on the west) / 1,284,000 square kilometers (495,000 square miles), slightly smaller than the combined size of California, Nevada, Arizona and New Mexico.

Major Cities: N'Djamena (capital), Sarh, Moundou, Abeche

Languages: French (official), Arabic (official), Sara, and various other Nilo-Saharan, Congo-Kordofanian and Afro-Asiatic languages (between 100-200)

Major Ethnic Groups: Muslim (located in central and northern Chad): Arab, Fulbe, Kotoko, Hausa, Kanembou, Bagirmi, Boulala, Zaghawa, Hadjerai, Maba; non-Muslim (located in southern Chad): Sara, Ngambaye, Mbaye, Goulaye, Moudang, Moussei, Massa.

Religion: Muslim 50 per cent, Christian 25 per cent, traditional beliefs 25 per cent

Economic Overview

Currency: Communaute Financiere Africaine (CFA) franc

Market Exchange Rate (9/25/02): US$1 = 671.47CFA

Real Gross Domestic Product (GDP) (2000E): $1.36 billion

Real GDP Growth Rate (2001E): 8.4 per cent *(2002E):* 9.8 per cent

Inflation Rate (2001E): 12.4 per cent *(2002E):* -5.1 per cent

Current Account Balance (2001E): -$660 million

Major Trading Partners: France, Cameroon, Nigeria, Portugal, Germany, South Africa

Merchandise Exports (2001E): $191 million

Merchandise Imports (2001E): $559 million

Merchandise Trade Balance (2001E): -$368 million

Major Export Products: Cotton, livestock/meat, textiles, fish

Major Import Products: Food, petroleum products, machinery, industrial goods

Total External Debt (2000E): $1.12 billion

Energy Overview

Recoverable Oil Reserves (1999E): 1 billion barrels

Oil Production (2002E): None

Oil Consumption (2002E): 1,000 barrels per day

Natural Gas Production (2000E): None

Natural Gas Consumption (2000E): None

Electric Generating Capacity (1/1/00): 0.021 million kilowatts

Electricity Generation (2000E): 0.092 billion kilowatthours, all of which is thermal

Environmental Overview

Total Energy Consumption (2000E): 0.002 quadrillion Btu (0.1 per cent of world total energy consumption)

Energy-Related Carbon Emissions (2000E): 0.046 million metric tons of carbon (<0.1 per cent of world carbon emissions)

Per Capita Energy Consumption (2000E): 0.3 million Btu (vs. U.S. value of 351.0 million Btu)

Per Capita Carbon Emissions (2000E): 0.006 metric tons of carbon (vs. U.S. value of 5.6 metric tons of carbon)

Fuel Share of Energy Consumption (2000E): Oil (100.0 per cent), Natural Gas (0.0 per cent), Coal (0.0 per cent)

Fuel Share of Carbon Emissions (2000E): Oil (100.0 per cent), Natural Gas (0.0 per cent), Coal (0.0 per cent)

Status in Climate Change Negotiations: Non-Annex I country under the United Nations Framework Convention on Climate Change (ratified June 7th, 1994). Not a signatory to the Kyoto Protocol.

Major Environmental Issues: Inadequate supplies of potable water; improper waste disposal in rural areas contributes to soil and water pollution; desertification.

Major International Environmental Agreements: A party to Conventions on Biodiversity, Climate Change, Desertification, Endangered Species, Nuclear Test Ban, Ozone Layer Protection and Wetlands. Has signed, but not ratified: Law of the Sea and Marine Dumping.

The total energy consumption statistic includes petroleum, dry natural gas, coal, net hydro, nuclear, geothermal, solar and wind electric power.

Oil and Gas Industries

Major Oil Fields: Doba Basin (Bolobo, Kome, Miandoun), Lake Chad Basin (Kanem, Kumia, Sedigi)

Major Foreign Oil Company Involvement: ChevronTexaco, ExxonMobil, Petronas

26

Chile

Chile has limited indigenous energy resources and relies on imports to meet its rapidly growing energy demand. In recent years, Chile has emerged as one of Latin America's most successful economies. Note: Information contained in this report is the best available as of August 2003 and can change.

General Background

In 2002, exports accounted for around two-fifths of Chile's gross domestic product (GDP). Mineral resources constitute most of the country's exports, with copper alone accounting for 34 per cent. Accordingly, Chile's economic performance is highly dependent on external demand and commodity prices. In the 1990s, the Chilean government established the Copper Stabilization Fund to offset the volatility of the copper market and the loss of fiscal revenues. Recent crises in neighboring countries and downturns in the world economy, particularly in the United States, also affected Chile's economy. Chile's economic growth rates were among the world's highest in the past decade but have slowed recently. In 2002, Chile had a real domestic product growth of 2.1 per cent. In 2003, the economy is forecasted to expand by 3.1 per cent. In the first quarter of 2003, the economy recorded a growth rate of 3.5 per cent, the highest quarterly rate since 2001. Chile is considered to have one of South America's most robust and open economies. The country is a party to several bilateral trade agreements, such as the Common Market of the Southern Cone (MERCOSUR). In April 2002, Chile entered into a trade agreement with the European Union whereby both sides will eliminate all import tariffs between them over the next ten years. Chile also has been in prolonged negotiations with the United States regarding inclusion in the North American Free Trade Agreement (NAFTA). On June 6, 2003, Chile and the United Stated signed a bilateral Free Trade Agreement. The agreement requires approval from legislatures in both countries.

Energy

Chile has limited indigenous energy resources and relies on imports for most of its hydrocarbon needs. Crude oil comes primarily from Argentina, Brazil, Nigeria, and Peru, and is processed by one

of the country's three state-owned refineries. Natural gas is imported via pipelines from Argentina. A significant amount of the country's electricity is supplied domestically by hydroelectric generators, but natural gas-fired power plants are playing an increasingly important role in Chile's power production sector. Chile's energy sector is mostly privatized. Energy policy decisions are the shared responsibility of the National Energy Commission, the Ministry of the Economy, the Superintendency of Electricity and Fuels, and the Chilean Commission on Nuclear Energy.

Oil

Chile is a small oil producer. Over the last two decades, production has declined steadily from 54,000 barrels per day (bbl/d) in 1982 to 14,000 bbl/d in 2002. In contrast, consumption has increased by about 135 per cent since 1982, to 240,000 bbl/d, resulting in net imports of 226,000 bbl/d. According to the National Energy Commission (CNE), Chile's main sources of oil imports in 2002 were Argentina (74 per cent), Brazil (8 per cent), Nigeria (7 per cent), Peru (6 per cent), Venezuela (3 per cent) and Malaysia (2 per cent). As of January 2003, Chile's proven oil reserves stood at 150 million barrels.

Upstream

Chile's domestic hydrocarbon reserves are concentrated in the Magallanes basin, located on the country's southern tip, known as Tierra del Fuego. Overall, oil production in Chile has been in decline for more than twenty years, as existing wells have matured and exploration efforts elsewhere have proven unsuccessful. Empresa Nacional de Petróleo (ENAP), Chile's national oil company, is the operator in Magallanes. Over the years, the oil company has sought to establish joint venture partners to reactivate its idle and declining fields. Since 1977, ENAP has entered into 15 joint ventures.

In February 2003, ENAP received approval from the Ministry of Economy to invest $264 million into projects. This investment is part of a larger plan to double the company's value by 2006. The investments will focus on modernizing domestic oil refineries and oil exploration and production outside of Chile. ENAP's foreign exploration subsidiary, Sipetrol, is responsible for international activities. Sipetrol, in conjunction with local firms, has operations in Argentina, Colombia, Ecuador, and Egypt.

Downstream Chile has three refineries, all of which are controlled by ENAP. The largest is the *Petrox SA* facility, which has an approximate crude throughput capacity of 100,640 bbl/d, and is located near the city of Talcahuano. Petrox is supplied with imported crude oil by the Trasandino pipeline, which extends from Argentina's Neuquén Basin westward to the refinery.

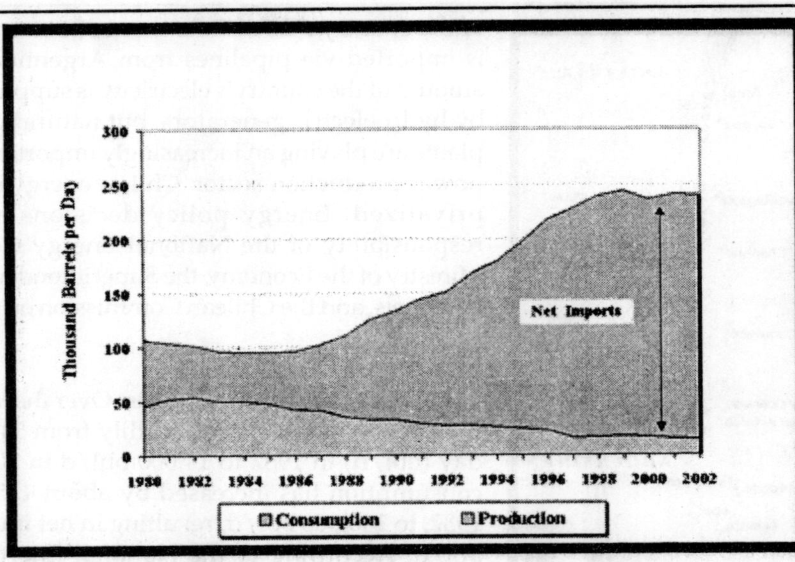

CHILEAN OIL PRODUCTION AND CONSUMPTION, 1980-2002

The country's second largest refinery is the *Refinería de Petróleo Concón,* which has a capacity of approximately 94,350 bbl/d, and is located north of Santiago. In August 2003, Brazil's state oil company, Petrobrás, announced that it was in negotiations with ENAP over potentially participating in a $300 million expansion of the Concón refinery.

Gregorio-Magallanes, the country's smallest refinery, has a capacity of 9,859 bbl/d. The refinery is located in the south and is connected to a separate pipeline system (Daniel-Gregorio) associated with the Magallanes basin.

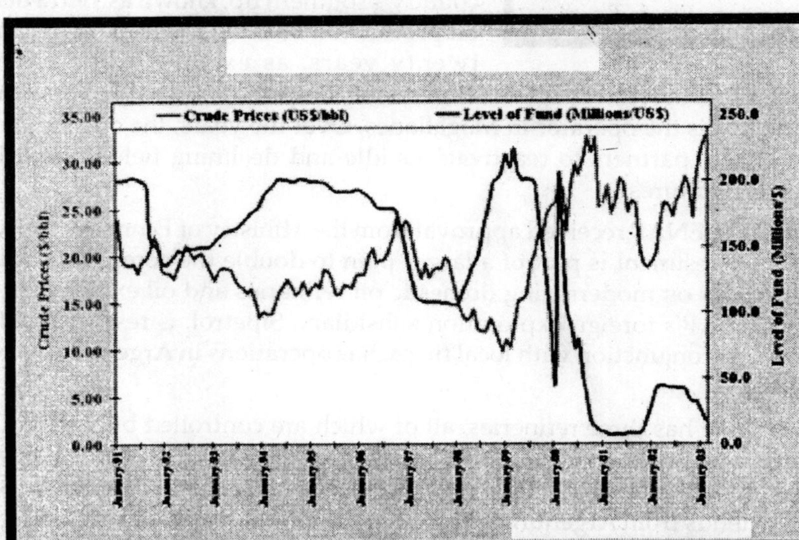

OIL STABILIZATION FUND, 1991-2003

Oil Transportation

Chile has an extensive domestic network to transport petroleum products. The country has two import pipelines, the *Trasandino* (Argentina) and the *Arica-Sica* (Bolivia) pipeline. The Trasandino pipeline is 268-miles long and has a 115,000 barrel per day capacity.

Because of Chile's growing oil import dependency, the government created the Oil Stabilization Fund (FEPP) in 1991. The fund works to stabilize prices for Chilean consumers, providing subsidies when oil prices are high and imposing taxes to bolster the fund when oil prices are low. High oil prices in past few years have stretched the fund such that fuel prices have risen several times since 2000, which is reflected in the graph above.

Natural Gas

Chile has natural gas reserves of about 3.5 trillion cubic feet (Tcf). As with oil, Chilean natural gas production has declined over the last two decades. In 2001, Chile produced 40 billion cubic feet (Bcf), down 44 per cent since 1997. Conversely, consumption has skyrocketed in the last ten years. In 1991, the country consumed 52 Bcf of natural gas, but consumption jumped to 228 Bcf in 2001, a 338 per cent increase. According to projections from Chile's National Energy Commission (CNE), consumption is expected to reach 571 Bcf by 2011. Dwindling production and rapidly expanding demand have resulted in Chile becoming a highly dependent net natural gas importer. The country imported an estimated 188 Bcf of natural gas in 2001, all of which came from Argentina.

In the 1990s, Chile began diversifying its energy mix away from hydroelectric power and towards natural gas-fired electric generation. A severe drought in 1998-1999, which resulted in blackouts, supported the government's change in energy policy. In April 2002, CNE foresaw ten new gas-fired plants and one new hydropower project by 2010. However, labor unrest in Argentina in 2002 temporarily disrupted natural gas supply to Chile. A rupture in the NorAndino pipeline also briefly

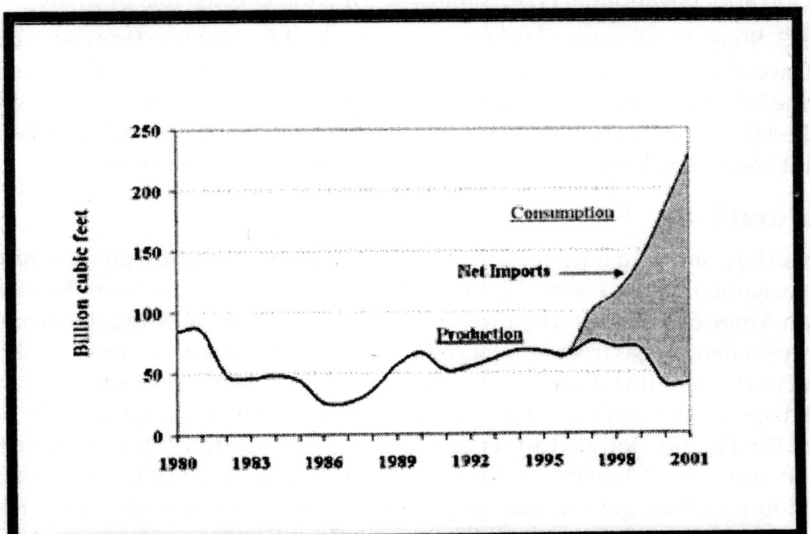

CHILEAN NATURAL GAS PRODUCTION AND CONSUMPTION, 1980-2001

interrupted natural gas supply. These events once again have prompted CNE to rethink the country's long-term energy strategy. One policy change under consideration is requiring new natural gas-fired power plants to be able to burn fuel oil.

Import Pipelines

Since 1997, four natural gas pipelines have come onstream, transporting natural gas from Argentina to Chile: *GasAtacama, Norandino, GasAndes,* and *Pacifico.* These pipelines supply Chile's urban centers, mines and power generators. Power generators are the country's largest consumers of natural gas and are projected to consume about 50 per cent of the country's supply by 2011, according to CNE.

Developments

In July 2003, the owners of *GasAtacama*, Endesa and CMS Energy, suspended plans to invest $20 million into the pipeline, citing lower than expected demand from power plants which the pipeline supplies. The pipeline began operation in June 1999. It runs 578 miles from Cornejo in Argentina to Mejillones in Chile.

In 2003, the *GasAndes* pipeline was extended 46 miles south from Santiago to a smelter in Caletones, which Chile's state copper company Codelco operates. The 280-mile pipeline was completed in 1997. It delivers natural gas from Argentina's Neuquén basin to three, 370-MW power plants around the country's capital. The GasAndes pipeline is controlled by TotalFinaElf in conjunction with Compania General de Combustibles, AES (which is looking to sell its stake), and Metrogas.

Since its inception, the *Pacifico* pipeline has been underused, mainly due to cancelled development of natural gas-fired power plant. The owners of the pipeline, El Paso, TransCanada Pipelines, Gasco, ENAP, and Repsol-YPF, have been attempting to attract power developers in order to boost throughput. As of July 2003, some investors had shown interest in building a power plant. The pipeline became operational in 1999. It stretches from the *Neuquén basin to region VIII in southern Chile*, supplying cities such as Concepción and Talcahuana. (The *Norandino* pipeline is 100 per cent owned by Tractebel).

In March 2003, Sipetrol announced that it planned to build a pipeline from Argentina in order to supply ENAP's Cabo Negro industrial complex. The complex is located near Punta Arenas in southern Chile. The pipeline is to be expected to have a capacity of 17.5 million cubic feet per day. The natural gas will be supplied from the *Poseidon project*, located offshore Tierra del Fuego, in Argentinean waters. The project is joint-owned by Repsol-YPF and Sipetrol, which is the operator.

Liquefied Natural Gas

Chile may also be a site for a liquefaction terminal for Bolivian natural gas en route to the United States and Mexico. Known as the Pacific LNG project, consortium members are Repsol-YPF, BG, and BP-controlled Pan American Energy. The project entails a $5 billion investment in transportation and port facilities to export natural gas from landlocked Bolivia. The Chilean port city of Patillos is competing with the Peruvian port city of Ilo for inclusion in the project. A decision has been postponed repeatedly, however. A land dispute between Bolivia and Chile has made the decision politically sensitive (in the 1879-1883 War of the Pacific, Bolivia lost its only outlet to the sea to Chile). A majority of Bolivians prefer a Peruvian port to a Chilean. In August 2003, Mining and Hydrocarbons Minister Jorge Berindoague announced that government is planning to hold a referendum in either October or November on whether to export natural gas through Chile or Peru.

Coal

Chile has total recoverable coal reserves of 1,302 million short tons (Mmst), and produced 0.63 Mmst in 2001. Consumption in 2001 totaled 3.53 Mmst, making Chile a net coal importer of 2.9 Mmst. Since 1997, consumption has tailed off, decreasing about 49 per cent. According to CNE, coal consumption is expected to increase. Over the years, the level of coal consumption has tended to fluctuate, as the power sector, the country's largest coal consumer, uses coal largely as a back-up to hydropower. For example, coal consumption increased significantly during the 1998-1999 droughts.

Most imported coal comes from Australia. Domestic coal production is located in the Lota/ Coronel area and in the extreme south on Tierra del Fuego. The country's largest coal mine closed in 1997, and only two small mines remain in operation.

Electricity

Chile consumed about 40.1 billion kilowatt hours (bkwh) of electricity in 2001, of which 21.5 bkwh (53 per cent) was generated by hydropower. Hydropower has historically been Chile's single largest electricity source, at times comprising over half of the country's installed electric generation capacity and production. In 1995, for example, hydropower supplied 72 per cent of electricity and represented 59 per cent of installed capacity.

From late 1997 to 1999, a severe drought crippled the country's electricity sector and caused rolling blackouts in Santiago. As a result, the Chilean government began diversifying its energy mix in order to become less reliant on hydropower. As of 2001, thermal-generated electricity represented 57 per cent of installed capacity and generated 49 per cent of the country's electricity, a significant increase from 1995 when thermal electricity sources represented 41 per cent and 26 per cent, respectively.

As mentioned in the natural gas section, the Chilean government has had to rethink its energy strategy after labor unrest in Argentina and pipeline damage resulted in blackouts in 2002. The

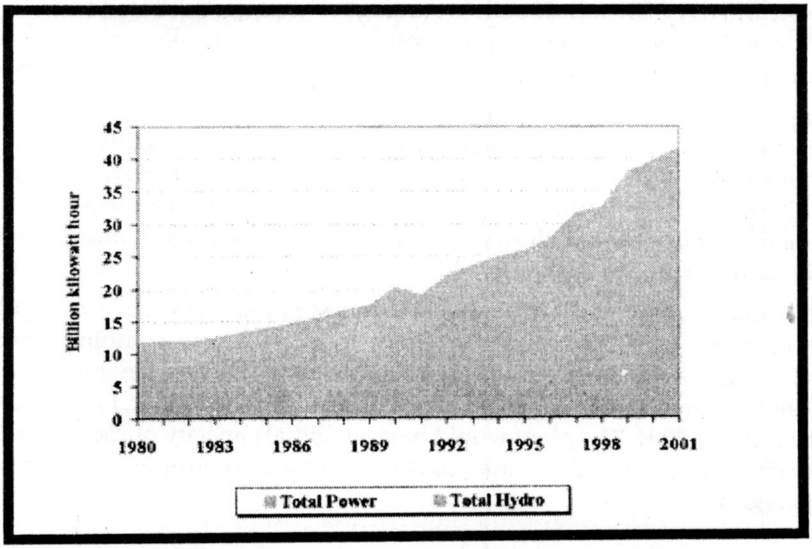

CHILEAN ELECTRICITY GENERATION, 1980-2001

government has indicated that it will seek legislation to force all gas-fired power generators to be equipped to burn oil, as well as to have oil stocks in proximity in order to avoid another future crisis.

Transmission Grids

Chile has four electric grids in operation: the Central Grid (*Sistema Interconectado Central*), which services over 90 per cent of Chile's population, generates 60 per cent of its electricity from hydropower and 40 per cent from thermal; the Northern Grid (*Sistema Interconectado del Norte Grande*), relies mainly on thermal and serves mostly mineral-processing centers in the North; and the *Asyen* and the *Magallanes* systems, which are both located in the southernmost region of the country. These latter regions are remote and sparsely populated. The Aysen system relies on a combination of thermal (64 per cent), hydro (28 per cent) and wind (8 per cent), whereas the Magallanes relies on only thermal generated electricity (83 per cent natural gas and 17 per cent diesel). The two systems' combined generation capacity was less than 1 per cent of the country's total in 2002, according to Chile's National Energy Commission.

The Economic Dispatching Center (CDEC) coordinates the systems. The CDEC is an autonomous entity composed of members from all utilities within each system to ensure efficiency and security of the electric system.

New Power Bill

In late July 2003, a new bill which would regulate electrical power transmission fees was introduced to Chile's Congress. The bill has resulted in disagreement on how much of the transmission fees the generators and consumers should pay. A new interconnector between the SIC and SING power grids, as well as a connect to Argentina is being debated.

Power Projects

With electricity demand forecasted to grow by 6 per cent annually, power companies have been developing several new projects. *Endesa* (Spain) is developing the 570-MW Ralco hydroelectric project, located on the Bio Bio river, 310 miles south of Santiago. The project will lead to an 18 per cent boost in production to the SIC's current power supply. Completion was originally expected in 2003, but protests from four Pehuenche Indian women who refused to be relocated have delayed completion.

AESGener postponed for one year (until December 2003) the construction of a proposed 740-MW, gas-fired project in Totihue in central Chile. The company cited difficulty in receiving regulatory approval, stemming from stiff environmental opposition to the project. However, many fear that if the plant is not built, central Chile could face power supply shortages in the near future. Only two other plants are being built the region: Endesa's 570-MW Ralco hydro project, and Colbun's expansion of the existing Nehuenco from 380 MW to 800 MW, which is expected to be completed in 2004.

In July 2003, *Colbun* announced that it plans to invest $1 billion in order to double the company's generation capacity by 2010. The company plans to build a new combined cycle power plant, Candelarai, in Codegue, which is south of Santiago. The plant will have an initial capacity of 250 MW, which would later be expanded to 750 MW. An additional gas-fired plant, Los Pinos, is planned, with construction to begin in 2004. The plant would have an initial capacity of 250 MW, which would later be expanded to 750 MW. Colbun also is intending to increase its hydropower capacity by 200-300 MW in the Santiago region.

Environment

The primary environmental threats to Chile are air pollution from vehicle and industrial emissions, water pollution from untreated industrial sewage, deforestation and soil erosion. Air pollution in Santiago is the most obvious and severe environmental problem in Chile.

Mitigating threats to the environment, however, is the increasing use of alternative fuels in Chile's industrial and energy sectors. Reliance on natural gas and hydroelectric generation to power the country has kept total carbon emissions in check over the past decade.

The continuing evolution of Chile's fuel mix away from petroleum and coal is key to the country's environmental future. With total energy demand expected to continue to grow by 6 per cent annually, and air pollution in Santiago already reaching critical levels, energy policies in Chile have commanded the attention of the country's current leadership, international donor organizations, and the private sector.

Country Overview

Independence: September 18, 1810 (from Spain)

Population (2002E): 15.7

Location/Size: Southern South America/757,000 sq km (292,000 sq mi), slightly smaller than twice the size of Montana

Major Cities: Santiago (capital), Concepción, Valparaíso, Antofagasta, Puerto Montt, Punta Arenas

Languages: Spanish

Ethnic Groups: White and White-Amerindian 95 per cent, Amerindian 3 per cent, other 2 per cent

Religions: Roman Catholic (89 per cent), Protestant (11 per cent)

Defense (8/98E): Army (51,000), Air Force (13,500), Navy (30,000), Paramilitary Security Forces (29,500)

Economic Overview

Central Bank President: Carlos Massad

Currency: Peso

Market Exchange Rate (08/13/03): US$1 = 723 Pesos

Nominal Gross Domestic Product (GDP, 2002E): $66.2 billion *(2003F):* $68 billion

Real GDP Growth Rate (2002E): 2.1 per cent *(2003F):* 3.1 per cent

Inflation Rate (consumer prices, 2002E): 2.5 per cent *(2003F):* 3.4 per cent

Major Export Products: Copper, cellulose, chemicals, salmon, wood products, mollusks, wine

Major Export Destinations: United States, Japan, United Kingdom, China

Major Import Products: Raw materials excluding petroleum, crude petroleum, capital goods, consumer goods

Major Import Origins: United States, Argentina, Brazil, China

Unemployment Rate (2002E): 7.7 per cent

Trade Balance: (2002E): \$2.5 million *(2003E)*: \$2.6 million

Total Foreign Debt (2002E): \$34.2 billion *(2003E)*: \$34.1 billion

Energy Overview

Proven Oil Reserves (1/1/03E): 150 million barrels

Oil Production (2002E): 14,000 barrels per day (bbl/d)

Oil Consumption (2002E): 240,000 bbl/d

Net Oil Imports (2002E): 226,000 bbl/d

Natural Gas Reserves (1/1/03E): 3.5 trillion cubic feet (Tcf)

Natural Gas Production (2001E): 40 billion cubic feet (Bcf)

Natural Gas Consumption (2001E): 228 Bcf

Net Natural Gas Imports (2001E): 188 Bcf

Recoverable Coal Reserves (2001E): 1.3 billion short tons

Coal Production (2001E): 0.63 million short tons (Mmst)

Coal Consumption (2001E): 3.53 Mmst

Electric Generation Capacity (2001E): 9.7 million kilowatts (4.13 of which is hydropower)

Electricity Generation (2001E): 41.6 billion kilowatthours (bkwh, of which 21.5 bkwh was hydropower)

Electricity Consumption (2001E): 40.1 Bkwh (of which 21.5 bkwh was hydropower)

Environmental Overview

Total Energy Consumption (2001E): 1.06 quadrillion Btu (0.26 per cent of world total energy consumption)

Energy-Related Carbon Emissions (2001E): 14.75 million metric tons of carbon (0.22 per cent of world carbon emissions)

Per Capita Energy Consumption (2001E): 68.9 million Btu (vs. U.S. value of 341.8 million Btu)

Per Capita Carbon Emissions (2001E): 0.96 metric tons of carbon (vs. U.S. value of 5.5 metric tons of carbon)

Energy Intensity (2001E): 12,942 Btu/ \$1995 (vs. U.S. value of 10,736 Btu/ \$1995)**

Carbon Intensity (2001E): 0.18 metric tons of carbon/thousand \$1995 (vs. U.S. value of 0.17 metric tons/thousand \$1995)

Fuel Share of Energy Consumption (2001E): Oil (46 per cent), Natural Gas (23 per cent), Coal (8.5 per cent)

Fuel Share of Carbon Emissions (2001E): Oil (61 per cent), Natural Gas (24 per cent), Coal (15 per cent)

Status in Climate Change Negotiations: Non-Annex I country under the United Nations Framework Convention on Climate Change (ratified December 22nd, 1994). Signatory to the Kyoto Protocol (June 17th, 1998 and ratified on August 26, 2002).

Major Environmental Issues: Air pollution from industrial and vehicle emissions; water pollution from raw sewage; deforestation contributing to loss of biodiversity; soil erosion; desertification.

Major International Environmental Agreements: A party to the Antarctic-Environmental Protocol, Antarctic Treaty, Biodiversity, Climate Change, Desertification, Endangered Species, Environmental Modification, Hazardous Wastes, Law of the Sea, Marine Dumping, Nuclear Test Ban, Ozone Layer Protection, Ship Pollution, Wetlands and Whaling.

The total energy consumption statistic includes petroleum, dry natural gas, coal, net hydro, nuclear, geothermal, solar, wind, wood and waste electric power. The renewable energy consumption statistic is based on International Energy Agency (IEA) data and includes hydropower, solar, wind, tide, geothermal, solid biomass and animal products, biomass gas and liquids, industrial and municipal wastes. Sectoral shares of energy consumption and carbon emissions are also based on IEA data.

Oil and Gas Industries

Organization: Largely privatized. Empresa Nacional de Petróleo (ENAP) is the national oil and gas company. The National Energy Commission (NEC) is responsible for overall energy planning and tariff regulation.

Major Ports: Santiago, Puerto Montt, Concepción, Valparaíso

Major Oil and Gas Fields: Posesión, Daniel Este-Dungeness, Skua, Spiteful

Major Refineries (crude oil capacity): Petrox–Talcahuano (100,640 bbl/d), Refinería de Petróleo–Concón (94,350 bbl/d), Gregorio-Magallanes (9,859 bbl/d)

27

China

The People's Republic of China (China) is the world's most populous country and the second largest energy consumer (after the United States). Production and consumption of coal, its dominant fuel, is the highest in the world. Rising oil demand and imports have made China a significant factor in world oil markets. China is expected to surpass Japan as the world's second-largest petroleum consumer for 2003.

General Background

China is the world's most populous country, with a rapidly growing economy. Economic development has proceeded unevenly, with urban coastal areas, particularly in the Southeast, experiencing more rapid economic development than other areas of the country. China has a mixed economy, with a combination of state-owned and private firms. A number of state-owned firms have undergone partial or full privatization in recent years. The Chinese government has encouraged foreign investment–in some sectors of the economy and subject to constraints–since the 1980s, offering several "special economic zones" in which foreign investors receive preferable tax, tariff, and investment treatment.

In March 2003, a long-expected transition in China's political leadership took place. Hu Jintao assumed the country's presidency, as well as chairmanship of the ruling Communist Party. Wen Jinbao became the new premier. Former president Jiang Zemin has retained the chairmanship of the Central Military Commission.

With China's entry into the World Trade Organization (WTO) in November 2001, the Chinese government made a number of specific commitments to trade and investment liberalization which, if fully implemented, will substantially open the Chinese economy to foreign firms. In the energy sector, this will mean the lifting or sharp reduction of tariffs associated with imports of some classes of capital goods, and the eventual opening to foreign competition of some areas such as retail sales of petroleum products.

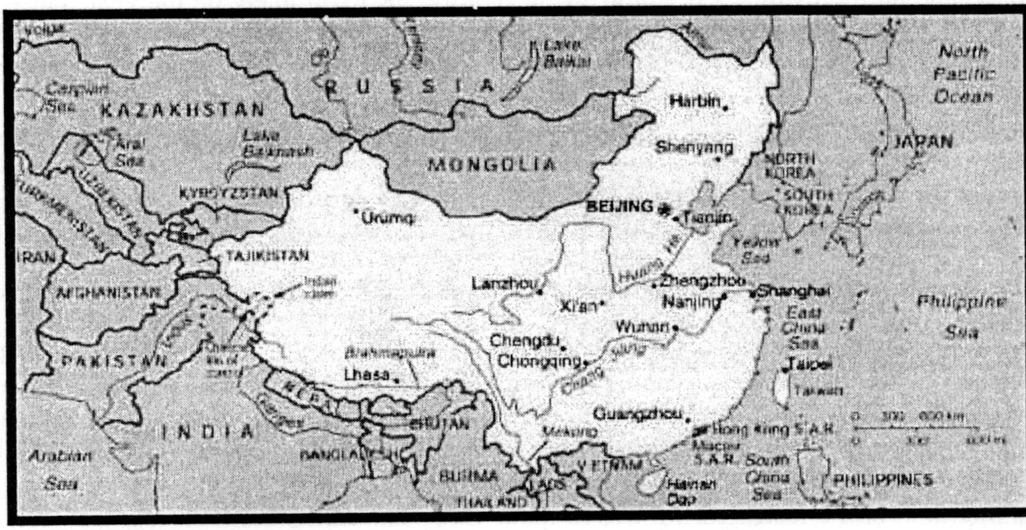

Despite moves toward privatization, much of China's economy remains controlled by large State Owned Enterprises (SOE's), many of which are inefficient and unprofitable. Restructuring of the SOE sector, including the privatization of some enterprises, is a major priority of the government, as is restructuring of the banking sector. Many Chinese banks have had to write off large amounts of delinquent debts from state-owned enterprises.

Layoffs have been part of the restructuring of the SOEs, as many were severely overstaffed. This has created unemployment, and also has been a burden on the government budget, as the government begins to provide social benefits which were previously the responsibility of the SOEs. The geographic concentration of privately-owned industry in the urban centers along the coast also has created social strains.

China's real GDP grew by 8.0 per cent in 2001, according to official Chinese figures, though some outside analysts have questioned the reliability of China's official economic data. Real GDP growth for 2003 is forecast at 7.1 per cent, with the decline taking into account the economic disruptions caused by the SARS epidemic in the second quarter. The Chinese government's current Five Year Plan (2001-2005) sets a target of 7.0 per cent real annual GDP growth.

Inflows of Foreign Direct Investment (FDI) into China in 2002 totalled $52.7 billion, a new record, and data from the first four months of 2003 shows continuing strength. Japan, Taiwan, and the United States are China's most important sources of FDI.

After three years of decline from 1999 to 2001, China's trade surplus has rebounded. The 2002 trade surplus was $50.0 billion, up from $30.3 billion in 2001. Imports have been increasing, largely capital goods being acquired to refurbish outdated industrial facilities, but it was offset by a strong 19.6 per cent surge in merchandise exports in 2002.

Oil

China was the world's third largest consumer of petroleum products in 2002, following the United States and Japan, with total demand of 5.26 million barrels per day (bbl/d). Since Japan's demand is stagnant and China's is still growing rapidly, 2003 will likely be the year that China surpasses Japan in petroleum consumption. China's oil demand is projected by EIA to reach 10.9 million bbl/d by 2025, with net imports of 7.5 million bbl/d, making it a major factor in the world oil market.

China's petroleum industry has undergone major changes in recent years. In 1998, the Chinese government reorganized most state owned oil and gas assets into two vertically integrated firms–the China National Petroleum Corporation (CNPC) and the China Petrochemical Corporation (Sinopec). Before the restructuring, CNPC had been engaged mainly in oil and gas exploration and production, while Sinopec had been engaged in refining and distribution. This reorganization created two regionally-focused firms, CNPC in the north and west, and Sinopec in the south, though CNPC is still tilted toward crude oil production and Sinopec toward refining. Other major state sector firms in China include the China National Offshore Oil Corporation (CNOOC), which handles offshore exploration and production and accounts for more than 10 per cent of China's domestic crude production, and China National Star Petroleum, a new company which was created in 1997. Regulatory oversight of the industry now is the responsibility of the State Energy Administration (SEA) which was created in early 2003.

The intention of the restructuring was to make these state firms more like similar vertically integrated corporate entities elsewhere. In connection with this process, the firms have been spinning off or eliminating many unprofitable ancillary activities such as running housing units, hospitals, and other services near company facilities. Massive layoffs also have been undertaken, as like many other Chinese SOEs, they were severely overstaffed.

The three largest Chinese oil and gas firms–Sinopec, CNPC, and CNOOC–all have successfully carried out initial public offerings (IPOs) of stock between 2000 and 2002, bringing in billions of dollars in foreign capital. CNPC separated out most of its high quality assets into a subsidiary called PetroChina in early 2000, and carried out its IPO of a minority interest on both the Hong Kong and New York stock exchanges in April 2000. The IPO raised over $3 billion, with BP the largest purchaser at 20 per cent of the shares offered. Sinopec carried out its IPO in New York and Hong Kong in October 2000, raising about $3.5 billion. Like the PetroChina IPO, only a minority stake of 15 per cent was

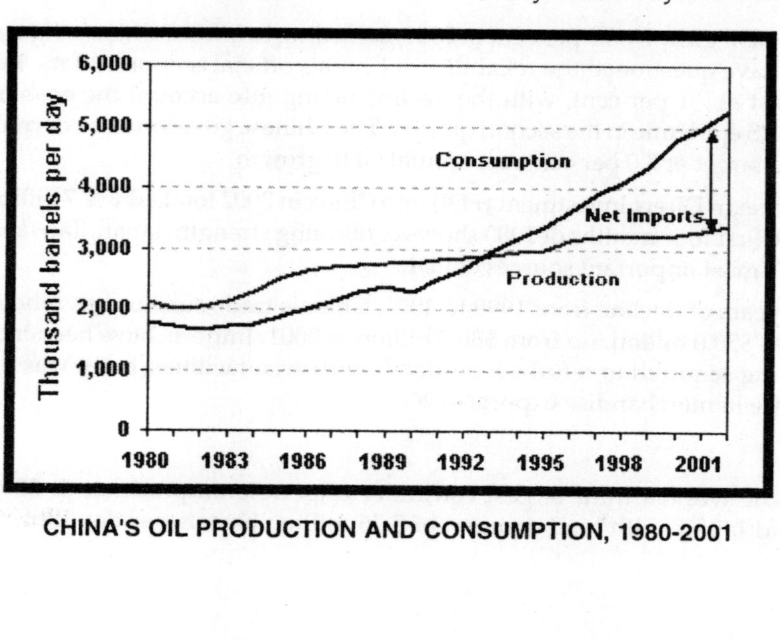

CHINA'S OIL PRODUCTION AND CONSUMPTION, 1980-2001

offered. About $2 billion of this amount was purchased by the three global super-majors–ExxonMobil, BP, and Shell. CNOOC held its IPO of a 27.5 per cent stake in February 2001, after an earlier attempt in September 1999 was canceled. Shell bought a large block of shares valued at around $200 million. In 2002, Chinese oil companies began to look at separating out some of their business units into subsidiaries. CNPC has set up subsidiaries for drilling services and geological survey work, and plans to eventually spin them off through international IPOs. CNOOC also has created an oilfield services unit–China Oilfield Service, Ltd. (COSL)–which was listed on the Hong Kong stock exchange in November 2002.

Several aspects of these stock offerings were very atypical. First, they all involved only minority stakes. Second, they have not given the foreign investors a major voice in corporate governace. The Chinese government still holds majority stakes in all three firms, and the foreign investors have not received seats on their boards of directors. Analysts have generally seen these investments as attempts by the supermajors to gain a foothold in China, which will necessarily involve partnerships with the Chinese majors. Even with the opening to foreign investment envisioned in China's commitments for membership in the WTO, it is still likely that almost all major oil and gas projects in China will involve one of the Chinese majors. The Chinese government stipulated in July 2001 that only CNPC and Sinopec will be allowed to open new retail filling stations prior to fulfillment of China's market-opening commitment in 2004. This is seen as an attempt to strengthen their control of retail sales of petroleum products and ensure that foreign firms will have to partner with one or the other of the Chinese majors to enter the retail market, even after 2004. All three of the global supermajors, BP, ExxonMobil, and Shell, are planning to enter the Chinese retail market in partnership with CNPC, Sinopec, or both.

As a net oil importer since 1993, China's petroleum industry is focused on meeting domestic demand, but it does still export a modest amount of crude oil. The largest export customer by far is Japan, which imports Daqing crude oil to burn directly in electric power plants. As of early 2003, China's exports of Daqing crude oil to Japan were around 65,000 bbl/d, down substantially from export levels during the 1990s, but up from the previous year due to problems with Japan's nuclear power sector.

Most Chinese oil production capacity, close to 90 per cent, is located onshore. One field alone, Daqing in northeastern China, accounts for about 1.0 million bbl/d of China's production, out of a total crude oil production of around 3.3 million bbl/d. Daqing is a mature field, however, having begun production in 1963, and production fell by nearly 3 per cent in 2002. At China's second-largest producing field, Liaohe in northeastern China, CNPC has solicited proposals from potential foreign partners to help it enhance recovery rates and extend production, though no contracts have yet been signed. In December 2000, regulatory changes were announced which will remove some of the barriers to foreign firms forming partnerships with Chinese oil majors. Government priorities focus on stabilizing production in the eastern regions of the country at current levels, increasing production in new fields in the West, and developing the infrastructure required to deliver western oil and gas to consumers in the East. Offshore development also is a high priority. Chinese officials have said that they expect production in Xinjiang to reach 1 million bbl/d by 2008, but that seems ambitious, given that transportation of that oil to consumers in the East remains a major obstacle.

Recent offshore oil exploration interest has centered on the Bohai Sea area, east of Tianjin, believed to hold more than 1.5 billion barrels in reserves, and the Pearl River Mouth area. Phillips Petroleum announced in March 2000 that it had completed its appraisal drilling of the Peng Lai find in Block 11/05, and would proceed with development. Commercial production began in December 2002, and the field is expected to reach its full output of 100,000 bbl/d in 2004. Shell and CNOOC currently are negotiating over a possible production charing agreement for the Bonan project in the Bohai Sea, after a successful exporation effort which began in 2001. CNOOC also signed a production sharing contract with Canadian independent Husky Oil in July 2001 for Block 39-05 in the Pearl River Mouth, near the Wenchang 13-1/13-2 blocks, where Husky Oil and CNOOC currently are producing about 50,000 bbl/d. Another major offshore oilfield has been developed in the Pearl River Mouth area by a consortium including ChevronTexaco, ENI, and CNOOC. The field began production in February 1999. ChevronTexaco also concluded an agreement with CNOOC in October 2002 for the development of the Bozhong field in the Bohai Sea, which has reserves estimated at 1.3 billion barrels. Meanwhile, improvement in Sino-Vietnamese relations has opened the way for oil and gas exploration in the Beibu Gulf (known in Vietnam as the Gulf of Tonkin). China and Vietnam signed an agreement in December 2000 which settled their outstanding disputes over sovereignty and economic rights in offshore areas near their border.

With China's expectation of growing future dependence on oil imports, China has been acquiring interests in exploration and production abroad. CNPC has acquired oil concessions in Kazakhstan, Venezuela, Sudan, Iraq, Iran, and Peru, and Azerbaijan. Sinopec also has begun seeking to purchase overseas upstream assets. The most significant deal thus far is CNPC's acquisition of a 60 per cent stake in the Kazakh oil firm Aktobemunaigaz, which came with a pledge to invest significantly in the company's future development over the next twenty years. While there had been some discussion of a possible oil pipeline from Kazakhstan to China, CNPC has said that it would only be considered if reserves were sufficient and it was economical, which looks doubtful. CNPC's position in Kazakhstan also suffered a major blow in May 2003, when the consortium partners in the Kashagan oilfield in the Caspian exercised their rights to block the sale of a 16.7 per cent interest in the project from BG to CNPC. The Greater Nile Petroleum Operating Company (GNPOC), the Sudanese oil project in which CNPC owns a stake, began exports in August 1999. CNOOC also has purchased an upsteam equity stake in the small Malacca Strait oilfield in Indonesia.

Russia's Far East is seen as a potential source of Chinese crude oil imports. The Russian and Chinese governments have been holding regular discussions on the feasibility of pipelines to make such exports possible. One proposed plan is a pipeline which would carry 600,000 bbl/d of crude oil from Anagarsk in Russia to join the existing Chinese pipeline network at Daqing. Yukos Oil of Russia and CNPC signed a memorandum of understanding in June 2003 for sales of oil via the pipeline, contingent on the pipeline being built. An alternative plan, proposed by Russian pipeline operator Transneft, would take Russian crude from both West Siberia and East Siberia via a 1 million bbl/d pipeline to an export terminal at the Pacific coast port of Nakhodka. China presumably would be one of the major consumers of oil from such a project, but it would also give Russia increased access to the Japanese, South Korean, and other East Asian markets. Both options, or possibly both eventually if oil reserves are sufficient, are still officially under consideration, according to recent Russian government statements. Japan has been actively promoting the Nakhodka option, offering to assist with financing, but the line to Daqing appears to be more likely to be built in the near future.

Downstream infrastructure development in China centers primarily on upgrading existing refineries rather than building new ones, due to overcapacity. In the late 1990s, the Chinese government shut down 110 small refineries, which generally made inferior quality petroleum products. Dozens of other small refineries owned by provincial and local governments have been merged into CNPC and Sinopec. Another major issue in the Chinese downstream sector is the lack of adequate refining capacity suitable for heavier Middle Eastern crude oil, which will become a necessity as Chinese import demand rises in the mid-term future. Several existing refineries are being upgraded to handle heavier and more sour grades of crude oil.

Chinese officials have spoken of their intention to build a national strategic petroleum reserve, and Chinese officials announced a policy decision in February 2003 to support the creation of a strategic petroleum reserve, and have reportedly been studying several options for the development of storage capacity. In the meantime, anecdotal evidence has suggested that China may have built up its petroleum stocks earlier this year in anticipation of possible war in Iraq.

Natural Gas

Historically, natural gas has not been a major fuel in China, but given China's domestic reserves of natural gas, which stood at 53.3 trillion cubic feet (Tcf) at the beginning of 2003, and the environmental benefits of using natural gas, China has embarked on a major expansion of its gas infrastructure. Until the 1990s, natural gas was used largely as a feedstock for fertilizer plants, with little use for electricity generation. Natural gas currently accounts for only around 3 per cent of total energy consumption in China, but consumption is expected to more than double by 2010. This will involve increases in domestic production, and imports, by pipeline and in the form of liquefied natural gas (LNG).

The country's largest reserves of natural gas are located in western and north-central China, necessitating a significant further investment in pipeline infrastructure to carry it to eastern cities. China has a pipeline under construction, the "West-to-East Pipeline," from natural gas deposits in the western Xinjiang province to Shanghai, picking up additional gas in the Ordos Basin along the way. Shell was chosen in February 2002 as the lead firm for the project, and Gazprom and ExxonMobil hold significant stakes. Sinopec also has been added as an equity partner. Construction began in July 2002, and a section of the pipeline east of the Ordos Basin is scheduled to begin operation in early 2004. The segment connecting to Xinjiang will be completed by early 2005. While it is unlikely to happen in the near future, the West-to-East Pipeline eventually could serve as a trunkline which could be extended to receive natural gas from Central Asia.

China announced the discovery of a major gas field at Sulige in the Ordos Basin in the Inner Mongolia Autonomous Region, adjacent to the Changqing oilfield, in 2001. While the field is still under evaluation, unofficial reserve estimates cited in the trade press put reserves in the range of 16-21 Tcf, substantially more than was assumed when the discovery was first announced. Some natural gas from the Ordos Basin is likely to be put into the West-to-East Pipeline, which was to run through the area in any case, to help make it economically viable. A pipeline was completed in 1997 between the Ordos Basin and Beijing, and a second pipeline may become necessary, as demand for natural gas in Beijing, Tianjin, and nearby Hebei province already is outstripping the capacity of the original pipeline.

Another proposed pipeline project would link the Russian natural gas grid in Siberia to China and possibly South Korea via a pipeline from the Kovykta gas fields near Irkutsk, which hold reserves of more than 50 Tcf. The cost of the project has been estimated at $12 billion, and a feasibility study is underway, due to be completed at the end of July 2003. The pipeline would have a planned capacity of 2.9 billion cubic feet per day (Bcf/d), of which China would likely consume about 1.9 Bcf/d and South Korea 1 Bcf/d. The main South Korea gas company, Kogas, formally joined the feasibility study in November 2000. The main foreign backer of the project is BP, which owns a 30 per cent stake in Rusia Petroleum, the license holder for the Kovykta gas field. Due to tensions on the Korean peninsula, the route currently under consideration for the section of the pipeline to South Korea would bypass North

Korea by running undersea from the city of Dalian in China to the South Korean coast near Seoul. The new route also would bypass Mongolia.

Aside from these huge projects, other pipelines are being developed to link smaller natural gas deposits to other consumers. A pipeline was completed in early 2002 linking the Sebei natural gas field in the Qaidam Basin with consumers in the city of Lanzhou. Another planned project would link gas deposits in Sichuan province in the southwest to consumers in Hubei and Hunan provinces in central China at an estimated cost of $600 million.

One major hurdle for natural gas projects in China is the lack of a unified regulatory system. Currently, natural gas prices are governed by a patchwork of local regulations. The Chinese government is in the process of drafting a new legal framework for the natural gas sector, but the process has been slow, and there are still considerable uncertainties regarding price regulation and taxation issues dealing with natural gas sales.

Offshore gas projects also are becoming a significant part of China's gas supply. The Yacheng 13-1 field, developed in the mid-1990s, has been producing gas for Hong Kong and Hainan Island since 1996. The Chunxiao gas field in the East China Sea, being developed by China National Star Petroleum, is also expected to become a significant producer within the next decade. The company puts the field's reserves at more than 1.6 Tcf. Another area where where exploratory drilling is planned is the Xihu Trough, in the East China Sea about 250 miles east of Shanghai. Shell reportedly has been in negotiations with Sinopec and CNOOC for the development of natural gas reserves in the area, but no agreement has been concluded.

Imported liquefied natural gas (LNG) will be used primarily in China's southeastern coastal region. Guangdong province already has launched a project to build six, 320-megawatt (MW) gas-fired power plants, and to convert existing oil fired plants with a capacity of 1.8 gigawatts (GW) to LNG. In March 2001, it was announced that BP had been selected to build China's first LNG import terminal, to be located near the city of Guangdong. BP will take a 30 per cent equity stake in the project, with CNOOC holding 31 per cent and the rest held by local firms from Guangdong and Hong Kong. A supply contract has been signed for LNG from Australia's North West Shelf LNG terminal. The project has been delayed somewhat, due to slow progress in concluding sales agreements with end-users of the natural gas. It is likely that the commercial operation of the project will be delayed until early 2007. A second LNG terminal is planned for Zhangzhou, in Fujian province farther up the coast. A supply agreement has been concluded with BP for LNG from its Tangguh project in Indonesia. A third LNG import project is under consideration for a startup date around 2010, but it is in the preliminary stages and has not secured government approval. If built, it would likely be located somewhere near the Yangtze River Delta.

Coal

Coal makes up the bulk, 64 per cent, of China's primary energy consumption, and China is both the largest consumer and producer of coal in the world. China's coal consumption in 2001 was 1.38 billion short tons, or over 26 per cent of the world total. The Chinese government has made major upward revisions to coal production and consumption figures covering the last several years. The new figures show coal consumption rising sharply in 2001, reversing the decline seen from 1997 to 2000. The decline during that period also is much less than the previously reported data.

China's coal industry has had a serious oversupply problem in recent years, particularly in the late 1990s, and the government has begun implementing major reforms aimed at reducing the oversupply, returning large state-owned mines to profitability as a prelude to possible future privatization, and reducing mine accidents. Large state-owned coal mines had experienced buildups of unused inventories in the mid-to-late-1990s, and many were operating at a financial loss. A large number of small, unlicensed mines also have added to the oversupply. In 1998, the government launched a large-scale effort to close down the small mines. Many small coal mines were ordered closed. It has become clear, however, through much anecdotal evidence, that not all of the "closed" mines have actually ceased operation, and the upward revisions to the Chinese State Statistical Bureau's production and consumption figures appears to reflect this. China also is increasingly seeking export markets for its coal as a way of dealing with its surplus production, and as of 2002 it was the world's second-largest coal exporter. Japan and South Korea are the primary markets, and China is beginning to emerge as a serious competitor to Australia for Japanese coal imports. India also has been importing modest quantities of Chinese coal.

Over the longer term, China's coal demand is projected to rise significantly. While coal's share of overall Chinese energy consumption is projected to fall, coal consumption will still be increasing in absolute terms. Several projects exist for the development of coal-fired power plants co-located with large mines, so called "coal by wire" projects. Other technological improvements also are being undertaken, including the first small-scale projects for coal gasification, and a coal slurry pipeline to transport coal to the port of Qingdao. Coalbed methane production also is being developed, with recent American investors in this effort including BP, Texaco, and Virgin Oil, which was awarded a concession for exploration in Ningxia province in January 2001. ChevronTexaco is the largest foreign investor in coalbed methane, with activities in several provinces.

In contrast to the past, China is becoming more open to foreign investment in the coal sector, particularly in modernization of existing large-scale mines and the development of new ones. The China National Coal Import and Export Corporation is the primary Chinese partner for foreign investors in the coal sector. Areas of interest in foreign invesment concentrate on new technologies only recently introduced in China or with environmental benefit, including coal liquefaction, coal bed methane production, and slurry pipeline transportation projects. Over the longer term, China plans to aggregate the large state coal mines into seven corporations by the end of 2005, in a process similar to the creation of CNPC and Sinopec out of state assets. Such firms might then seek to pursue foreign capital through international stock offerings.

China has expressed a strong interest in coal liquefaction technology, and would like see liquid fuels based on coal substitute for some of its petroleum demand for transportation. A coal liqufaction facility is under construction by the Shenhua Group in Inner Mongolia, with a projected startup date of 2005.

Electricity

As with coal, China's electric power industry experienced a serious oversupply problem in the late 1990s, due largely to demand reductions from closures of inefficient state-owned industrial units, which were major consumers of electricity. The Chinese government responded to the short-term oversupply in part by implementing a drive to close down small thermal power plants and by imposing a moratorium (with a few exceptions) on approval of new power plant construction, which ran through January 1, 2002. Until recently, the backlog of projects approved in the mid-1990s had kept pace with demand increases. In the first half of 2003, however, the Chinese government has approved 30 major new electric power projects, with a total of around 22 gigawatts (GW) of capacity. Construction has begun on 17 of these projects. A total of 18.5 GW of new capacity is scheduled to be completed this year.

The largest project under construction, by far, is the Three Gorges Dam, which, when fully completed in 2009, will include 26 separate 700-MW generators, for a total of 18.2 GW. Plans were announced in March 2002 to reorganize the Three Gorges project into the China Yangtze Three Gorges Electric Power Corporation. The corporation is expected to seek capital through an equity offering open to foreign and domestic investors, similar to those already carried out by the major Chinese oil companies. The IPO has been scheduled for September 2003. The reservoir created by the dam began to fill in June 2003, and the first test runs of the initial group of electric turbines is set for August 2003.

Another large hydropower project involves a series of dams on the upper portion of the Yellow River. Shaanxi, Qinghai, and Gansu provinces have joined to create the Yellow River Hydroelectric Development Corporation, with plans for the eventual construction of 25 generating stations with a combined installed capacity of 15.8 GW.

Many of the major developments taking place in the Chinese electricity sector recently involve nuclear power. China's total installed capacity for nuclear power generation increased from 2.1 GW at the beginning of of 2002 to 5.4 GW at the beginning of 2003. The first generation unit of the Lingao nuclear power plant in Guangdong province began commercial operation in May 2002, with a capacity of 1-GW. The second 1-GW generating unit began operating in January 2003. An additional 600-MW generating unit at the Qinshan nuclear power plant in Zhejiang province began operation in February 2002, and another 600-MW unit at the same site came online in December 2002.

A major issue for China's electric power industry is the distribution of generation among power plants. China's stated intention eventually is to create a unified national power grid, and to have a modern power market in which plants sell power to the grid at market-determined rates. In the short term, though, traditional arrangements still hold sway, and state-owned power plants which have government connections tend to have a higher priority than independent private plants. Additionally, some private plants with "take-or-pay" contracts, which provide for guaranteed minimum sales amounts, have had trouble getting the provincial authorities running the local grids to honor those terms.

Growth in Chinese electricity consumption is projected at an average of 4.3 per cent per year through 2025. The largest gainer in terms of fuel share in the future is expected to be natural gas, due largely to environmental concerns in China's rapidly industrializing coastal provinces, though the largest increase in absolute terms is likely to be coal. If a truly competitive market for electric power develops as planned, the Chinese market may once again become attractive to foreign investment. At present, foreign direct investment is allowed only in power generation, but loan financing has been obtained for some power transmission projects.

The Chinese government is in the early stages of formulating a fundamental long-term restructuring of their electric power sector, embodies in the National Power Industry Framework Reform Plan promulgated by the State Council in April 2002. As with many other countries reform programs, generating assets are being largely separated from transmission and distribution. The State Power Corporation (SPC) divested most of its generating assets and was split into 11 regional transmission and distribution companies in December 2002. Electricity prices will still be regulated, but there are likely to be major changes in tarriffs and the overall regulatory structure for electricity pricing. The process is at an early stage, and many of the details remain to be worked out. A new electricity law, superseding the one established in 1995, is expected to be probulgated within the next year.

Environment

China suffers from major energy-related environmental problems. According to a report by the World Health Organization (WHO), seven of the world's ten most polluted cities are in China. The country's heavy use of unwashed coal leads to large emissions of sulfur dioxide and particulate matter. China also is important to any effort to curb emissions of greenhouse gases, as it is projected to experience the largest absolute growth in carbon dioxide emissions between now and the year 2020.

China is a non-Annex I country under the United Nations Framework Convention on Climate Change, meaning that it has not agreed to binding targets for reduction of carbon dioxide emissions under the Kyoto Protocol. While the Chinese government is concerned with its environmental problems, it tends to be more concerned with local problems, such as particulate matter and sulfur dioxide emissions. Thus, it is undertaking efforts to lessen emissions of pollutants such as sulfur dioxide and nitrogen oxide, through improved pollution controls on power plants as well as policies designed to increase the share of natural gas in the country's fuel mix, particularly around major metropolitan areas.

Country Overview

Population (July 2002E): 1.3 billion

Location/Size: Eastern Asia/3.7 million square miles (9.6 million square kilometers, slightly smaller than the United States)

Major Cities: Beijing (capital), Shanghai, Tianjin, Guangzhou, Shenyang, Wuhan, Chengdu, Hong Kong

Languages: Mandarin (official), many local dialects

Ethnic Groups: Han Chinese (92 per cent); Zhuang, Uygur, Hui, Yi, Tibetan, Miao, Manchu, Mongol, Buyi, Korean, others (8 per cent)

Religion: Officially atheist; Daoism, Buddhism, Muslim (2-3 per cent); Christian (1 per cent)

Defense (8/98): Army (2.1 million), Navy (260,000), Air Force (470,000), reserves (1.2 million), People's Armed Police (1 million)

Economic Overview

Currency: Yuan

Exchange Rate (6/25/03): US$1 = 8.3 Yuan/Renminbi

Gross Domestic Product (2002E): $1.24 trillion *(2003F):* $1.34 trillion

Real GDP Growth Rate (2002E): 8.0 per cent *(2003F):* 7.1 per cent

Inflation Rate (2003F): 0.6 per cent

Current Account Surplus (2003F): $8.0 billion

Major Trading Partners: Japan, United States, European Union, South Korea, Taiwan

Merchandise Exports (2003F): $442.1 billion

Merchandise Imports (2003F): $399.2 billion

Merchandise Trade Surplus (2003F): $42.9 billion

Major Export Products: Light industrial and textile products, mineral fuels, heavy manufactures, agricultural goods

Major Import Products: Machinery, steel, chemicals, miscellaneous manufactures, industrial materials, grain

External Debt (2003F): $185.3 billion

Energy Overview

Proven Oil Reserves (1/1/03E): 18.3 billion barrels

Oil Production (2002E): 3.39 million barrels per day (bbl/d)

Oil Consumption (2002E): 5.26 million bbl/d

Net Oil Imports (2002E): 1.87 million bbl/d

Crude Oil Refining Capacity (1/1/03E): 4.5 million bbl/d

Natural Gas Reserves (1/1/03E): 53.3 trillion cubic feet (Tcf)

Natural Gas Production (2001E): 1.07 Tcf

Natural Gas Consumption (2001E): 1.07 Tcf

Recoverable Coal Reserves (1/1/96E): 126.2 billion short tons

Coal Production (2001E): 1.49 billion short tons

Coal Consumption (2001E): 1.38 billion short tons

Electric Generation Capacity (1/1/01E): 318 GW (237 GW thermal; 79 GW hydro; 2 GW nuclear)

Electricity Generation (2001E): 1,420 billion kilowatthours (1,139 conventional thermal; 263 hydro; 17 nuclear)

Statistical note: All data reported here exclude Hong Kong, a former British colony which reverted to China on July 1, 1997.

Environmental Overview

Total Energy Consumption (2001E): 39.7 quadrillion Btu (9.8 per cent of world total energy consumption)

Energy-Related Carbon Emissions (2001E): 831.7 million metric tons of carbon (12.7 per cent of world carbon emissions)

Per Capita Energy Consumption (2001E): 28.8 million Btu (vs. U.S. value of 341.8 million Btu)

Per Capita Carbon Emissions (2001E): 0.65 metric tons of carbon (vs. U.S. value of 5.5 metric tons of carbon)

Energy Intensity (2001E): 35,469 Btu/$1995 (vs. U.S. value of 10,736 Btu/$1995)**

Carbon Intensity (2001E): 0.75 metric tons of carbon/thousand $1995 (vs. U.S. value of 0.17 metric tons/thousand $1995)**

Fuel Share of Energy Consumption (2001E): Oil (25.8 per cent), Natural Gas (3.1 per cent), Coal (64.0 per cent)

Fuel Share of Carbon Emissions (2001E): Oil (21.1 per cent), Natural Gas (2.2 per cent), Coal (76.8 per cent)

Status in Climate Change Negotiations: Non-Annex I country under the United Nations Framework Convention on Climate Change (ratified January 5th, 1993). Signatory to the Kyoto Protocol (signed May 29th, 1998–not yet ratified).

Major Environmental Issues: Air pollution (greenhouse gases, sulfur dioxide particulates) from the overwhelming use of high-sulfur coal as a fuel, producing acid rain which is damaging forests; water shortages experienced throughout the country, particularly in urban areas and in the north; future growth in water usage threatens to outpace supplies; water pollution from industrial effluents; much of the population does not have access to potable water; less than 10 per cent of sewage receives treatment; deforestation; estimated loss of one-fifth of agricultural land since 1949 to soil erosion and economic development; desertification; trade in endangered species.

Major International Environmental Agreements: A party to the Antarctic-Environmental Protocol, Antarctic Treaty, Biodiversity, Climate Change, Desertification, Endangered Species, Hazardous Wastes, Law of the Sea, Marine Dumping, Nuclear Test Ban, Ozone Layer Protection, Ship Pollution, Tropical Timber 83, Tropical Timber 94 and Wetlands. Has signed but not ratified: Nuclear Test Ban.

The total energy consumption statistic includes petroleum, dry natural gas, coal, net hydro, nuclear, geothermal, solar, wind, wood and waste electric power. The renewable energy consumption statistic is based on International Energy Agency (IEA) data and includes hydropower, solar, wind, tide, geothermal, solid biomass and animal products, biomass gas and liquids, industrial and municipal wastes. Sectoral shares of energy consumption and carbon emissions are also based on IEA data.

Energy Industry

Organization: Coal–China National Local Coal Mines Development Corp., China Northeast & NEI-Mongolia United Coal Co., numerous local state-owned mines and rural collectives; *Coal import/exports*–China Coal Import and Export Group; *Petroleum*–China National Petroleum Corp. (CNPC, PetroChina is its publicly traded subsidiary), China National Offshore Oil Corp. (CNOOC), China National Oil & Gas Exploration & Development Corp. (CNODC), China National Star Petroleum (Star); China National Petrochemical Corp. (SINOPEC); *Oil imports/exports*–China National Chemicals Import and Export Corporation (SINOCHEM), China United Petroleum Corporation (China Oil), China United Petrochemical Corp. (UNIPEC); *Electric power*–China State Power Corp., Huaneng Group, Inc., China National Power Industry Corp. (CNPIC), regional electric power corporations, China National Nuclear Industry Corp., China International Water and Electric Corp. (CWE).; *Energy Finance*–China National Energy Investment Corp.

Major Producing Oil Fields (2001 Production): Daqing (1.1 MMBD), Shengli (0.5 MMBD), Liaohe (0.3 MMBD)

Major Refineries (1/1/03 Capacity): Fushun (184,800 bbl/d), Maoming (170,700 bbl/d), Qilu (160,700 bbl/d), Gaoqiao (150,600 bbl/d), Dalian (142,600 bbl/d), Yanshan (190,800 bbl/d), Jinling (140,600 bbl/d); Zhenlai (160,700 bbl/d)

28

Colombia

Colombia is an important petroleum and coal producer, although political unrest has led to decreased exports. Colombia aims to boost exploration to preserve its status as a net oil exporter in the future.

General Background

Colombia continues to weather difficult political and economic conditions. According to Colombia's statistics office, the economy grew 1.5 per cent in 2002, a slight increase over 2001. Unemployment also remained high in 2002 at an estimated 17.7 per cent.

Since taking office, Colombia's new president, Alvaro Uribe, has undertaken measures to spark the slow-growing economy. In January 2003, Congress approved five reform bills, focusing on tax, pension, labor, public administration and banking. Still pending approval is a government cost-cutting referendum, which seeks to reduce the country's fiscal burden by reducing the size of Congress, freezing most public-sector salaries for two years, and limiting pensions for state employees. The IMF backed Colombia's reforms with funding, with a 2-year, $2.1 billion stand-by agreement last January.

During the presidential campaign, President Uribe vowed to regain authority over the country's territory, of which rebel groups control 40 per cent. After former President Pastrana broke off three-years of peace talks with the largest rebel group, the *Fuerzas Armada Revolucionarias de Colombia* (FARC) in February 2002, full-scale violence erupted. Previously conflicts between left-wing rebel groups, such as the FARC and the Army of National Liberation (ENL) and right-wing paramilitary group, the United Self-Defense Forces of Colombia, had been confined primarily to remote jungle regions, with oil industry infrastructure as a common target. However, since then, the violence has moved into the country's urban centers. In addition to bombings aimed at the general civilian population, rebel groups target politicians. In 2002, a reportedly 3,500 Colombians were killed and an additional 2,000 kidnapped by guerrilla forces, including former presidential candidate Ingrid Beancourt. Recently in

May 2003, rebels shot hostages, including the former defense minister Gilberto Echeverri and a state governor, when the government attempted to rescue them.

The conflict has taken its toll on the Colombian energy infrastructure, particularly the oil industry. Rebels have targeted oil pipelines, electricity transmission infrastructure, and trains transporting coal. They have also taken employees associated with these operations hostage. As oil reserves decline, the government has been boosting its security efforts in order to attract new investment in replacement fields. Colombian official have repeatedly warned that if no significant new fields are found, the country soon will become a net importer.

Oil

As of January 2003, Colombia had about 1.84 billion barrels of proven oil reserves, up from 2002 estimates of 1.75 billion barrels. Colombia's oil production declined in 2002 to 591,250 barrels per day (bbl/d), after reaching an all-time high of 830,000 bbl/d in 1999. Colombia exported 256,000 bbl/d to the United States during 2002, down from 296,000 bbl/d (13.5 per cent) in 2001. Estimates indicate that, without new discoveries, Colombia could become a net oil importer in the medium term.

Sector Organization

The Colombian government owns the country's hydrocarbon reserves. Oil and natural gas development is regulated by state oil company *Empresa Colombiana de Petroleos* (Ecopetrol) and the Energy and Mines Ministry. Private companies operate in joint ventures with Ecopetrol, with Ecopetrol legally required to hold at least a 30 per cent share. Britain's BP and U.S. Occidental are the most active foreign companies in the Colombian oil sector.

Since 1999, the government has taken measures to make the investment climate more attractive to foreign oil companies in Colombia. The first measure was a reduction in Ecopetrol's mandatory share in joint ventures from 50 per cent to 30 per cent. The second measure was a new law passed in June 2002, which eliminated the flat 20 per cent royalty in favor of a sliding scale from 8 per cent to 25 per cent. The rate is based on production levels, e.g., royalties of 8 per cent will be assessed on fields producing 5,000 bbl/d or less and 20 per cent to 25 per cent for production between 400,000 bbl/d and 600,000 bbl/d. The sliding scale aims to improve the financial viability of smaller producing fields. Since its introduction, the Colombian government has issued an increased number of exploration and production contracts.

Exploration and Production

Vast unexplored and potentially hydrocarbon-rich territories remain in Colombia. The country also shares many of the geological features of its oil-rich neighbor Venezuela. In particular, the Llanos and Magdalena basins are thought to have great potential for hydrocarbon development. Much of Colombia's crude oil is lighter and sweeter than that of other major Latin American oil producers, with its three export crude oils ranging between 20° and 36° API.

Colombia's largest production fields are the Cusiana and Cupiagua fields in the eastern Andes foothills, and the Cano Limón field in the Arauca province, near the Venezuelan border. The Cusiana and Cupiagua fields are thought to have reached peak production, while Cano Limón, the largest producing field, is nearing depletion. BP operates Cusiana/Cupiagua, while U.S.-based Occidental operates Cano Limón. Much of the fields' crude oil is exported to the U.S. Gulf Coast, with about 90 per cent of light sweet Cusiana and Cupiagua crude bound for the U.S. market.

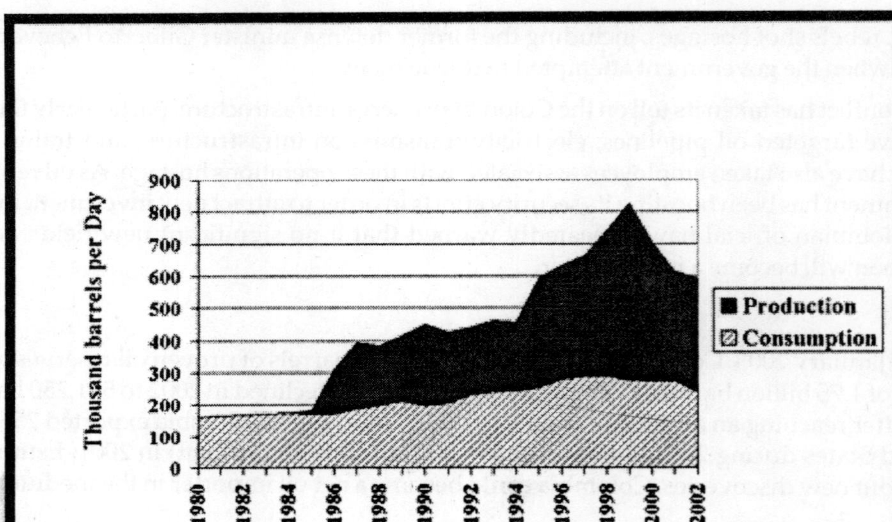

COLOMBIAN OIL PRODUCTION AND CONSUMPTION, 1980-2002

Recent Developments

Ecopetrol and Colombian Energy and Mines officials hope that new regulations will lead to more active exploration and production (E&P) in Colombia, as well as result in increasing the country's oil reserves by one billion barrels over the next four years. During the last year, the following companies have signed E&P contracts with Ecopetrol: Chile's Sipetrol; Colombian-based Argosy Energy International; Petrominerales, a subsidiary of Canada's Petrobank; Petrobras; Nexen; US-based Mercantile Oil & Gas; Occidental; and Russia's Lukoil.

In 2002, some new production has been brought online, notably at the Suroriente and Guando fields. In June, a Petrobras and Nexen joint venture with Ecopetrol began oil production at one of Colombia's largest new fields–Guando, located in the upper Magdalena Valley. The oil production at the field is expected to reach 10,000 bbl/d in 2003, ramping up to a peak of between 20,000 bbl/d and 30,000 bbl/d between 2005-2007.

In December 2002, the Colombia Energy Consortium began pumping oil at the Suroriente field. In March 2003, the Consortium signed a contract to sell oil produced at the field to Ecopetrol. Production is currently limited, as oil is being transported to the Mansolia-Orito pipeline by truck. After a pipeline linking the field to the Mansolia-Orito line is constructed, production will increase to 180,000 barrels per month. The Consortium, consisting of Russia's Rosneft (45 per cent) and of Colombian companies Petrotesting Colombia (27.5 per cent) and Holsan Chemicals (27.5 per cent), won a tender for development of this field in 2001.

In 2003, Canada's Petrobank Energy and Resources plans to invest $45 million in Colombia primarily developing the Orito block. Petrobank also has operations at the Neiva block. BP also plans to begin exploration at the Niscota block.

Setback in Exploration

The Colombian government's plans to increase oil production and reserves experienced a recent setback. In March 2003, Ecopetrol officials announced that they had discovered one of the largest oil deposits in over a decade, the Gibraltar well in the Siriri block. They estimated that the well contained 200 million barrels of reserves. A preliminary study, however, showed that the Gibraltar well contained mostly water. Irrespective of the study, Colombia's Mines and Energy Minister Luis Ernesto Mejia, confirmed in mid-April that the well is actually producing both natural gas and crude oil.

Occidental originally drilled the well, but the company turned it over to the Colombian government in May 2002 after it said it drilled 12,000 feet and found only some water and gases. Occidental explained that its decision was based solely on the economics of that particular field, as it was not showing the projected commercial quantities of oil. The company went on to say that its decision was unrelated to conflicts with the indigenous U'wa tribe or attacks on company investments elsewhere in Colombia. The U'wa continue to resist the Colombian government's further exploration of the region.

Oil Pipelines

Colombia has five major oil pipelines, four of which connect production fields to the Caribbean port town of Coveñas. These include the 500-mile Ocensa pipeline, which transports 615,000 bbl/d from the Cusiana and Cupiagua fields; the 490-mile Cano Limón pipeline; and the smaller Alto Magdalena and Colombia Oil pipelines. The fifth pipeline, the TransAndino, transports crude from Colombia's Orito field in the Putumayo basin (as well as Ecuadorian crude) to Colombia's Pacific port at Tumaco.

Security Issues

Security problems continue to plague the Colombian oil industry. Oil production has declined not only due to the mature nature of the country's largest producing fields, but also as result of repeated attacks on the oil industry infrastructure.

Left-wing rebel groups continue to target the oil industry. In April 2002, Colombia's second-largest rebel group, the National Liberation Army (ELN), declared that oil companies working in the country were military targets, specifically naming Occidental, Repsol-YPF, and Ecopetrol. The ELN and the FARC bombed the Cano Limón pipeline a record 170 times (about 14 times per month) in 2001, holding in more than 24 million barrels of crude oil, according to Ecopetrol estimates. Bombings in 2002 fell to 41, allowing for increased exports of Cano Limón crude to the U.S. Gulf Coast. Aid in the form of money and military training from the U.S. government has helped decrease the number of attacks on pipeline this year.

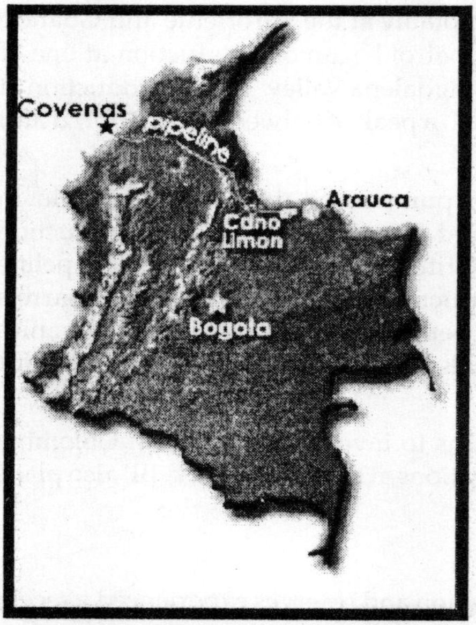

The BP Cusiana/Cupiagua complex's Ocensa pipeline is by far a less frequent target for bombings. While it carries more oil than the Cano Limón pipeline, more of it is underground. The company's wells are attacked occasionally, but the pipeline has fared far better than the Cano Limón. The TransAndino is subject to more bombings than the Ocensa, but fewer than the Cano Limón.

In addition to pipeline bombings, security problems also are responsible for labor unrest. Ecopetrol strikes have followed disappearances of oil union officials. An estimated 165 union members were killed in Colombia in 2001. Right-wing paramilitary organizations have claimed responsibility for abductions of oil union (USO) members, charging those abducted with ELN affiliations. Strikes in February, March, and April protested abduction and killings of oil union workers.

Oil disputes have not been limited to the domestic arena. In July 2002, Nicaragua opened a tender process for offshore oil exploration contracts in the Caribbean. Colombia claimed that some of area included acreage near the islands of San Andres and Providencia, which have belonged to Colombia since 1928. As of May 2003, Nicaragua has failed to attract bidders for exploration acreage in offshore territories. Colombia has also stated that it would not rule out military intervention if oil exploration takes place in its territory.

Refining

Colombia's current refining capacity is about 285,850 bbl/d, and all refineries are 100 per cent owned and operated by Ecopetrol. Although Colombia is a net oil exporter, it imports gasoline to meet domestic product demand. The government has stated that it will not construct new refineries, and any new construction must be undertaken by the private sector.

Natural Gas

Colombia's proven natural gas reserves stood at an estimated at 4.5 trillion cubic feet (Tcf), as of January 2003, up slightly from the January 2002 estimate of 4.3 Tcf. Both production and consumption in 2002 totaled 201 billion cubic feet (Bcf). Consumption is expected to grow significantly over the next decade. The government's *Plan de Masificacion de Gas Natural* (Natural Gas Mass Consumption Plan) aims to increase natural gas use, especially for electricity generation and public transportation.

Sector Organization

Colombia's natural gas industry is controlled by three key players: Ecopetrol, which manages exploration and development of natural gas reserves; the Energy and Gas Regulatory Commission (CREG), which oversees regulation and distribution markets; and *Empresa Colombiana de Gas* (Ecogas), a state-owned company tied to the Ministry of Energy and controlled by Ecopetrol, in charge of managing the country's transmission system and operating most transportation pipelines. CREG has segmented the industry into production, transmission, distribution, and marketing sectors. Some distribution in major urban areas has been privatized.

Production

Natural gas production is centered along the Northern Coast and Barranca regions; other areas of production are in the south and in basins to the east of Bogotá. The Guajira basin is thought to have the most potential for large new discoveries. Despite government efforts to galvanize the Colombian natural gas sector, development has been slow.

Most of Colombia's natural gas is produced at the Chuchupa offshore field and the Ballena onshore field by ChevronTexaco Colombia subsidiary, Texas Petroleum Company, which has an association contract with Ecopetrol through 2004. On February 8, 2003, both companies signed the Catalina project agreement, which extends the current agreement until 2016. Without the contract, ChevronTexaco's rights in Colombia would expire at the end of next year. The company would be reduced to a contractor, operating the fields for a fee, with Ecopetrol receiving 100 per cent of the output. Currently, a local court is examining the agreement to determine whether ChevronTexaco should pay higher royalty fees. ChevronTexaco claims that it would rule out exploration and upstream investment in Colombia if the government cannot resolve the dispute over royalties.

Natural Gas Pipelines

Ecogas owns and operates Colombia's domestic pipeline network. In April 2003, Colombia and Venezuela agreed to build a $120 million pipeline, allowing Colombia to export natural gas from the Guajira basin to the Maracaibo region of Venezuela. Eventually Venezuela will deliver natural gas through the same pipeline.

Coal

Colombia is the largest coal producer in Latin America, and its proven reserves–7,328 million short tons in 2001–consist of high-quality bituminous coal and a small amount of metallurgical coal. Colombia's coal is relatively clean-burning, with a sulfur content of less than 1 per cent. Colombia ranks as one of the largest coal exporters in the world.

The Colombian government has recently increased its investment in and promotion of coal exportation. In January 2003, the Colombian Mining Association (Asomineros) and the Colombian government launched a campaign to promote coal exports to other Central and South American countries. In April 2003, the government also earmarked $320 million for infrastructure development that would facilitate coal output in reaching the country's ports. Part of the investment will go to the construction of a second rail line linking Cesar region with Santa Marta port on the Atlantic coast.

Minercol, Colombia's state mining company, plans to tender exploitation concessions for 15 mineral properties across the country by the middle of 2003, of which 6 are coal properties.

Sector Organization

Colombia's coal mines are in private hands. The Carbones del Cerrejon consortium of South Africa's Anglo-American, the UK/Australian BHP Billiton, and Switzerland's Glencore owns and operates the massive Cerrejón Zona Norte project, which is the largest coal mining operation in Latin America and the largest open-cast coal mine in the world. The consortium plans to invest $130 million to increase production. U.S.-based Drummond is active in Colombia, operating the country's second-largest mine, La Loma. The company also owns El Descanso, another coal property located in the same area, and is planning to invest $1 billion over the next five years in order to increase coal exports.

Electricity

In 2001, Colombia's electricity generation capacity was 12.7 gigawatts. The country consumed 39.8 billion kilowatthours (Bkwh) and produced 43 Bkwh. Typically 70 per cent of generation is hydroelectric. Colombia's power sector, however, has endured some difficult years in the 1990s, as a severe drought left hydroelectric plants unable to meet electricity demand and forced power rationing. As a result, the government began to encourage the construction of coal and natural gas-fired power plants over the medium to long term. Hydropower's share of generation is therefore expected to decline.

Colombia has electricity interconnections with Venezuela and Ecuador. In January 2003, a new 131-mile interconnection line between Ecuador and Colombia began operation. This interconnection follows electricity power agreements which the five Andean Community (CAN) members agreed on in December 2002. Bolivia, Colombia, Ecuador, Peru, and Venezuela agreed to integrate their power

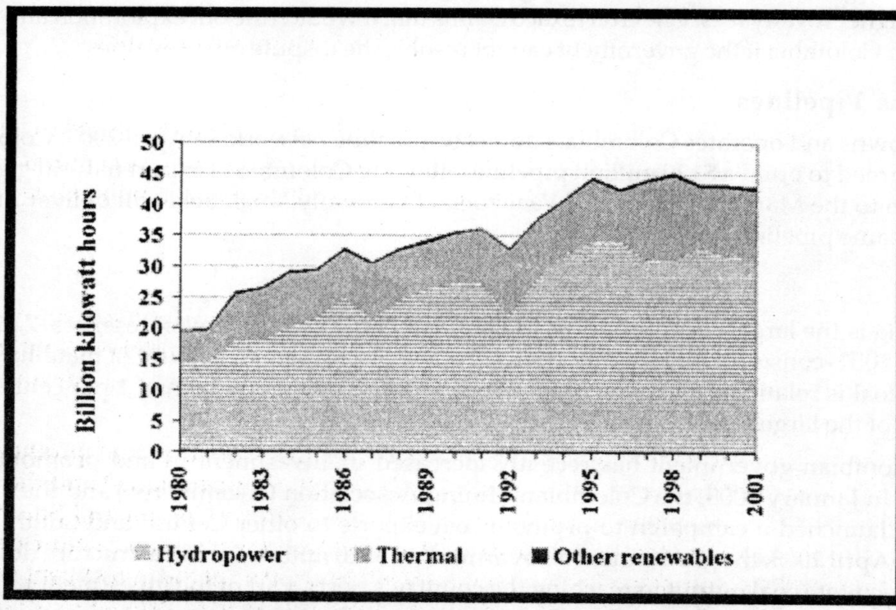

COLOMBIAN ELECTRICITY GENERATION, 1980-2001

markets to create a regional power market. This is part of CAN's Regional Infrastructure Integration Initiative.

Colombia and Panama also signed in April 2003 a memorandum of understanding on energy integration. The goal of the agreement is to increase energy integration in order to interconnect South America with Central America.

Sector Organization

The Colombian electricity sector continues to face serious challenges. Repeated attacks on electric infrastructure by rebel separatist groups are straining the system. Grid interconnections have been blown up, leaving the country divided into several, smaller grids and preventing repair crews from reconnecting the fragments. Plans to connect Colombia's Atlantic Coast and its capital city, Bogota, by transmission cable have been put on hold.

The current problems have delayed privatization in the sector. A 1994 electricity reform law deregulated the industry, and privatization progressed throughout the late 1990s. However, the increase in guerrilla attacks on the system has put privatization plans on hold. Of the 28 generation companies on the national interconnected grid, most of the larger companies remain in publicly owned.

Country Overview

Independence: July 20, 1810 (from Spain)

Population (2002E): 43.8 million

Location/Size: NW South America / 1.1 million square kilometers (440,000 square miles), about the size of New Mexico, Texas, and Louisiana combined

Major Cities: Bogotá (capital), Cali, Medellin, Barranquilla

Language: Spanish

Ethnic Groups: Mestizo (58 per cent), White (20 per cent), Mulatto (14 per cent), Black (4 per cent), Black-Amerindian (3 per cent), Amerindian (1 per cent)

Religion: Roman Catholic (90 per cent)

Defense (8/98): Army (121,000); Navy (18,000); Air Force (7,300); Paramilitary Police (87,000)

Economic Overview

Exchange Rate (5/08/03): US$1 = 2,926.4 Colombian pesos

Gross Domestic Product (GDP, nominal, 2002E): $81.4billion

Real GDP Growth Rate (2002E): 1.5 per cent *(2003F)*: 2.1 per cent

Inflation Rate (consumer prices) (2002E): 6.3 per cent *(2003F):* 7.5 per cent

Unemployment Rate (2002E): 17.7 per cent *(2003F):* 16.1 per cent

Total Foreign Debt (2002E): $35 billion

Current Account Balance (2002E): -2.0 per cent of GDP

Major Trading Partners: United States, Venezuela, Germany, Japan, Peru

Major Exports: Petroleum, coal, coffee

Major Imports: Capital goods, industrial inputs, consumer goods

Energy Overview

Proven Oil Reserves (1/1/03E): 1.84 billion barrels

Oil Production (2002E): 591,250 barrels/day (bbl/d), of which 577,250 bbl/d was crude oil

Oil Consumption (2002E): 239,000 bbl/d

Net Oil Exports (2002E): 352,250 bbl/d

Oil Exports to the U.S. (2002E): 256,000 bbl/d

Crude Refining Capacity (1/1/03E): 285,850 bbl/d

Natural Gas Reserves (1/1/03E): 4.5 trillion cubic feet

Natural Gas Production (2001E): 201 billion cubic feet (Bcf)

Natural Gas Consumption (2001E): 201 Bcf

Recoverable Coal Reserves (2001E): 7.3 billion short tons

Coal Production (2001E): 47.9 million short tons (MMST)

Coal Consumption (2001E): 4.5 MMST

Electric Generating Capacity (2001E): 12.7 gigawatts

Electric Generation (2001E): 43.0 billion kilowatthours (31.3 Bkwh hydroelectric, 11.2 Bkwh thermal, 0.6 Bkwh other renewables)

Environmental Overview

Total Energy Consumption (2001E): 1.1 quadrillion Btu (0.3 per cent of world total energy consumption)

Energy-Related Carbon Emissions (2001E): 15.0 million metric tons of carbon (0.2 per cent of world carbon emissions)

Per Capita Energy Consumption (2001E): 26.5 million Btu (vs. U.S. value of 341.8 million Btu)

Per Capita Carbon Emissions (2001E): 0.4 metric tons of carbon (vs. U.S. value of 5.5 metric tons of carbon)

Energy Intensity (2001E): 11,568 Btu/ $1995 (vs. U.S. value of 10,736 Btu/ $1995)**

Carbon Intensity (2001E): 0.15 metric tons of carbon/thousand $1995 (vs. U.S. value of 0.17 metric tons/thousand $1995)

Fuel Share of Energy Consumption (2001E): Oil (44.0 per cent), Natural Gas (17.0 per cent), Coal (9.7 per cent)

Fuel Share of Carbon Emissions (2001E): Oil (62.0 per cent), Natural Gas (19.0 per cent), Coal (19.0 per cent)

Status in Climate Change Negotiations: Non-Annex I country under the United Nations Framework Convention on Climate Change (ratified March 22, 1995). Party to the Kyoto Protocol, but not yet ratified.

Major Environmental Issues: Deforestation; soil damage from overuse of pesticides; air pollution, especially in Bogota, from vehicle emissions

Major International Environmental Agreements: A party to the Antarctic Treaty, Biodiversity, Climate Change, Endangered Species, Hazardous Wastes, Marine Life Conservation, Nuclear Test Ban, Ozone Layer Protection, Ship Pollution, Tropical Timber 83 and Tropical Timber 94. Has signed, but not ratified, the Antarctic-Environmental Protocol, Desertification, Law of the Sea and Marine Dumping

The total energy consumption statistic includes petroleum, dry natural gas, coal, net hydro, nuclear, geothermal, solar, wind, wood and waste electric power. The renewable energy consumption statistic is based on International Energy Agency (IEA) data and includes hydropower, solar, wind, tide, geothermal, solid biomass and animal products, biomass gas and liquids, industrial and municipal wastes. Sectoral shares of energy consumption and carbon emissions are also based on IEA data.

Energy Overview

Organization: Oil: Empresa Colombiana de Petroleos (Ecopetrol); *Natural gas*: Ecopetrol, Energy and Gas Regulatory Commission (CREG), Empresa Colombiana de Gas (Ecogas)

Major Ports: Barranquilla, Buenaventura, Cartagena, Leticia, Puerto Bolívar, San Andres, Santa Marta, Tumaca, Turbo

Major Oil-Producing Fields: Cusiana (BP Amoco); Cano Limón (Occidental)

Major Oil Pipelines: Oleoducto Central (Ocensa, from Cusiana/Cupiagua to Covenas); Cano Limón (to Covenas); Transandino (OTA, from Ecuador to Tumaco)

Major Natural Gas Pipelines: Mariquita-Cali (TransGas Occidente); Ballena-Barrancabermeja; Barrancabermeja-Neiva-Bogotá (Centro Oriente)

Refineries (1/1/03E Capacity): 285,850 bbl/d, all run by state-run Ecopetrol. Two main refineries, Barrancabermeja–Santander (205,000 bbl/d) and Cartagena–Bolívar (75,000 bbl/d). Other refineries: Tibu–N. de Santander (1,800 bbl/d), Orito–Putumayo (1,800 bbl/d), and Apiay (2,250 bbl/d).

29

Congo-Brazzaville

With increased interest in offshore West Africa and the development of new offshore oil fields, Congo-Brazzaville is becoming increasingly important to world energy markets.

General Background

Denis Sassou-Nguesso was elected to a seven-year term as president of Congo in March 2002. In the two rounds of elections held in May and June 2002, the parties that supported President Sassou-Nguesso, the Parti Congolais du Travail (PCT) and the Forces Démocratiques Unies (FDU) won 83 of the 153 seats in the Assemblée Nationale (National Assembly). In July 2002, it was announced that the PCT and its allies had won 56 of the 72 seats in the Senate. It was reported in April 2003 that over 2,000 rebel Ninja soldiers had surrendered with their weapons in Congo's Pool region. This followed a peace agreement reached on March 17 between the government and Ninja leader, Rev. Frederic Bitsangou (Pastor Ntoumi). The disarmed Ninjas, who have been guaranteed amnesty by the government, are now awaiting either integration into the military or assistance in returning to civilian life. Conflict erupted in Pool between government forces and Ninja rebels loyal to Ntoumi in late March 2002. An additional 600 Ninja rebels were reported to have surrendered in May 2003.

Congo's economy consists mainly of village agriculture, an urban informal sector (*i.e.*, unregulated business, commerce, and service activities), and an industrial sector dominated by oil and oil-related services with few linkages to the rest of the economy. Since the 1980s, the oil industry has provided the major share of government revenues and exports, replacing timber production and exports as the principle growth sector. Oil accounts for over 50 per cent of Congo's real gross domestic product (GDP), 60 per cent -80 per cent of the government budget and about 95 per cent of Congo's export earnings. Oil exports grew sharply, from approximately $820 million in 1994 to around $2.3 billion in 2002.

In spite of its oil wealth, Congo has experienced budgetary shortfalls as a result of public sector expenditures, slumps in world oil prices (1998-1999), and armed conflicts (1997, 1999 & 2002).

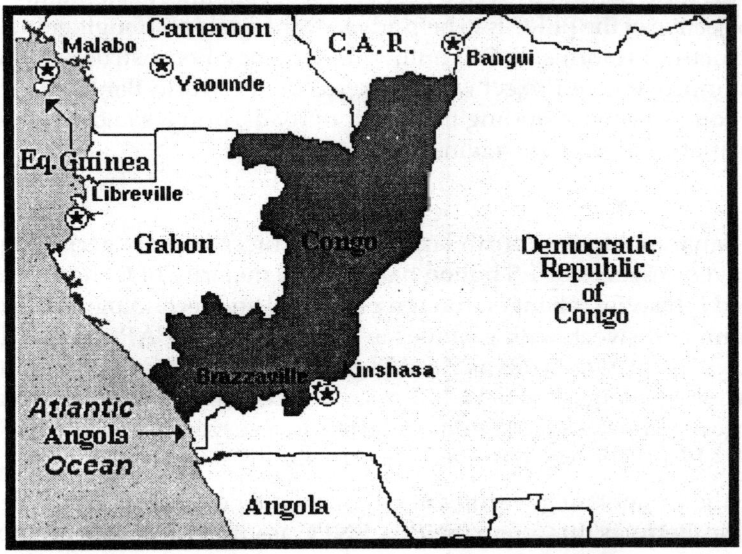

Mirroring the pattern of armed conflict in the country, Congo's real GDP growth rate has fluctuated over the past several years, from a decline of 3 per cent in 1999 to 7.5 per cent growth in 2000, following a cessation of internal hostilities (plus higher oil prices). In 2001, real GDP growth was an estimated 2.9 per cent and real growth for 2002, originally forecast at 4.4 per cent, was flat following a return to armed conflict. Inflation in 2002 was about 3.6 per cent.

Congo's business and administrative infrastructure was badly damaged during the 1997 and 1998-1999 civil wars, increasing the petroleum sector's dominance of the economy (since oil production was not directly harmed by the fighting). Overall, Congo's non-oil sector shrank by 20 per cent from 1996 to 1999. Rebel attacks on the CFCO railway, which runs from the port of Pointe Noire to the capital Brazzaville and the interior, have hindered the flow of goods. Passenger and cargo trains were suspended as a result of frequent attacks, which led to several deaths and extensive material damage in Pool. According to official figures, the CFCO incurred losses estimated at between 70-80 million CFA per day as a result of insecurity. Economic activity was further hampered by the fact that over 800,000 Congolese, nearly 30 per cent of the population, fled their homes during the conflicts.

In November 2000, The International Monetary Fund (IMF) approved a loan of $14 million for emergency post-conflict assistance to support the Congo government's reconstruction and economic recovery program. Congo later signed an agreement with the IMF on a "staff-monitored program" (SMP) as part of post-conflict recovery plans. The aim of the SMP is to lead to a formal "poverty reduction and growth facility" (PRGF) IMF lending program for the country. Implementation of the PRGF would permit Congo to benefit from relief on its external debt under the Initiative for the Heavily Indebted Poor Countries (HIPC Initiative). In exchange, Congo is expected to increase its financial and fiscal transparency (including at the state oil company), to improve the management of its fiscal balance and external arrears (*i.e.*, by reducing fraud in the customs area), and to reduce cost over-runs for civil service salaries.

In July 2001, the World Bank approved a Post-conflict Economic Rehabilitation Credit (PERC) for Congo. The main objective of the PERC is to help advance the goals of strengthening economic stability and supporting structural reforms. A $40 million Emergency Reconstruction, Rehabilitation, and Living Conditions Improvement Project was approved by the World Bank in July 2002. The project comprises several components, including the repair of roads, bridges and schools; rehabilitation of the railway; job creation; and the fight against HIV/AIDS.

Oil

Congo is sub-Saharan Africa's fourth largest oil producer (after Nigeria, Angola, and Gabon), with estimated proven reserves of 1.5 billion barrels. The majority of Congo's crude production is located offshore and is heavily reliant on foreign personnel and technology. Congo's crude oil types are typically medium and sweet, with gravities in the 22°API to 33°API range. The country's main crude export blend is Djeno, which has a 27.6°API gravity and a 0.23 per cent sulfur content. Recent production on the country's largest oil field–N'Kossa, with 500 million barrels in recoverable reserves–and the second largest–Kitina, with 145 million barrels in reserves–has yielded lighter, higher value crude types of 40.7°API and 38°API gravities, respectively.

In April 1998, the Congolese government established a new national petroleum company, the Société Nationale des Pétroles du Congo (SNPC). SNPC assumed all upstream functions of Hydro-Congo (the older state-owned oil firm retained its downstream operations), when exploration and production assets were formally transferred to SNPC in February 2000. In 1994, Congo enacted new hydrocarbon legislation offering production-sharing agreements (PSAs) to foreign oil companies. The PSAs were designed to replace joint venture (JV) arrangements in existence since 1968. Under the PSAs, foreign oil companies act as contractors for SNPC. The PSAs are intended to ensure a constant minimum flow of revenue to the government, commensurate with similar agreements between other regional oil exporting states and major oil companies. Congo's PSAs commit foreign partners to carry out exploration and development within a pre-determined time (usually three years for each phase). Contractors (foreign oil companies) finance all investment costs, and recover their investments when production begins. In 1995, foreign companies were given the option of converting existing exploration and production JVs to PSAs. All major operators in Congo have signed PSAs for their respective field developments.

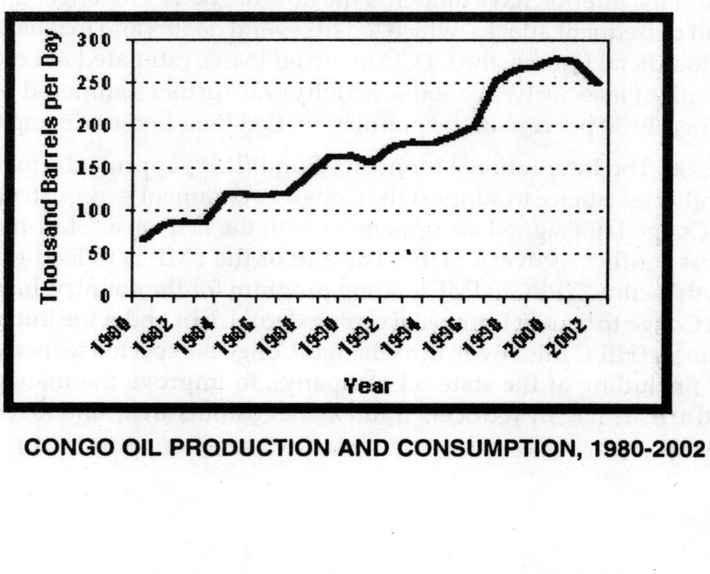

CONGO OIL PRODUCTION AND CONSUMPTION, 1980-2002

Production

Oil production in Congo commenced in 1957 from the onshore Pointe Indienne field, reaching a peak output of about 2,500 barrels per day (bbl/d) in the mid-1960s. Over the past twenty years, Congo's crude production has quadrupled, from 65,000 bbl/d in 1980 to an average of 280,000 bbl/d in 2000. Production has been declining recently, largely due to drop-off in production at mature fields, and delays in bringing several fields online. Crude production averaged 275,000 bbl/d in 2001, and 249,000 bbl/d in 2002. For the first quarter of 2003, production averaged 240,000 bbl/d, and Congo will likely lose its position as the fourth largest sub-Saharan African oil producer to Equatorial Guinea by the end of 2003. Overall, Congo's oil production is expected to decline through 2004 before rebounding as new field production (including the Libondo, Tchibeli, Litanzi, and Yanga-Sud fields) begins to come online and offset reduced output at more mature fields. Approximately one-third of the oil produced goes directly to the government and it is sold by SNPC on the state's behalf.

Total (formally TotalFinaElf) operates the majority of Congo's crude oil producing fields, about two-thirds of the country's total output. Total operates Congo's largest field, N'Kossa, located on the Haute Mer permit. Discovered in 1984, N'kossa (two fixed platforms in 623 feet of water) began production in June 1996 at 5,000 bbl/d, but it has remained less than the originally projected 100,000 bbl/d-110,000 bbl/d. N'Kossa production averaged approximately 34,000 bbl/d of crude for the first quarter of 2003, with an additional 10,000 bbl/d of LPG (liquefied petroleum gas). Total's (51 per cent interest) partners on the Haute Mer Permit are ChevronTexaco (30 per cent), SNPC (15 per cent), and Cape Town-based Energy Africa (EA) holds a 4 per cent interest.

Many of the smaller Total-operated fields are located on the Pointe Noire Grande Fonds (PNGF) concession. PNGF fields that Total has brought online in the past few years include: Tchibouela, Sendji, Yanga, Tchendo, and Likouala. Total brought the Tchibouela-Est field (three miles east of the Tchibouela hub) into production by employing a "hubs and clusters" development strategy designed to boost productivity without heavy capital outlays. Tchibouela-Est holds an estimated 40 million barrels of recoverable reserves. Total's partner on the PNGF concession is Italy's ENI-Agip (Agip) which has a 35 per cent interest.

Agip, the second largest producer, averaged around 72,000 bbl/d for the first two months of 2003. Agip is the operator of the following fields (interest per cent): Djambala (65 per cent), Foukanda (100 per cent), Kitina (55 per cent), Loango (50 per cent), Mwafi (100 per cent) and Zatchi (65 per cent). Total is Agip's partner on both Loango and Zatchi, while ChevronTexaco and SNPC are the other shareholders with Agip on Katina. Agip began production on the offshore Kitina field (southern Marine VII license) in December 1997. Production began from a single well at a rate of 5,300 bbl/d. Kitina development included the construction of a production platform, the drilling of 16 wells and the installation of a 39-mile (65-kilometer) underwater pipeline to the Djeno oil terminal. Kitana production currently averages 60,000 bbl/d.

Discovered in 1998, Agip began production at the Foukanda and Mwafi fields in 2001. Peak production of 21,000 bbl/d is expected to be reached by the end of 2003. Foukanda, with estimated reserves of 40 million barrels, was developed with a five-well platform located on the central portion of the Marine VI Block. Production from the Mwafi fields comes from six wells on a satellite platform. The oil will be piped to the Loango platform and then on to Djeno terminal. Mwafi reserves are estimated at 26 million barrels.

European-independent Perenco (50 per cent and Operator) currently has production of nearly 20,000 bbl/d from two offshore fields. The Yombo field produces some 14,000 bbl/d of heavy crude

(16.2°API) which is processed in the field to a fuel oil product. The Emeraude field lies in relatively shallow waters 264 feet (80 meters). Low reservoir pressure and the viscous crude (22.3°API) complicate production operations. Emeraude development has involved many platforms and deviated wells. In 2001, there were 82 producing wells on the field. US-based Nuevo Energy announced in May 2003 that it has put its 50 per cent stake in Yombo up for sale.

Calgary-based Heritage Oil began production from the onshore Kouakouala field in May 2000. Initial production of about 1,000 bbl/d was trucked from the fields to the Congolaise de Raffinage (CORAF) refinery in Pointe Noire. Production doubled when a six-inch, 18 mile (30-kilometer)-pipeline was completed in April 2002. The pipeline runs from the Kouakouala field to the Djeno terminal. Production at Kouakouala is expected to reach over 4,000 bbl/d by the end of 2004. Heritage Oil's (30 per cent interest) partners on Kouakouala are: France's Maurel and Prom (35 per cent) and Tacoma Resources (35 per cent). Heritage sold its 30 per cent interest in the Kouilou License to partner Maurel and Prom in April 2002. The Kouilou License includes the M'Boundi discovery.

Exports

While most of Congo's crude oil exports are destined for Western Europe (mainly France) and the United States, Congo also has sought to increase its sales to Asian markets. Congo exported 23,000 bbl/d of crude to the United States during 2002, ranking Congo as the 22nd largest foreign supplier of crude oil to the United States. This is a dramatic decline from the 38,000 bbl/d Congo exported to the United States (18th largest) in 2001.

Field Development and Exploration

Total and its partners are continuing development and exploration on the Haute Mer permit, the location of the N'Kossa field. The Moho field, also in the Haute Mer area, was discovered in 1995. Moho has estimated reserves of 400 million barrels of oil. The $1 billion Moho/Bilondo oil project will develop three separate fields: Moho North, Moho South and Bilondo. The different crude oil and reservoir characteristics will make the task of designing a transport system suited to all fields difficult. Total initially looked at two options for the fields development–a subsea production system (tied back to existing infrastructure) or a Floating Production Unit (FPU–a converted oil tanker)–but both technologies will be utilized on Haute Mer. Contracts for both the FPU and subsea production system packages are set to be awarded by mid-2003. First oil is set to be produced in 2005 with output to hit 75,000 bbl/d of oil and 45 million cubic feet per day (Mmcf/d) of associated gas. Gas will be sent to Nkossa with oil piped to shore to the Djeno terminal via the Likouala platform.

Total also plans to develop two N'Kossa satellite fields, N'Kossa Sud Sud and N'Soko. N'Kossa Sud Sud lies south of the main N'Kossa field, and it is likely that it will be developed via subsea wells. Four wells have been drilled on the field to date. N'Soko lies three miles (five kilometers) southeast of N'Kossa and was discovered 10 years ago. Two production and one water injection well will be needed for the development. Nkossa Sud Sud development is scheduled to start in the third quarter of 2004. Nsoko has been approved for development to start in the second quarter of 2004.

Total is in the process of developing several smaller fields on the PNGF permit. Production on the Kombi and Likalala fields, with an estimated output of 40,000 bbl/d, began in December 1999. Other fields scheduled for development on the PNGF permit include Libondo, Tchibeli, Litanzi and Yanga-Sud. Total expects to begin production on all its PNGF fields by 2004. Total reserves are estimated to be 180 million barrels, and the total development costs are estimated at $135 million.

Total is also the operator of one of Congo's two ultra-deepwater blocks, Mer Tres Profonde Sud (MTPS). Total's (40 per cent) partners on MTPS are Agip (30 per cent) and ExxonMobil (30 per cent). In May 2000, Total announced an oil discovery on MTPS. The Andromede Marine-1 well, drilled in 6,200 feet (1,893 meters) of water, tested at 7,000 bbl/d. Agip (55 per cent interest) operates the Mer Tres Profonde Nord (MTPN) block. Total holds the remaining interest on MTPN.

In November 2001, Agip announced that it had struck oil at the deepwater Awa Marine-1 well on the Marine X permit. Oil flowed at a rate of 3,800 bbl/d with a density of 32° API gravity. Production from the field could be transported through existing infrastructures to the nearby Djeno terminal. Agip's (72 per cent interest) partners are South Africa's Sasol (18 per cent)and SNPC (10 per cent).

In July 2001, Heritage Oil and its partners announced an oil discovery at M'Boundi on the Kouilou License in Eastern Congo near the border with Cabinda (Angola). The M'Boundi well flowed at 1,476 bbl/d of 39° API crude oil. Heritage Oil estimates that M'Boundi could contain 100 million barrels of recoverable oil reserves. UK-independent, Burren Energy announced in January 2002 that it had purchased a stake in the PSA to develop the onshore oilfield containing the M'Boundi discovery. Burren acquired 51 per cent of the 35 per cent interest in the concession held by Swiss-based Tacoma Resources, and has an option on the remaining 49 per cent. Maurel and Prom holds the remaining 65 per cent, as it acquired Heritage's share in April 2002. Maurel and Prom was awarded the La Noumbi License located in northwestern Congo near the Gabonese border in April 2003. Initial exploratory drilling is planned for early 2004.

In May 2003, U.S.-based Murphy Oil signed PSAs for two adjoining blocks Mer Profonde Sud (MPS) and Mer Profonde Nord (MPN). MPS and MPN were relinquished by Total (MPS) and ExxonMobil (MPN) in 2001. Murphy Oil holds an 85 per cent interest in both blocks with the SNPC holding the balance. Water depths on the MPS and MPN range from 450 to 6,900 feet and total acreage is 1.8 million acres. Murphy plans to acquire additional seismic data and drill one exploratory well on MPS in 2004. Drilling on MPN is not anticipated to begin before 2005.

Perenco also plans to expand its position in Congo. Perenco is reportedly interested in acquiring Nuevo's per cent 0 per cent interest in the Yombo concession. Perenco is also interested in the Marine IV offshore block, adjacent to Gabonese waters. ChevronTexaco relinquished Marine IV in 1999.

Congo and Angola have set up a joint development area ending a border dispute that had blocked deepwater oil exploration. Framework establishing the ZIC (Zone d'Interet Commun) was signed in March 2003. ZIC, similar to Nigeria and Sao Tome and Principe's JDZ (Joint Development Zone),will initially concentrate on development of fields straddling Angola's Block 14 and Congo's Haute Mer field. Bruno Itoua, president of SNPC, says the two governments will split their share of oil revenues 50-50. Haute Mer partners are Total, EA, and ChevronTexaco. Block 14 partners are ChevronTexaco, Agip, Total, and Petrogal. ChevronTexaco will operate the acreage in the joint zone, and all companies active on both concessions will be entitled to take equity. An exploratory well on the joint-area is planned for the summer of 2003. An oil-bearing structure with the potential to hold 500 million–1 billion barrels of oil lies within the joint-area.

In an effort to reverse its declines in oil production, Congo plans to offer several blocks for bidding in the third quarter of 2003. Several designated blocks as well as onshore acreage are expected to be tendered. The offshore blocks lie in waters of depths up to 1,980 feet (600 meters). Congo plans to promote exploitation of its natural gas reserves, which several of the newly-offered zones are expected to contain.

Refining and Downstream

Congo's sole refinery, CORAF, was commissioned in 1974. The nameplate capacity of the refinery is 21,000 bbl/d, but the refinery often operates at less than half-capacity. CORAF, which was facing serious technical and financial problems, ceased refining activities in August 1997, forcing the country to import oil products, primarily from the Democratic Republic of Congo. In March 2000, CORAF resumed operations after nearly four years of inactivity. The government, in December 1999, provided 750 million CFA for restoration and maintenance of CORAF's facilities. In September 2000, fuel from the refinery was again able to reach Brazzaville due to the reopening of the CFCO rail line. The rail line had been closed for two years because of repeated sabotage by rebels, and sporadic attacks by Ninja rebels have continued.

In June 2000, a South Africa-brokered financial arrangement was announced to pay for the dredging of the port of Pointe Noire. The port, originally at a depth of 43 feet (13 meters), had silted up to a depth of only 23 feet (7 meters), effectively blocking large carrier ships. The last thorough dredging of Pointe Noire took place in 1987. The oil companies that use the port agreed to make direct payments to the South African bank which financed the dredging.

In May 2001, a tentative agreement was reached to divide oil distribution and marketing in Congo between Shell and Total, effectively privatizing the downstream oil sector. State-owned Hydro-Congo had held a monopoly in the sale and distribution of oil products in Congo for over 25 years. In August 2001, the Congolese government approved a bill privatizing the distribution and marketing of petroleum products. Oil Minister Jean-Baptiste Tati Loutard stated that the privatization would eventually lead to the closure of Hydro-Congo. Total and Shell have agreed to retain nearly two-thirds of the Hydro-Congo's staff.

In May 2002, it was announced that ChevronTexaco had stepped in to replace Shell in Congo's downstream oil sector, where it would operate alongside Total and the British consortium of Tacoma/ Puma-Energy. In July 2002, the three oil firms and the government formed a new company to transport oil products to Brazzaville. Congo's privatization committee stated that SNPC, Total, ChevronTexaco and Tacoma/Puma Energy would each hold 25 per cent in the new company, SCLOG (Societe Commune de Logistique Petroliere). ChevronTexaco also acquired 40 of Hydro-Congo's service stations, which will operate under the Texaco brand. Terminals and storage facilities at Pointe-Noire and Brazzaville will be upgraded.

Natural Gas

Congo contains an estimated 3.2 trillion cubic feet of natural gas reserves, the third largest gas resource base in sub-Saharan Africa behind Nigeria and Cameroon. Although the majority of Congo's natural gas reserves are associated (found alongside oil deposits), the country contains some major non-associated fields, including the offshore Banga Marine and Litchendjili Marine fields. The majority of Congo's natural gas output is currently vented or flared because of a lack of infrastructure, but the government plans to reduce this by utilizing the natural gas for electric power production over the next several years.

In December 2002, Agip and ChevronTexaco inaugurated a natural gas-fired power plant in Djeno near Pointe Noire. The plant, with a capacity of 25 megawatts (MW), was completed in 13 months at a cost of $32 million. The power plant is fired from associated gas from the Kitina, Djambala and Foukanda offshore fields. The gas is separated onshore at the Djeno oil terminal and piped to the nearby power plant.

Electricity

Congo currently has an electrical generation capacity, primarily hydropower, of 118 MW. The potential for hydroelectric power generation in Congo is large (3,000 MW), but it has been largely unexploited. The 74-MW Bouenza (Moukoukoulou) and the 15-MW Djoué hydroelectric plants have been the mainstays of Congo's power generating capability since the early 1980s.

Congo's electricity consumption currently is low, as most people in rural areas rely on wood as their primary source of fuel. Moreover, electricity transmission links are non-existent in many parts of the country, and fighting during the civil war periodically disrupted power supplies. Moukoukoulou, repeatedly a target for rebels during the civil war, has had its productive capacity reduced to 55 MW. In 1996, the Djoué hydropower station on the outskirts of Brazzaville was refurbished by Rotek, a branch of the South African power company, Eskom; but it was subsequently damaged in the fighting. Officials from Congo's parastatal electric utility, Societe Nationale d'Electricite (SNE) estimate that the cost of necessary repairs to the generating facilities at 15 billion CFA ($20.4 million). The generation shortage has forced Congo to import increasing amounts of power from the Democratic Republic of Congo. Congo plans to reduce its reliance on electricity imports by expanding current facilities and constructing additional generation facilities.

In November 2001, Congo and two Chinese companies signed a $220 million contract to build a 120-MW hydroelectric power plant on the Congo River. Construction of the facility is expected to take five years, and when completed it will double Congo's power generating capacity.

Development of the $925 million, 1-gigawatt (GW) Sounda Gorge hydroelectric project has been postponed. Sounda Gorge, at the confluence of the Niari and Kouilou Rivers, is located approximately 85 miles north of Pointe Noire. Before independence, the French built access roads, a cement-processing factory and a deviation tunnel at Sounda Gorge. Post-independence saw Electricite de France (EdF) unsuccessfully try to complete the generating facilities. The project, as envisioned by South African financiers, called for the implementation of a three-phase construction plan. The first phase would have involved the installation of two turbines in the deviation tunnel. These turbines would have had a capacity of 10 MW and would have been be used to generate cash flow for subsequent phases. The second phase would have entailed the construction of a 130-foot high dam, which would have boosted the plant's generating capacity to 240 MW. A third phase was to increase the dam's height to over 300 feet and its generating capacity to 1 GW.

Canadian-based Magnesium Alloy Corporation (MAC) originally proposed establishing a 180-MW hydroelectric facility on the Kouilou River. MAC says that the facility would primarily supply power for its magnesium extraction project, but an approximate 50 MW of power also would be available for sale to domestic and regional consumers. The facility plans were seemingly mooted when MAC announced, in September 2002, that it had signed a Memorandum of Understanding (MOU) with Eskom Enterprises. The MOU defined the steps to be undertaken jointly to establish a long-term power contract, with regional impacts, for the delivery of low-cost electricity to MAC's planned Kouilou Magnesium Project. Eskom will evaluate production and transmission assets to be built or rehabilitated, including the establishment of new wheeling routes that would link Congo to the Southern Africa Power Pool.

Prior to the 1997 civil war, SNE was one of several government entities considered for privatization. Subsequent damage inflicted upon SNE's infrastructure in the Brazzaville area during the civil war decreased the likelihood of immediate privatization. EdF had expressed interest in SNE, and offered assistance in restoring services to Brazzaville. In February 2002, British-based Bi-Water was chosen as the concessionaire for Congo's national water distribution company, the Societe Nationale de Distribution d'Eau (SNDE). Bi-Water was chosen over French firms Saur International and Vivendi Water. The three firms have submitted bids for the concession to run SNE.

Country Overview

Independence: August 15, 1960 (from France)

Population (2002E): 2.96 million

Location/Size: Central West Africa; borders the Atlantic Ocean, the Democratic Republic of the Congo (to the east and south), Gabon (to the west), Cameroon and the Central African Republic (to the north), and the Angolan enclave of Cabinda (to the south) / 342,000 square kilometers (132,000 square miles), slightly smaller than Montana

Major Cities: Brazzaville (capital), Pointe Noire

Languages: French (official), Kikongo, Lingala and Monokutuba, plus other African languages also spoken

Ethnic Groups: Kongo (48 per cent), Sangha (20 per cent), Batéké (17 per cent), M'Bochi (12 per cent), others including Europeans (3 per cent)

Religion: Christian–predominantly Roman Catholic (50 per cent), Traditional Beliefs (48 per cent), Muslim (2 per cent)

Defense (1998E): Army (8,000), Navy (800), Air Force (1,200), Paramilitary Forces (5,000)

Economic Overview

Currency: Communauté Financière Africaine (CFA) franc

Market Exchange Rate (7/10/03): US$1 = 600.21 CFA

Gross Domestic Product (GDP) (2002E): $3 billion

Real GDP Growth Rate (2002E): 3.1 per cent *(2003F):* 3.6 per cent

Inflation Rate (2002E): 2.6 per cent *(2003F):* 2.6 per cent

Merchandise Exports (2002E): $2.3 billion *(2003F):* $2.3 billion

Merchandise Imports (2002E): $0.7 billion *(2003F):* $0.7 billion

Major Export Products: Crude oil, timber, sugar, cocoa, coffee, diamonds

Major Import Products: Capital goods, foodstuffs, petroleum products, machinery, vehicles and spare parts

Major Trading Partners: United States, France, South Korea, China, Germany, Italy

Total External Debt (2000E): $4.9 billion

Reserves excluding gold (2000E): $222 million

Energy Overview

Proven Oil Reserves (1/1/03E): 1.5 billion barrels

Oil Production (2002E): 249,000 barrels per day (bbl/d), of which all is crude oil *(2003Q1E):* 240,000 bbl/d

Oil Consumption (2002E): 5,000 bbl/d *(2003F):* 5,000 bbl/d

Net Oil Exports (2002E): 244,000 bbl/d

Refining Capacity (1/1/03E): 21,000 bbl/d

Natural Gas Reserves (1/1/03E): 3.2 trillion cubic feet (Tcf)

Natural Gas Consumption (2001E): None

Natural Gas Vented/Flared (2000E): 125.8 million cubic feet per day

Coal Reserves: None

Coal Production/Consumption (2001E): None

Electric Generation Capacity (1/1/01): 118 megawatts

Electricity Generation (2001E): 358 million kilowatthours (MMkwh)

Electricity Consumption (2001E): 633 MMkwh

Environmental Overview

Total Energy Consumption (2001E): 0.017 quadrillion Btu* (0.004 per cent of world total energy consumption)

Energy-Related Carbon Emissions (2001E): 0.844 million metric tons of carbon (0.129 per cent of world carbon emissions)

Per Capita Energy Consumption (2001E): 5.5 million Btu (vs U.S. value of 341.8 million Btu)

Per Capita Carbon Emissions (2001E): 0.3 metric tons of carbon (vs U.S. value of 5.5 metric tons of carbon)

Fuel Share of Energy Consumption (2001E): Oil (60.4 per cent), Hydroelectric Power (21.5 per cent)

Fuel Share of Carbon Emissions (2001E): Natural Gas (76.5 per cent), Oil (23.5 per cent)

Status in Climate Change Negotiations: Non-Annex I country under the United Nations Framework Convention on Climate Change (ratified October 14th, 1996). Not a signatory to the Kyoto Protocol.

Major Environmental Issues: Air pollution from vehicle emissions; water pollution from the dumping of raw sewage; tap water is not potable; deforestation.

Major International Environmental Agreements: A party to Conventions on Biodiversity, Climate Change, Endangered Species, Ozone Layer Protection, Tropical Timber 83, Tropical Timber 94. Has signed, but not ratified: Desertification and Law of the Sea.

The total energy consumption statistic includes petroleum, dry natural gas, coal, net hydro, nuclear, geothermal, solar and wind electric power. The renewable energy consumption statistic is based on International Energy Agency (IEA) data and includes hydropower, solar, wind, tide, geothermal, solid biomass and animal products, biomass gas and liquids, industrial and municipal wastes. Sectoral shares of energy consumption and carbon emissions are also based on IEA data.

Oil and Gas Industries

Organization: State-owned Société Nationale des Pétroles du Congo (SNPC) oversees offshore and onshore oil exploration and production: state-owned Hydro-Congo participates in downstream oil sector (marketing); Societe Commune de Logistique Petroliere (SCLOG) responsible for petroleum product distribution (SNPC, Total, ChevronTexaco and Tacoma/Puma Energy each hold 25 percent).

Major Producing Fields: N'Kossa, Loango, Zatchi, Kitina

Major Refineries (1/1/03 Capacity bbl/d): Congolaise de Raffinage (CORAF)–Pointe Noire (21,000 bbl/d)

Major Oil Terminals: Djeno

Major Foreign Oil Company Involvement: Agip, Anadarko, ChevronTexaco, Energy Africa, ExxonMobil, Heritage Oil, Maurel and Prom, Murphy Oil, Perenco, Sasol, Shell, Tacoma, Total

30

Cote d'Ivoire

Natural gas reserves and excess electricity generating capacity could lead to Cote d'Ivoire becoming a significant regional energy supplier in coming years. Recent offshore discoveries in the Gulf of Guinea, including natural gas finds in its territorial waters, make Cote d'Ivoire a leading area for hydrocarbon exploration in sub-Saharan Africa.

General Background

Cote d'Ivoire is currently in a state of political upheaval, following a failed coup attempt in September 2002. Laurent Gbagbo is still president, although his government controls only the southern half of thecountry. In 2000, Gbagbo defeated General Robert Guei in a controversial election that saw candidates from Cote d'Ivoire's two main parties–Cote d'Ivoire Democratic Party (PDCI) and Rally of the Republicans (RDR)–excluded from the race. Guei had assumed the presidency following the Christmas Eve 1999 coup that toppled President Henri Konan Bedie. Although peace had been restored for two years after Gbagbo's election, violence erupted in September 2002 during a military coup that ultimately failed. Since then, the rebel group Patriotic Movement for Cote d'Ivoire (MPCI) has taken over cities in the northern half of the country, while a second group has taken over cities in the west, where many of the nation's cocoa plantations are located. So far there is no sign that oil production has been affected by the turmoil.

Cote d'Ivoire's economy is heavily reliant on agriculture (primarily cocoa, coffee and timber. which accounts for approximately 40 per cent of Cote d'Ivoire's gross domestic product (GDP) and approximately 70 per cent of the total export earnings. The country's real GDP shrank by 0.9 per cent in 2001, due to low cocoa prices, but was expected to increase by 3.5 per cent in 2002. Inflation declined from 4.3 per cent in 2001 to an estimated 3.0 per cent in 2002. Since 1990, the government of Cote d'Ivoire (GOC) has privatized 44 of the 61 entities it had scheduled for privatization. In 1990, the GOC privatized the management of the Compagnie Electricite Ivoirienne (CIE), the state electric utility.

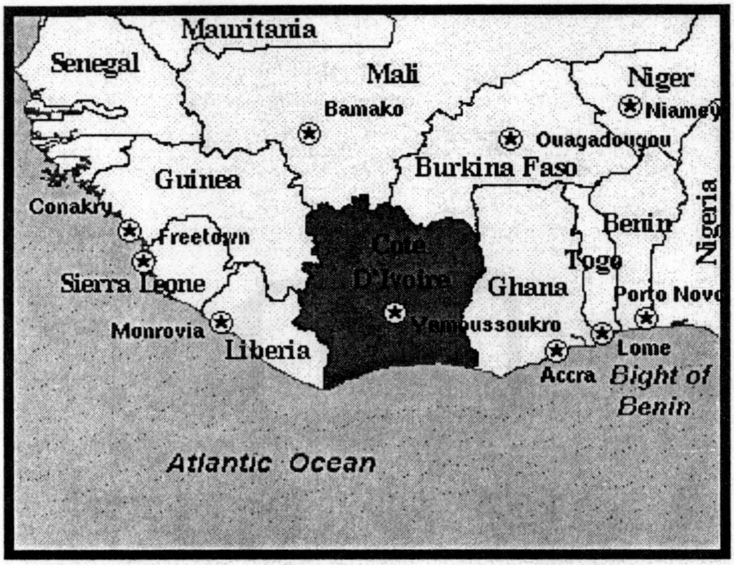

Other enterprises scheduled for privatization include the state's portion of the Societe Ivoirienne Raffinage (SIR) oil refinery, and segments of Petrolieres de la Cote d'Ivoire (Petroci), the national oil company.

Between 1999 and 2002, the World Bank did not provide any budgetary assistance to Cote d'Ivoire, citing the country's poor progress in economic management and governance. Although cocoa and coffee production increased, secondary sector industries, such as telecommunications, mechanical and electrical industries, textiles, garments, and public works all experienced negative growth. In January 2002, Cote d'Ivoire cleared its arrears with the World Bank, while the country's successful economic performance led to a staff-monitored program (SMP) and later to the approval of a Poverty Reduction and Growth Facilty (PRGF) by the International Monetary Fund (IMF). The nation's Poverty Reduction Strategy Paper was due to be finalized in December 2002, although it may be delayed due to political and economic problems stemming from the failed coup attempt in September 2002 and the violence that has followed.

Cote d'Ivoire is an important country in West Africa. It is the leading member (accounting for 41 per cent of the group's combined GDP) of the West African Economic and Monetary Union (UEMOA). UEMOA is composed of eight West African states (Benin, Burkina Faso, Cote d'Ivoire, Guinea Bissau, Mali, Niger, Senegal and Togo) that share the same currency (the CFA franc). Cote d'Ivoire is also a member of the 16-nation Economic Community of West African States (ECOWAS). In September 1998, the Abidjan Stock Exchange was replaced by the Bourse Regionale des Valeurs Mobilieres (BRVM), a regional stock exchange. The BRVM serves as the regional exchange for the member nations of the UEMOA.

Oil

Cote d'Ivoire contains an estimated 100 million barrels of recoverable oil reserves, with offshore reserves first discovered in the 1970's. The national oil company, Petroci, was established in 1975.

Petroci was restructured in 1998 and four new entities were created: Petroci Holding, a fully state-owned company that is responsible for the state's portfolio management in the oil sector and the three subsidiaries; Petroci Exploration-Production, responsible for upstream hydrocarbon activities; Petroci-Gaz, responsible for development of the gas sector; and Petroci Industries-Services, responsible for all other related services. Up to 49 per cent interest in the three subsidiaries is available to private sector investors.

In June 2002, Calgary-based Canadian Natural Resources (CNR), through its wholly-owned subsidiary Ranger Oil Cote d'Ivoire, signed a production-sharing contract (PSC) for the CI-400 Block. Block CI-400 is comprised of parts of two other blocks, and is located in deep waters of up to 2,000 meters. CNR also holds interests in several other offshore blocks, including a 58.67 per cent interest and the role of operator in the PSC for Block CI-26, which contains the previously abandoned Espoir field. CNR's partners on CI-26 are Ireland's Tullow Oil (21.33 per cent) and Petroci (20 per cent).

Re-development of the Espoir field, which was closed down in 1988 due to high production costs, has been estimated at $265 million. The field includes two unmanned wellhead platforms with hydrocarbon processing on a Floating, Production, Storage and Offloading (FPSO) vessel. The FPSO has crude oil production capacity of 40,000 barrels per day (bbl/d) and water injection capacity of 60,000 bbl/d. Espoir's oil production is exported by shuttle tanker, while the natural gas is piped to shore where it is used to generate electricity. First production occurred in January 2002, and the field has a life expectancy of 20-25 years. Initial production began at 12,000 bbl/d of oil and 34 mmcf/d of natural gas and is expected to peak at 28,000 bbl/d of oil and 35 million cubic feet per day (mmcf/d) of natural gas. In March 2002, the new FPSO, Espoir Ivoirien, tested successfully and began production. The FPSO is contracted to produce for an initial ten-year term, with ten one-year renewal options after that. Espoir's recoverable reserves are estimated at 93 million barrels of oil and 180 billion cubic feet (bcf) of gas.

CNR operates Block CI-40 with a 61 per cent interest, and is joined by Svenska Petroleum (29 per cent) and Petroci (10 per cent). In March 2001, CNR announced a discovery on Block CI-40, Cote d'Ivoire's first deepwater oil find. The Baobab discovery had a flow rate of 6,700 bbl/d of heavy (22°-23° API) oil and the well is expected to come online by 2005. A second successful well was completed in February 2002 and tested at a rate of 10,000 bbl/d. Block CI-40 lies 5 miles (8 kilometers) south of the CNR-operated Espoir field and is now estimated to have around 150 million barrels of recoverable oil reserves.

U.S.-based Ocean Energy (Ocean) operates the Lion and Panthere fields on Block CI-11. Current production from the block is 20,000 bbl/d of oil and nearly 70 million cubic feet per day (mmcf/d) of natural gas. Ocean's partners on Block CI-11 are Petroci, Pluspetrol of Argentina, and International Finance Corporation. Ocean has continued development work on Block CI-11 and is now extracting oil from the Lion A5 deep-sea oil well. Recoverable reserves on Block CI-11 are estimated to be 210 million barrels of oil and 495 bcf of natural gas.

Ocean holds interests (ranging from 35 per cent -80 per cent) in several other blocks in Cote d'Ivoire. Development on offshore block CI-01, where several small fields have been discovered, is continuing. Ocean is the operator of that block with an 80 per cent interest. Although primarily natural gas discoveries have been made so far, the Gazelle field has recoverable oil reserves estimated at 8 million barrels. Ocean is the majority holder on Blocks CI-02, CI-12 and CI-104. Ocean holds a 35 per cent interest and is the operator on Cote d'Ivoire's first deep-water offshore block, CI-105. Ocean's partners are Shell (55 per cent) and Petroci (10 per cent).

The U.K.-based Dana Petroleum has an exploration agreement with Cote d'Ivoire for Block CI-100, which is located directly west of acreage held by Dana in Ghana. Dana made an oil discovery on the Ghanaian block in March 2000, and hopes that the oil-bearing structures continue into Ivorian waters. CI-100 is adjacent Block CI-01 where hydrocarbon discoveries were also made.

Refining and Downstream Oil Activities

Cote d'Ivoire's refining facilities are composed of the 65,200-bbl/d SIR refinery and an adjacent 10,000-bbl/d asphalt plant in Abidjan. An oil pipeline connects the SIR refinery to the Lion and Panthere fields. The refinery also receives crude oil from Nigeria for refining. SIR is scheduled to be almost fully privatized by the end of 2003, after several long delays due to political turmoil. The state currently owns approximately 47 per cent of SIR, and expects to retain a 10 per cent interest after privatization. Other shareholders in SIR are: TotalFinaElf (25.3 per cent), Shell (10.3 per cent), ExxonMobil (8.1 per cent), the government of Burkina Faso (5.4 per cent), ChevronTexaco (3.7 per cent) and the GOC (1.5 per cent). The GOC's interest in SMB also has been scheduled for privatization. SMB's industry partners are SIR (53 per cent), Petroci (20 per cent) and Shell (6 per cent). An additional 21 per cent is held privately (SMB stock is traded on the BRVM).

A petroleum products depot adjacent to SIR stores petroleum products for domestic use as well as for export to other countries in the West African region. The depot, owned by the Societe de Gestion des Stocks Petroliers de Cote d'Ivoire (Gestoci) supplies products to Mali, Burkina Faso, Niger and Chad. Nigeria has received shipments of refined products from Cote d'Ivoire to help ease fuel shortages created by problems with its refineries. Gestoci operates fuel depots in Bouake and Yamoussoukro.

Gestoci, currently joint-owned by the GOC and the petroleum marketers, is set to be restructured. The plan calls for the GOC to have its stake in Gestoci reduced to 34 per cent. The GOC anticipates that refinery expansion coupled with upgrades to Gestoci's Abidjan storage facilities will transform the country into the "Rotterdam of Africa", and make Cote d'Ivoire the main supplier of refined products on Africa's Atlantic seaboard. A products pipeline from Abidjan to Bouake is being considered. Currently, products are shipped by rail and truck to the Bouake depot. The line, if constructed, may be extended to Burkina Faso, Guinea and Mali. Several foreign oil companies (and one local firm, PetroIvoire) are involved in the distribution and marketing of refined products in Cote d'Ivoire. ENI-Agip, ExxonMobil, ChevronTexaco, TotalFinaElf and Shell control over 90 per cent of the downstream retail sector.

Natural Gas and LPG

Natural gas reserves in Cote d'Ivoire, first discovered in the 1980's, recently have begun to be developed and utilized. Current estimates of Cote d'Ivoire's recoverable natural gas reserves stand at 1.1 trillion cubic feet. Over the next four years, the GOC estimates that natural gas consumption will grow by 50 per cent. Natural gas production from Espoir is expected to average 35 mmcf/d over 20 years.

Ocean and its partners are currently producing natural gas from the Panthere field on Block CI-11. The natural gas is transported to a site near Abidjan via pipeline, where it is used primarily to generate electricity. Future industrial uses are planned. Ocean announced in April 1997 that it had signed a ten-year, take-or-pay agreement to supply 170 bcf of gas for electricity generation. Under a take-or-pay agreement the buyer agrees to pay a certain price for a specified amount of gas, whether the buyer actually uses (takes) the gas or not. The gas would be supplied from fields operated by Ocean located offshore eastern Cote d'Ivoire in Block CI-01 (Kudu, Eland and Ibex fields) and Block CI-02

(Gazelle field). The contract calls for initial deliveries, to begin in early 1999, of 30 mmcf/d for the first two years increasing to 50 mmcf/d for the remainder of the contract.

In April 1997, U.S.-based Apache was granted a ten-year take-or-pay contract for Block CI-27. Apache's partners on the field are Electricite de France (EDF), Petroci and Saur–Bouygues (SAUR). The offshore block, which contains the Phillips-discovered Foxtrot natural gas field, is primarily seen as having the greatest potential as a gas producer. The original estimate of recoverable reserves on Block CI-27 included over six million barrels of condensate. Foxtrot is now the country's largest natural gas deposit, with an estimated 650 bcf of recoverable gas. The partners initiated production in June 1999. In September 1999, Apache reduced and restructured its holdings on Block CI-27 by selling its stake in Foxtrot field to Mondoil. Mondoil and SAUR assumed joint management of the operation. Apache continues to provide technical services to the project, and it retains an interest through its participation in Mondoil (Apache is a shareholder in Mondoil).

Cote d'Ivoire is poised to become a regional natural gas exporter. Negotiations between Cote d'Ivoire and Ghana to adopt a memorandum of understanding (MOU) on the sale of gas began in 1997. In April 1999, an agreement was signed by the governments of Cote d'Ivoire and Ghana with the UK government and Penspen for a feasibility study to build a gas pipeline between Cote d'Ivoire and Ghana. The UK Department of Trade and Industry has said it will share the cost of the study with Penspen. The pipeline would run from Abidjan to Takoradi in Ghana, which is the location of a power plant currently fueled by light oil. Construction of the pipeline is estimated to take 15-18 months. The possibility of connecting the line to the proposed West Africa Gas Pipeline (WAGP) is being studied.

The GOC estimates that domestic consumption of butane will more than double over the next five years. In 1998, Ocean began operation of a LPG extraction plant it built near Abidjan. The facility produces 40,000 metric tons of liquefied petroleum gas (LPG) and gasoline annually, from 70 mmcf/d of natural gas feedstock. The GOC and Morocco signed an agreement in September 1998 for the establishment of an industrial plant in Cote d'Ivoire, Simgaz, to produce gas cylinders. Estimated production from the plant is 350,000 gas cylinders per year. A pilot LPG-fuelled vehicle program has been proposed. Two filling and conversion stations will be constructed in Abidjan.

Electricity

Cote d'Ivoire currently has installed electric generation capacity of 0.89 gigawatts. Approximately three-fourths of Cote d'Ivoire's installed generating capacity is hydroelectric, although hydroelectricity accounts for only 20 per cent of generated electricity. The remaining 80 per cent is provided by thermal generating facilities powered primarily by oil. The major hydroelectric facilities are: Ayame I, Ayame II, Buyo, Grah, Kossou and Taabo. An additional hydroelectric facility at Soubre is being considered. A feasibility study for the facility, which would be constructed under the build, own, operate transfer (BOOT) scheme, was delayed by Cote d'Ivoire's political upheavals. Construction of the facility will take six to ten years.

Cote d'Ivoire is a leading proponent of the development of the West African Power Pool (WAPP). The WAPP was formed by an agreement between ECOWAS Energy Ministers in November 1999. Although the formal structure and operational agreements for the pool have yet to be determined, it is envisioned that the WAPP will interconnect the power networks of all the mainland ECOWAS nations. Cote d'Ivoire is currently tied to Ghana, Togo and Benin through one interconnection, and a link to Burkina Faso is currently under construction as well. Additional connections to Mali and Guinea are being studied. At a summit in Ghana in April 2002, it was announced that the ambitious project still needs to secure at least $10 billion in financing over the next fifteen years if the project is to be completed. It is hoped that most of the funding will come form the New Partnership for African Development (NEPAD). The U.S. Federal Energy Regulatory Commission (FERC) also signed a deal that links the Indiana Public Utilities Regulatory Commission to the WAPP project.

A new thermal power plant is being built at Azito outside of Abidjan by a consortium named Azitoenergie, which includes: Swiss-based Asea Brown Boveri (ABB), Industrial Promotion Services (IPS)–an affiliate of the Aga Khan Fund for Economic Development–and EDF. The plan called for a $225 million, 420-megawatt (MW) facility, with operation of the first 144-MW phase beginning in early 1999. Construction of the plant began in the last quarter of 1997. The first phase of the Azito power plant was inaugurated on January 23, 1999. A second, 144-MW natural gas turbine was commissioned in February 2000. A steam-powered turbine and two recovery boilers, which will transform the facility into the combined-cycle format, will be added to boost the facility's capacity to the planned 420 MW. The project has received financing from the International Finance Corporation (IFC), part of the World Bank group, announced that it had entered into a 14-year, $32 million interest rate swap with the Azito project. The agreement followed the IFC's earlier $60 million loan to the project. The Commonwealth Development Corporation (CDC) and the AfDB also provided financing for the Azito project.

CIE handles the management of the government-owned generation facilities as well as transmission and distribution of electricity in Cote d'Ivoire. SAUR and EDF are the majority owners of CIE (51 per cent share). Other investors in CIE included the GOC (20 per cent) and private and employee investors (29 per cent). CIE has nearly doubled the number of its customers in the ten years following its privatization. The number of customers has increased from 400,000 (in 1991) to 750,000 (as of January 2001).

Country Overview

Prime Minister: Affi N'Guessan

Independence: August 7, 1960 (from France)

Population (July 2002): 16.8 million

Location/Size: Western Africa, bordering the Atlantic Ocean, between Liberia and Guinea (on the west); Ghana (on the east); Burkina Faso and Mali (on the north)/322,460 square kilometers (124,500 square miles), slightly larger than New Mexico

Major Cities: Yamoussoukro (capital), Abidjan, Bouake

Languages: French (official), over 60 native dialects

Ethnic Groups: Akan (the Baoule subgroup accounts for 23 per cent of the population), Kru (the Bete subgroup accounts for 18 per cent), Senoufo 15 per cent, Malinke 11 per cent, Agni, Mande. There are nearly 3 million foreign Africans (mainly from Burkina Faso and Mali), and an estimated 130,000-330,000 non-Africans (French 30,000 and Lebanese 100,000-300,000)

Religion: Muslim 60 per cent, Christian 22 per cent (predominantly Roman Catholic), traditional beliefs 18 per cent

Defense (1999E): Army (6,800), Navy (900), Air Force (700), Republican Guard (includes Presidential Guard) (1,100), Gendarmerie (4,400).

Economic Overview

Currency: Communaute Financiere Africaine (CFA) franc

Market Exchange Rate (12/09/02): US$1 = 650.0 CFA

Gross Domestic Product (GDP at market exchange rate) (2002E): $10.2 billion

Real GDP Growth Rate (2001E): -0.9 per cent *(2002E):* 3.5 per cent

GDP per Capita (2002E): $609

Inflation Rate (2001): 4.3 per cent *(2002E):* 3.0 per cent

Current Account Balance (2001E): -$2.9 billion

Major Trading Partners: European Union, Nigeria, United States, Ghana, Burkina Faso, Mali

Merchandise Exports (2001E): $3.69 billion *(2002E):* $3.92 billion

Merchandise Imports (2001E): $1.86 billion *(2002E):* $1.93 billion

Merchandise Trade Balance (2001E): $1.83 billion *(2002E):* $1.99 billion

Major Export Products: Cocoa, coffee, timber, petroleum products

Major Import Products: Food, consumer goods, industrial goods, machinery

Total External Debt (2000E): $13.2 billion

Energy Overview

Proven Oil Reserves (1/1/02E): 100 million barrels

Oil Production (2002E): 11,400 barrels per day (bbl/d), of which 10,400 bbl/d is crude oil

Oil Consumption (2001E): 30,000 bbl/d *(2002E):* 30,000 bbl/d

Net Oil Imports (2002E): 18,600 bbl/d

Refining Capacity (1/1/02E): 65,200 bbl/d

Natural Gas Reserves (1/1/02E): 1.05 trillion cubic feet (tcf)

Natural Gas Production (2000E): 47.3 bcf

Natural Gas Consumption (2000E): 47.3 billion cubic feet

Electric Generation Capacity (1/1/00E): 0.89 million kilowatts

Electricity Generation (2000E): 5.0 billion kilowatthours (of which 80 per cent is thermal, 20 per cent is hydroelectric)

Environmental Overview

Total Energy Consumption (2000E): 0.11 quadrillion Btu (0.027 per cent of the world total)

Energy-Related Carbon Emissions (2000E): 1.86 million metric tons of carbon (0.050 per cent of world carbon emissions)

Per Capita Energy Consumption (2000E): 6.6 million Btu (vs. U.S. value of 351.0 million Btu)

Per Capita Carbon Emissions (2000E): 0.1 metric tons of carbon (vs. U.S. value of 5.6 metric tons of carbon)

Energy Intensity (2000E): 8,745 Btu/ $1995 (vs. U.S. value of 10,918 Btu/ $1995)**

Carbon Intensity (2000E): 0.15 metric tons of carbon/thousand $1995(vs. U.S. value of 0.17 metric tons/thousand $1995)**

Fuel Share of Energy Consumption (2000E): Oil (57.2 per cent), Natural Gas (43.8 per cent)

Fuel Share of Carbon Emissions (2000E): Oil (63.4 per cent), Natural Gas (36.6 per cent)

Status in Climate Change Negotiations: Non-Annex I country under the United Nations Framework Convention on Climate Change (ratified November 29th, 1994). Not a signatory to the Kyoto Protocol.

Major Environmental Issues: Deforestation (most of the country's forests, once the largest in West Africa, have been cleared by the timber industry); water pollution from sewage and industrial and agricultural effluents

Major International Environmental Agreements: A party to Conventions on Biodiversity, Climate Change, Desertification, Endangered Species, Hazardous Wastes, Law of the Sea, Marine Dumping, Nuclear Test Ban, Ozone Layer Protection, Ship Pollution, Tropical Timber 83, Tropical Timber 94 and Wetlands

The total energy consumption statistic includes petroleum, dry natural gas, coal, net hydro, nuclear, geothermal, solar and wind electric power. The renewable energy consumption statistic is based on International Energy Agency (IEA) data and includes hydropower, solar, wind, tide, geothermal, solid biomass and animal products, biomass gas and liquids, industrial and municipal wastes. Sectoral shares of energy consumption and carbon emissions are also based on IEA data

Oil and Gas Industries

Organization: Petrolieres de la Cote d'Ivoire (Petroci) has four operating subsidiaries (Petroci-Holding, Petroci-Exploration/Production, Petroci-Gaz, and Petroci-Industries/Services

Major Oil Fields: Lion, Panthere

Major Gas Fields: Foxtrot, Kudu, Panthere

Major Refineries (1/1/02 Capacity): Societe Ivoirienne de Raffinage (SIR)–Abidjan (65,200 bbl/d), Societe Multinationale des Bitumes (SMB)–Abidjan (10,000 bbl/d)

Foreign Oil Company Involvement: Addax, Apache, Canadian Natural Resources, Dana Petroleum, Energy Africa, ENI-Agip, ExxonMobil, Gentry, Gulf Canada, Ocean Energy, PanCanadian Petroleum, Pluspetrol, Royal Dutch/Shell, Santa Fe Snyder, Stratic Energy, Texaco, TotalFinaElf, Tullow, Vanco Energy

31

Economic Community of West African States (ECOWAS)

The following provides a brief economic and energy sector overview of the fifteen countries that make up the Economic Community of West African States (ECOWAS). Information on non-ECOWAS member Mauritania is also included. ECOWAS members include: Benin, Burkina Faso, Cape Verde, Cote d'Ivoire , The Gambia, Ghana, Guinea, Guinea-Bissau, Liberia, Mali, Niger, Nigeria, Senegal, Sierra Leone, and Togo.

General Background

The Economic Community of West African States (ECOWAS) was created on May 28, 1975 in Lagos, Nigeria. ECOWAS was established to promote cooperation and integration in order to create an economic and monetary union for promoting economic growth and development in West Africa. ECOWAS has encountered many problems in the process of regionally integrating West Africa, including: political instability and lack of good governance that has plagued many member countries; the insufficient diversification of national economies; the absence of reliable infrastructure; and the multiplicity of organizations for regional integration with the same objectives. Several ECOWAS-member countries are currently part of the West African Monetary Union (UEMOA), a regional economic and monetary union which shares a common currency (the CFA Franc). The Francophone-countries of Benin, Burkina Faso, Cote d'Ivoire, Mali, Niger, Senegal and Togo, with Guinea Bissau (Lusophone), comprise UEMOA.

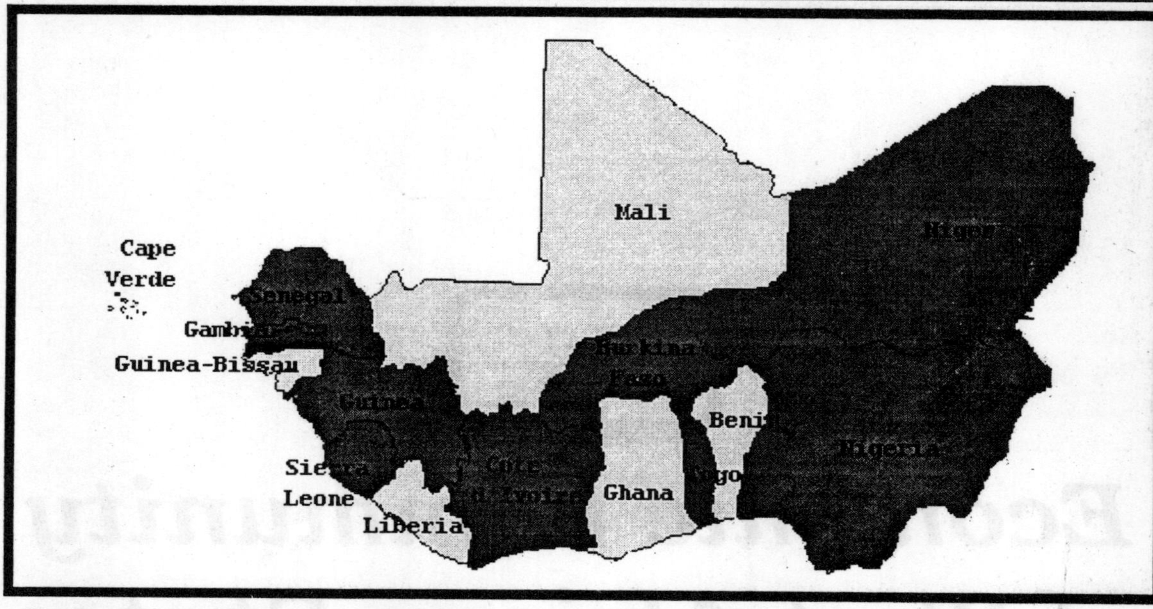

Senegal's President, Abdoulaye Wade, is the current ECOWAS Chairman. President Wade is continuing the process of focusing on regional economic integration. In April 2002, the ECOWAS Council of Ministers (COM) approved a new procedure for the ECOWAS Trade Liberalization Scheme (TLS). The TLS entitles the manufacturers of approved products to customs duty exemption within ECOWAS member states. The new procedure will consist of National Approval Committees, which are to be set up by member states, to handle the approval of products to be granted exemption under TLS. The decision by COM, abrogates the decision of 1988 granting the COM the monopoly for approving applications for such exemptions. It followed the harmonization of the ECOWAS procedure with that of the UEMOA, which already operates under this procedure.

ECOWAS is seeking international support to enable it train and equip the 15 battalions of troops pledged by member states as standby units for its peacekeeping force, ECOMOG. The training of the composite units will facilitate their effectiveness in peacekeeping, humanitarian assistance and other missions for which they could be deployed. ECOMOG forces have been deployed previously in civil conflicts in Sierra Leone and Liberia. Senegal announced in November 2002 that it is to boost its contribution to the ECOWAS military mission to Cote d'Ivoire and provide the Force Commander. Five countries–Benin, Ghana, Niger, Senegal and Togo–are contributing the 1,264 troops for the first phase of the mission. The force will take over from French troops who have been monitoring the October 17, 2002 agreement for the cessation of hostilities under an agreement involving ECOWAS, France and the government of Cote d'Ivoire. Nigeria will also participate in the mission and it has pledged to provide medical and signal teams as its contribution. Talks continue with representatives of the Patriotic Movement for Cote d'Ivoire (MPCI), the rebel group which has taken control of several cities in the northern portion of the country.

Economic Overview

In 2001, the combined Gross Domestic Product (GDP) for the ECOWAS region was estimated at $75.1 billion (see Table 1.). ECOWAS economies are at varying stages of development. Nigeria's economy, the region's largest with a GDP of $39.5 billion, is larger than the combined GDP of the other ECOWAS countries. While the region's economies grew at a combined rate of 3.4 per cent in 2001, the substantial external debt of individual states remains one of the region's greatest challenges. Internal strife has adversely affected economic performance in several states.

Table 31.1: Economic and Demographic Indicators

Country	*Gross Domestic Product (Nominal GDP), 2001E (Billions of U.S. $)*	*Real GDP Growth Rate, 2001 Estimate*	*Real GDP Growth Rate, 2002 Projection*	*Per Capita GDP, 2001E*	*Population 2001E (Millions)*
Benin	$2.4	4.9%	4.8%	$366	6.8
Burkina Faso	$2.4	5.1%	5.2%	$205	12.6
Cape Verde	$0.6	2.8%	3.4%	$1,251	0.4
Cote d'Ivoire	$9.2	-0.9%	3.5%	$562	16.8
Gambia	$0.4	5.0%	4.1%	$275	1.5
Ghana	$5.3	3.9%	5.1%	$269	20.2
Guinea	$2.9	2.9%	2.4%	$380	7.8
Guinea-Bissau	$0.2	7.2%	6.2%	$186	1.3
Liberia	$0.5	5.3%	4.2%	$188	3.3
Mali	$2.6	1.5%	7.9%	$237	11.3
Mauritania	$0.9	4.3%	3.7%	$344	2.8
Niger	$1.6	4.6%	2.9%	$139	10.6
Nigeria	$39.5	3.9%	3.3%	$302	129.9
Senegal	$4.6	5.7%	4.9%	$472	10.6
Sierra Leone	$0.7	5.5%	5.2%	$146	5.6
Togo	$1.3	2.2%	2.0%	$274	5.3
Regional Total/Average	**$75.1**	**3.4%**	**3.8%**	**$304**	**246.8**

Total regional exports, including intra-regional exports, were $26.1 billion in 2001. ECOWAS had a $2.5 billion trade surplus for 2001. The region's major export commodities were energy products (crude oil and refined petroleum products), minerals (gold, diamonds, and bauxite) and agricultural products (cocoa, coffee, groundnuts, and cotton). Overall trade between the United States and ECOWAS countries fell by 11.2 per cent in 2001. U.S. exports to the region grew 23 per cent in 2001, to nearly $1.7 billion. U.S. imports, with Nigerian crude oil accounting for the vast majority, from the region fell by 15.3 per cent in 2001, to $9.6 billion. On December 31, 2002, President Bush approved the designation of 38 sub-Saharan African countries as eligible for tariff preferences under the African Growth and Opportunity Act (AGOA). As required by the legislation, this annual determination signifies which countries are making continued progress toward a market-based economy, the rule of law, free trade, economic policies that will reduce poverty, and protection of worker's rights. Burkina Faso, Liberia and Togo were the only countries in the region not covered by AGOA.

It was announced in November 2002, that the five member states of the proposed second monetary zone have postponed the date for the launching of their common currency, the Eco, from January 1, 2003 to July 1, 2005. The decision to postpone the launching was taken following the inability of the countries to satisfy the four convergence criteria for their economies prior to the introduction. The five states–The Gambia, Ghana, Guinea, Nigeria and Sierra Leone,–which signed the 2000 Accra Declaration for the creation of the second monetary zone agreed to reform their economies to meet specific targets prior to the introduction of the Eco. These include a restriction on budget deficit to no more than 4 per cent by 2002, reducing inflation to 5 per cent by 2003, a ceiling on central bank financing of budget deficit to 10 per cent of the previous year's revenue and a minimum foreign reserve that would support at least six months of imports by 2003. The Eco will circulate simultaneously with the CFA Franc for a specific period prior to the creation of a single currency for West Africa. Liberia, which attended some meetings of the new zone, indicated interest in the membership but it has not signed the Accra Declaration formalizing membership. Cape Verde has attended some meetings of the zone as observer.

Energy Overview

Nigeria is ECOWAS's only net energy exporter, and the size of its exports alone makes the region a net energy exporter. In 2001, the countries of West Africa collectively consumed. 1.48 quadrillion British thermal units (Btu) of commercial energy (0.4 per cent of total world consumption) and produced 5.70 quadrillion Btu (1.4 per cent of total world production). Also in 2001, the region generated 31.7 million metric tons of carbon emissions (0.5 per cent of the world total). Nigeria accounted for 61.9 per cent (0.92 quadrillion Btu) of energy consumption, 96.3 per cent (5.49 quadrillion Btu) of energy production, and 74.2 per cent (23.52 million metric tons) of the region's carbon emissions.

Table 31.2: Total Energy and Carbon Dioxide Emissions, 2001

Country	*Total Energy Consumption, (Quadrillion Btu)*	*Total Energy Production, (Quadrillion Btu)*	*Net Energy Exports, (Quadrillion Btu)*	*Carbon Dioxide Emissions (Million Metric Tons of Carbon)*
Benin	0.030	0.004	-0.009	0.45
Burkina Faso	0.018	0.001	-0.010	0.31
Cape Verde	0.004	0.000	-0.002	0.06
Cote d'Ivoire	0.120	0.083	-0.025	1.92
Gambia	0.004	0.000	-0.003	0.08
Ghana	0.171	0.058	-0.050	1.48
Guinea	0.023	0.004	-0.017	0.37
Guinea-Bissau	0.005	0.000	-0.004	0.10
Liberia	0.007	0.000	-0.006	0.13
Mali	0.011	0.003	-0.008	0.16
Mauritania	0.050	0.000	-0.048	0.89
Niger	0.015	0.004	-0.012	0.30
Nigeria	0.916	5.136	4.305	23.52
Senegal	0.066	0.001	-0.063	1.25
Sierra Leone	0.014	0.086	-0.010	0.26
Togo	0.026	0.000	-0.019	0.40
Regional Total	**1.480**	**5.294**	**4.016**	**31.68**

Commercial energy resources in the region, primarily petroleum and natural gas, are concentrated in coastal and offshore regions. Electricity in West Africa is generated through thermal (58.8 per cent of installed capacity) or hydroelectric (41.2 per cent) resources. Natural gas has the potential to take a more significant role in the region's energy sector as fields in Nigeria, Cote d'Ivoire and Senegal are developed. Due to the region's relatively small urban population (approximately 33.9 per cent) and the lack of infrastructure, access to commercial energy sources is limited.

Petroleum

Nigeria, West Africa's only significant oil producer, had oil production averaging 2.118 million barrels per day (bbl/d) in 2002 (see Table 3). Nigerian oil production has averaged 2.219 million bbl/d for the first three months of 2003. OPEC (Organization of Petroleum Exporting Countries) statistics list Nigeria's estimated proven crude reserves at 31.5 billion barrels, and this constitutes 96 per cent of the region's estimated proven crude reserves. Smaller reserves are located in the Gulf of Guinea (offshore Benin, Cote d'Ivoire and Ghana), in the Atlantic Ocean (offshore Mauritania and Senegal) and in landlocked Niger. Nigeria is the region's largest oil consumer (nearly 60 per cent of the region's total).

Table 31.3: Petroleum Overview

Country	*Petroleum Production, 2002 (Thousand Barrels Per Day)*	*Petroleum Consumption, 2002 (Thousand Barrels Per Day)*	*Petroleum Net Exports, 2002 (Thousand Barrels Per Day)*	*Crude Oil Reserves, 1/1/2003 (Million Barrels)*	*Crude Oil Refining Capacity, 1/1/2003 (Thousand Barrels Per Day)*
Benin	1.00	13.05	-12.05	8.2	0.0
Burkina Faso	0.00	8.87	-8.87	0.0	0.0
Cape Verde	0.00	2.34	-2.34	0.0	0.0
Cote d'Ivoire	19.39	32.0	-12.61	100.0	65.2
Gambia	0.00	1.98	-1.98	0.0	0.0
Ghana	7.00	40.49	-33.49	16.2	45.0
Guinea	0.00	8.73	-8.73	0.0	0.0
Guinea-Bissau	0.00	2.59	-2.59	0.0	0.0
Liberia	0.00	3.20	-3.20	0.0	15.0
Mali	0.00	4.07	-4.07	0.0	0.0
Mauritania	0.00	24.13	-24.13	200.0	0.0
Niger	0.00	5.03	-5.03	300.0	0.0
Nigeria	2,122.86	272.96	1,849.90	31,506.0	438.8
Senegal	0.00	32.20	-32.20	700.0	27.0
Sierra Leone	0.00	6.66	-6.66	0.0	10.0
Togo	0.00	11.87	-11.87	0.0	0.0
Regional Total/Average	**2150.25**	**470.17**	**1,680.07**	**32,830.4**	**601.0**

Exploration and Production

Nigeria has the potential to significantly increase its crude oil production in the next few years as recent discoveries come on stream. ExxonMobil has launched its construction work on Nigeria's deep-water Erha field, which is estimated to contain 500 million barrels of reserves. At maturity Erha should yield 150,000 bbl/d, with first oil expected in 2005. ExxonMobil holds a 56.25 per cent share and operates Block OPL-209 (the block where Erha is located), under a production-sharing contract (PSC) with Nigerian National Petroleum Corporation (NNPC). Shell is the other partner on OPL-209. Erha's $1.1-billion field development is expected to include a Floating Production Storage and Offloading vessel (FPSO), 15 producing wells, five water injection wells, and four gas injection wells.

Work is continuing on Shell's massive Bonga deep-water development, which has the potential to produce 225,000 bbl/d. Bonga is estimated to contain 700 million barrels and it is scheduled to come on stream in early 2004. Shell is operator (55 per cent), partnered with ExxonMobil (20 per cent), Agip (12.5 per cent), and TotalFinaElf (12.5 per cent). The field will be serviced by a 2.0-million-bbl FPSO.

ExxonMobil is developing the 400-million-bbl Yoho field in the shallow waters of Block OML-104. ExxonMobil holds 40 per cent and NNPC 60 per cent of the $1.2 billion project. First oil is planned for early 2003, producing at 90,000 bbl/d. Yoho will utilize an FPSO to allow oil production almost two years ahead of the full-field production capacity. When full-field output of 150,000 bbl/d is achieved in late 2004, the FPSO will be replaced by a floating storage and offloading (FSO) vessel. Associated gas will be re-injected to maintain field pressures and eliminate flaring.

U.S.-based independent Kerr-McGee is currently exploring for oil offshore Benin. The wildcat well, Zifa-1 is expected to be completed in early 2003 on the eastern half of Block 4. Kerr-McGee holds 70 per cent and is operator, with Malaysia's Petronas having the remaining 30 per cent interest. A second well on the western half of the block, is likely before the end of 2003. Canada's Archean Energy Ltd holds a 100 per cent interest on Benin's shallow-water Blocks 2 and 3. In neighboring Togo, Petronas and U.S.-independent Hunt Oil are planning to commence an exploration drilling program before the end of the first quarter of 2003. Hunt Oil signed a PSC with Togo in late July 2002 for the country's first deepwater well. Hunt Oil was awarded the exclusive right to Togo's entire offshore area last May. The contract area, previously divided into 15 blocks, covers 1,570 square miles (4,067 square kilometers).

Saltpond Offshore Producing Ltd (SOPL), which is 60 per cent owned by U.S.-registered independent Lushann-Eternit and 40 per cent by the state-owned Ghanaian National Petroleum Company (GNPC), has recently begun production on a very modest scale from the Saltpond oil and gas field offshore Ghana. Production from the Saltpond field was originally curtailed in 1985. Redevelopment of Saltpond may include the building a pipeline to take associated gas from the field to shore for domestic use, a spokesman from SOPL stated.

In March 2000, Scottish-based Dana Petroleum's WT-1X exploration well flowed at 1,000 bbl/d on Ghana's West Tano Block. Dana Petroleum announced on October 18, 2002 that a second well on the West Tano Block had been successful. The WT-2X well was drilled to a total depth of 10,910 feet (3,300 meters) below sea level and encountered an oil-bearing reservoir, similar to that production-tested in the earlier WT-1X well. Several prospects in the deepwater portions of the Block, where the geology is different, are being evaluated with the intent of commencing deepwater drilling in late 2003.

It was announced in August 2002 that, Oklahoma-based Devon Energy and Canadian independent EnCana have entered into an agreement with the GNPC to explore for hydrocarbons offshore southeastern Ghana in the Keta Basin. The companies will invest $56 million in seismic research and exploration drilling. Devon intends to test-drill on the Keta Block by mid-2003. Devon has been active in Ghana since 1997 when it acquired the Keta concession. An earlier wildcat well (Dolphin) in shallow waters on the Keta block proved dry in the first quarter of 2001. Houston-based Vanco Energy also signed an exploration agreement with the Ghanaian government in August 2002. Vanco Energy said it plans to invest $30 million to explore for oil on the Cape Three Points Deepwater Block.

In June 2002, Calgary-based Canadian Natural Resources (CNR), signed a PSC for Cote d'Ivoire's CI-400 Block. Block CI-400 is located in deepwater of depths up to 6,600 feet (2,000 meters). CNR also holds interests in several other offshore blocks, including a 66 per cent interest and the role of operator for Block CI-26, which contains the previously abandoned Espoir field. CNR's partners on CI-26 are Ireland's Tullow Oil (24 per cent) and Cote d'Ivoire's state petroleum firm, Petroci (10 per cent).

Re-development of the Espoir field, which was closed down in 1988 due to high production costs, has been estimated at $265 million. The field includes two unmanned wellhead platforms with hydrocarbon processing on an FPSO. The FPSO has crude oil production capacity of 40,000 bbl/d and water injection capacity of 60,000 bbl/d. Espoir's oil production is exported by shuttle tanker, while the natural gas is piped to shore where it is used to generate electricity. First production occurred in January 2002, and the field has a life expectancy of 20-25 years. Initial production began at 12,000 bbl/d of oil and 34 million cubic feet per day (Mmcf/d) of natural gas and is expected to peak at 28,000 bbl/d of oil and 35 Mmcf/d of natural gas. In March 2002, a new FPSO, Espoir Ivoirien, tested successfully and began production. The FPSO is contracted to produce for an initial ten-year term, with ten one-year renewal options after that. Espoir's recoverable reserves are estimated at 93 million barrels of oil and 180 billion cubic feet (Bcf) of gas.

In May 2003, the consortium partners announced results of the Acajou South oil discovery on Block CI-26, 5 miles (9 kilometers) southeast of the Espoir field. The well, approximately 15 miles (24 kilometers) off the coast of Cote d'Ivoire was drilled in waters with depth of 3,050 feet. It reached a total depth of 8,027 feet and encountered a gross oil column of over 250 feet. The well had test flows of 3,500 bbl/d of 33° API oil.

CNR operates Block CI-40 with a 61 per cent interest, and is joined by Svenska Petroleum (29 per cent) and Petroci (10 per cent). In March 2001, CNR announced a discovery on Block CI-40, Cote d'Ivoire's first deepwater oil find. The Baobab discovery had a flow rate of 6,700 bbl/d of heavy (22°-23° API) oil and the well is expected to come online by 2005. A second successful well was completed in February 2002 and tested at a rate of 10,000 bbl/d. Block CI-40 lies 5 miles (8 kilometers) south of the Espoir field and is now estimated to have around 150 million barrels of recoverable oil reserves. CNR will test the Kossipo structure, which lies approximately 5 miles (8 kilometers) southeast of the similarly-structured Baobab find. The results from the Kossipo well will be used to determine the optimum development plan for the Baobab pool. First oil from the Baobab discovery could come online as soon as 2004 at rates of up to 95,000 bbl/d.

U.S.-based Ocean Energy (Ocean) operates the Lion and Panthere fields on Block CI-11. Current production from the block is 20,000 bbl/d of oil and nearly 70 mmcf/d of natural gas. Ocean's partners on Block CI-11 are Petroci, Pluspetrol of Argentina, and International Finance Corporation. Ocean has continued development work on Block CI-11 and is now extracting oil from the Lion A5 deep-sea oil well. Recoverable reserves on Block CI-11 are estimated to be 210 million barrels of oil and 495 bcf of natural gas.

Although the Mano River States (Guinea, Liberia and Sierra Leone) currently produce no hydrocarbons, sporadic exploration activity is taking place. Houston's SCS Corp., a wholly owned subsidiary of HyperDynamics, and partner U.S. Oil Corp. have obtained exclusive rights to explore and market the results for the offshore area of Guinea. The venture has total exploration and data marketing rights to the entire continental margin of Guinea, which covers 210 miles of coastline and up to 150 miles offshore. Under the agreement with U.S. Oil, SCS will operate the exploration operations for an area of approximately 31,000 sq. miles. Phase one, an initial seismic survey, which commenced in April 2002, was budgeted at $2.5 million. It was announced in July 2002, that the government of Sierra Leone was offering companies the chance to bid for leases on approximately 10,800 square miles (28,000 square kilometers of offshore area to develop the country's petroleum reserves. The concession areas are being split into seven blocks of approximately 1,550 square miles (4,000 square kilometers) each, and bids on the concessions were originally due by the end of March 2003. Oil companies interested in the offshore acreage petioned the government to extend the deadline so that they could have more time to evaluate seismic data. The blocks are likely to be awarded by the end of 2003. Several test wells drilled in the 1970's on Sierra Leone's continental shelf produced "shows" of oil.

In March 2002, U. K.- independent Premier Oil announced the results of its first exploratory well on the Sinapa prospect (Block 2) offshore Guinea-Bissau. The Sinapa-1 exploratory well has been designated plugged and abandoned with oil shows. Despite the disappointing results, it was reported that Premier Oil plans to retain its acreage in Guinea-Bissau, and is in the process of reviewing seismic data on its holdings. Premier Oil intends to drill two exploration wells on its Guinea-Bissau acreage during the second half of 2003. In August 2002, the national oil company of Guinea-Bissau, Petroguin, was near completing the arrangements to offer the country's new deep-water acreage to prospective investors. Exploration activity in the region has sparked interest in the remaining 11 offshore blocks.

In June 2002, U.S.-independent Amerada Hess (Hess) announced it had acquired a 68 per cent interest in the Croix du Sud Block. Croix du Sud is located in jointly-administered waters offshore Guinea-Bissau and Senegal. In 1995 Guinea-Bissau and Senegal established the Agence de Gestion et de Cooperation (AGC) for the joint-development of maritime resources located in the area. Hess' partners on the block are the AGC (12 per cent interest) and Australia's Fusion Oil (20 per cent). Fusion Oil also holds a 10 per cent interest in the AGC-administered Cheval Marin Block. Agip (75 per cent and operator) and the AGC (15 per cent) are Fusion Oil's partners on Cheval Marin. Seismic surveys were completed on Cheval Marin in April 2002, and additional seismic surveys are planned for both AGC blocks. In 1960, the French-firm Total (now TotalFinaElf) discovered the Dome Flore field, which contains estimated reserves of 700 million barrels of heavy crude, in waters now-administered by the AGC.

Houston-based Vanco Energy Company (Vanco) signed a PSC with the government of Senegal in October 1999. The PSC is for the Dakar Offshore Profond permit, a 7.9-million acre (32,000 sq km), deepwater block that extends from Senegal's offshore boundary with Gambia to the Mauritanian border. The license is the largest concession to date in offshore Senegal, and the country's first deepwater award. Vanco is operator of the concession with a 90 per cent interest and Senegal's state-owned oil company, Petrosen, holds the remaining 10 per cent. Seismic surveys on the block were conducted in 2000 and 2001, and Vanco plans to drill an exploratory well on the southern portion of the block by the end of 2003.

Two Australian independents, Roc Oil and Woodside Petroleum (Woodside), have decided to let the PSCs on Senegal's Casamance Blocks I, II, and III lapse in December 2002. In its announcement, Roc Oil said the decision was reached after reprocessing seismic data and carrying out seismic survey during the past three years. The blocks, located offshore southern Senegal, cover a total area of 2 million acres (8,187 square kilometers).

Hess (80 per cent interest) and Fusion Oil (20 per cent) hold the rights to Gambia's Deepwater PPL Block. The license has been issued for six years during which time an extensive exploration campaign involving seismic acquisition and the drilling of two wells will be undertaken. In December 1999, a 2D seismic survey over the western half of the Gambian licence was conducted, and in February 2001 a second 2D seismic survey was completed. The companies are currently finalizing interpretation studies and have been evaluating options for 3D seismic data acquisition in 2003.

There is currently little hydrocarbon exploration activities in the landlocked countries of Mali, Niger,and Burkina Faso. Mali plans to offer 15 blocks in four basins, with the final awards being granted before the end of 2003. Algeria , through its state-owned oil firm Sonatrach, has already inked an exploration pact with Mali to jointly explore the Taoudeni basin. Taoudeni extends across the borders of both countries. In November 2002, Niger's Mining and Energy Minister stated that a consortium consisting of ExxonMobil and Petronas, will begin drilling three exploratory oil wells in the Agadem Basin (in eastern Niger) starting in April 2003. Previous exploration in the Agadem Basin had confirmed reserves of 300 million barrels, but the oil was not exploited because the reserves were not considered commercially viable. Oil exploration has been carried out for more than 20 years in Niger's Djado region on the border with Libya, but no commercial finds have been discovered.

Several finds have been made offshore non-ECOWAS member Mauritania. A consortium of firms, led by Woodside, made the Chinguetti discovery in May 2001 on Block 4. Studies indicate the most likely recoverable reserves on the Chinguetti field are in excess of 100 million barrels. Woodside and its consortium partners signed PSCs for three Areas (A, B and C) covering exploration blocks 2-6. Woodside is the operator and its partners include: Agip, Roc Oil, Fusion Oil and Hardman Resources (Hardman). Participants in PSC Block A are Woodside (35 per cent and operator), Agip (35 per cent), Hardman (24.3 per cent), Fusion Oil (3.0 per cent), and Roc Oil (2.7 per cent). On PSC Block B, Woodside and Agip each hold 35 per cent, Hardman 21.6 per cent, Fusion 6.0 per cent, and Roc Oil 2.4 per cent. In February 2002, South Africa's Energy Africa acquired a 20 per cent stake in Block 2 off the coast of Mauritania. Energy Africa bought the interest from Woodside, Hardman Resources and Roc Oil. Participating partner's interests in Block 2 are: Woodside (operator) 48 per cent, Hardman 28.8 per cent, Energy Africa 20 per cent and Roc Oil 3.2 per cent. In May 2003, Premier Oil announced that it had reached an agreement with Fusion Oil to acquire its interest in the Mauritanian PSCs for a cash consideration of $10 million and an overriding royalty. As part of this transaction, Premier Oil has granted Fusion Oil the right to receive a 5 per cent interest from Premier Oil's 100 per cent interest in Block 2 and in Blocks 4A and 5A in Guinea Bissau, with these rights to be exercised within 60 days of completion of the drilling of the next well on each block.

The consortium partners planned an extensive drilling program to evaluate and delineate the Chinguetti discovery. The Chinguetti 4-2 appraisal well was the first of five proposed test wells scheduled for 2002. Other wells to be drilled were: the Banda (Chinguetti 4-3) exploration well, the Chinguetti 4-4 exploration well (PSC Area A), the Thon prospect (PSC Area C) and the Tiof exploration prospect on PSC Area B. It was reported in September 2002, that the Chinguetti 4-2 well flowed at an initial rate of 1,560 bbl/d of oil and 650,000 cubic feet per day of natural gas. In October 2002, results from the Banda well were announced. The well drilled offshore Mauritania by the Woodside consortium intersected an oil column and a gas column, but it was not flow-tested. An appraisal well may be drilled early in 2003. Woodside reported on October 11, 2002 that the Chinguetti 4-4 appraisal well had been plugged and abandoned. In November 2002, the Thon well was plugged and abandoned as a dry hole. Earlier in November 2002, Petronas acquired a 35 per cent interest in PSC Area C. Woodside remains the operator, but reduces its interest from 59.5 per cent to 37.6 per cent, and Hardman cut its stake from 35.5 per cent to 22.4 per cent. Roc Oil's share remains at 5.0 per cent. Drilling on the Tiof prospect was pushed to 2003. Up to five wells will be drilled in PSC Areas A and B during 2003. These include two exploration wells and three contingent wells (including two Banda appraisal wells).

Exploration activities on other blocks offshore Mauritania are taking place. A consortium led by Dana Petroleum is currently acquiring seismic data on Block 1. In September 2002, Energy Africa acquired an interest in Block 1. Participating interests in Block 1 following the acquisition are: Dana Petroleum (Operator) 48 per cent, Energy Africa 32 per cent, Hardman 18 per cent and Roc Oil 2 per cent. Drilling activity on Block 1 is tentatively scheduled to begin in late 2003 or early 2004. Seismic data is also being gathered on two other Dana-operated blocks, 7 & 8. Dana Petroleum's (51.4 per cent) partners on Block 7 are: Woodside (35 per cent), Hardman (11.6 per cent) and Roc Oil (2 per cent). On Block 8, Dana Petroleum (80 per cent) is partnered with Hardman (18 per cent) and Roc Oil (2 per cent). The Mauritanian Ministry of Mines and Industry, has opened discussions with companies about entering into PSCs on five ultra-deep water blocks. The Blocks 13-17, lie west of Chinguetti and Banda discoveries.

Refining

West Africa's petroleum refining capacity is concentrated in Nigeria. Nigeria's four refineries have the capacity to process 438,750 bbl/d of crude. Other ECOWAS refineries are in Cote d'Ivoire (Abidjan, 65,200 bbl/d); Ghana (Tema, 45,000 bbl/d); Liberia (Monrovia, 15,000 bbl/d); Senegal (Dakar, 27,000 bbl/d) and Sierra Leone (Freetown, 10,000 bbl/d).

Nigeria refinery problems including sabotage, fire, poor management and lack of turnaround maintenance (TAM) have decreased output, and forced the nation to often import refined petroleum products to meet demand. In January 2003, The NNPC announced plans to undertake TAM at the country's four refineries, but gave no schedule of when the work was likely to be completed.

The capacity of Ghana's Tema Oil Refinery (TOR) was increased from 25,000 bbl/d to 45,000 bbl/d in 1997, when it underwent a major modernization program at a cost of $208 million. TORC is heavily in debt, as a result of its providing subsidized petroleum products during times of high crude prices. In November 2002, a residual catalytic cracking (RCC) unit was commissioned at TOR. The RCC unit will treat 14,000 bbl/d of resid from TOR's distillation unit to extract more gasoline and diesel fuel after the first stage of refining. The RCC would also allow TOR to produce surplus LPG and gasoline for export.Cote d'Ivoire's SIR is scheduled to be almost fully privatized by the end of 2003, after several long delays due to political turmoil. The state currently owns approximately 48 per cent of SIR, and expects to retain a 10 per cent interest after privatization. Other shareholders in SIR are: TotalFinaElf (25 per cent), Shell (10 per cent), ExxonMobil (8 per cent), the government of Burkina Faso (5 per cent), and ChevronTexaco (4 per cent).

The Nigerian government is seeking investors to purchase its 48 per cent stake in Sierra Leone's West African Refinery Company (WARC). Nigeria's privatization agency, Bureau for Public Enterprises, stated that potential bidders must provide proof of their ability to operate and manage the refinery. WARC was acquired by the government of Sani Abacha in 1994. Unipetrol, a Nigerian oil marketing company also holds a 24.2 per cent interest in WARC.

Consumption

Petroleum consumption in West Africa averaged 470,170 bbl/d in 2002. The overwhelmingly vast majority of petroleum consumed in the region is imported; Nigeria is the only net exporter of petroleum.

In January 2003, the Ghanaian government announced fuel price hikes that would bring them in line with world markets and to help pay off TOR's massive debt. Energy Minister Albert Kan Dapaah stated that premium gasoline, which used to sell at 10,500 cedis ($1.27) per gallon, will now cost 20,000 cedis ($2.43), and diesel will sell at 17,500 cedis ($2.12), up from 8,800 cedis ($1.07). Minister Kan Dapaah also said the increases will help deter fuel smuggling between Ghana and neighboring countries where prices are much higher. A similar price increase, initiated in May 2001, saw the price of gasoline rise by nearly 8 per cent in Burkina Faso.

Several countries in the region have recently experienced periodic, sometimes severe, petroleum shortages. Limited fuel storage capacity and a lack of foreign exchange have contributed to Gambia's fuel shortages. The situation was exacerbated by the implementation of fuel rationing and a lack of imports arriving from Cote d'Ivoire, due to the political instability.

In September 2002, the coup attempt in Cote d'Ivoire created acute shortages of gasoline and diesel fuel in portions of the country. Deliveries to interior regions of the country were halted due to fears that fuel trucks might be hijacked by rebel forces.

A strike by tanker drivers caused a fuel shortage in Ghana in May 2003. The strike, apparently in protest of the arrest of some drivers for illegally dealing in petroleum products, caused temporary shortages at petrol stations throughout the country.

Natural Gas

There are significant reserves of natural gas in West Africa. Field discoveries have been confirmed and reserves have been proven in Benin (43 Bcf); Cote d'Ivoire (1.1 trillion cubic feet–Tcf); Ghana (840 Bcf); Nigeria (124.0 Tcf); and Senegal(106 Bcf) (see Table 4). West Africa contains approximately 32 per cent of Africa's natural gas reserves. Although natural gas is still in early stages of use in the region, several projects for the expansion of its use are under way.Nigeria's reserves are the 9th largest in the world, but due to a lack of infrastructure, Nigeria flares 75 per cent of the natural gas it produces. Nigeria is developing several projects to utilize its vast reserves of associated and non-associated natural gas. Nigeria's most ambitious natural gas project, the $3.8-billion LNG (liquefied natural gas) facility on Bonny Island, was completed in September 1999. Initially, the facility is to be supplied from dedicated natural gas fields, but within a few years it is anticipated that half of the input gas will consist of associated (currently flared) natural gas. The consortium developing the project, Nigeria Liquefied Natural Gas Corporation (NLNG), is comprised of the NNPC (49 per cent), Shell (25.6 per cent), TotalFinaElf (15 per cent), and Agip (10.4 per cent). In December 2002, NLNG commissioned the third LNG train and shipped the first cargo from the new facility. NLNG is in the process of constructing trains four and five and they should be on stream in 2005. A sixth LNG train is being considered.

In August 2002, Ghana signed production agreements with Britain's First Oil for the development of its Tano (North and South) offshore oil and natural gas fields. Gas from the fields is to power turbines mounted on barges in the Effasu Lagoon. North Tano is to be developed first, supplying 26 Mmcf/d to Effasu by the end of 2003, where a gas processing plant will be built adjacent to the 134-megawatt (MW) power-generation barge. South Tano is farther offshore, in deeper water, and it will come on stream later. When on stream, gas deliveries will rise to 60 Mmcf/d and generating capacity at Effasu/Mangyea can be increased to 270 MW. The South Tano field should also boost oil production from both fields to 3,000 bbl/d.

Table 31.4: Natural Gas Overview (Billion Cubic Feet)

Country	Production, 2001	Consumption, 2001	Reserves, 1/1/2003
Benin	0.00	0.00	43
Burkina Faso	0.00	0.00	0
Cape Verde	0.00	0.00	0
Cote d'Ivoire	47.68	47.68	1,050
Gambia	0.00	0.00	0
Ghana	0.00	0.00	840
Guinea	0.00	0.00	0
Guinea-Bissau	0.00	0.00	0
Liberia	0.00	0.00	0
Mali	0.00	0.00	0
Mauritania	0.00	0.00	0
Niger	0.00	0.00	0
Nigeria	553.74	277.22	124,000
Senegal	1.77	1.77	106
Sierra Leone	0.00	0.00	0
Togo	0.00	0.00	0
Regional Total	**603.18**	**326.67**	**126,039**

Natural gas reserves in Cote d'Ivoire, first discovered in the 1980's, have begun to be developed and utilized. Over the next four years, the Ivorian government estimates that natural gas consumption will grow by 50 per cent. Initial production of 34 Mmcf/d of natural gas from the re-developed Espoir field began in January 2002. Espoir's recoverable reserves are estimated at 180 bcf of natural gas.

Senegal's natural gas reserves are primarily located onshore. In October 2002, U.S.-independent Fortesa has begun natural gas production from the onshore Gadiaga Development Area and is applying for a license for the adjoining Kayar offshore field. Gadiaga currently produces two mmcf/d and Fortesa executives stated that the company planned to carry out additional seismic studies. Fortesa holds a 70 per cent stake in Gadiaga, with the remaining 30 per cent held by Petrosen, the Senegalese national oil company. Production from the onshore Diam Niadio field, which took place from 1986 to 1999, totaled more than 7 bcf.

The most significant natural gas development project being considered is the West African Gas Pipeline (WAGP) project. The WAGP will traverse 620 miles (1,033 kilometers) both on and offshore from Nigeria's Niger Delta region to its final planned terminus in Ghana. The $500 million WAGP will initially transport 120 Mmcf/d of gas to Ghana, Benin and Togo beginning in June 2005. Gas deliveries are expected to increase to 150 Mmcf/d in 2007, to 210 MMcf/d in 7 years and be at 400 Mmcf/d when the pipeline is functioning at its capacity (approximately 15 years after construction). Speculation has the WAGP eventually terminating in Senegal, but the current regional stability problems of several countries (Cote d'Ivoire, Liberia, Sierra Leone) that lie on the way to Senegal, will hinder any further extension of the WAGP.

In February 2003, the four nations signed an agreement on the implementation of the WAGP. The treaty, which is for a 20-year period, provides for a comprehensive legal, fiscal and regulatory framework, as well as a single authority for the implementation of the project. The WAGP partners are ChevronTexaco with 36.7 per cent, NNPC with 25 per cent, Shell with 18 per cent, Ghana's Volta River Authority (VRA) with 16.3 per cent and Societe Beninoise de Gas and Societe Togolaise de Gas each with a 2 per cent interest.

Electricity

West Africa's total installed electric generating capacity was 9.4 gigawatts (GW) at the beginning of 2001, the majority of which is thermal (see Table 5). Total electricity generation for the region in 2001 was 33.8 billion kilowatthours (bkwh), with Nigeria (15.7 bkwh), Ghana (8.8 bkwh) and Cote d'Ivoire (4.6 bkwh) being the largest generators. In 2001, total regional electricity consumption was 31.8 bkwh, led by Nigeria's 14.6 bkwh (45.8 per cent). Ghana (8.8 bkwh, 27.8 per cent), Cote d'Ivoire (3.0 bkwh, 9.4 per cent) and Senegal (1.4 bkwh, 4.4 per cent) were the next largest electricity consumers.

Table 31.5: Electricity Overview, Billion Kilowatthours except where noted

Country	*Consumption, 2001*	*Generation, 2001*	*Installed Capacity, 1/1/2001 (Gigawatts)*	*Thermal Capacity (Per cent of Total)*	*Hydroelectric Capacity (Per cent of total)*
Benin	0.63	0.27	0.094	28.7%	71.3%
Burkina Faso	0.26	0.28	0.121	73.6%	26.4%
Cape Verde	0.04	0.04	0.007	100.0%	0.0%
Cote d'Ivoire	2.98	4.61	0.892	31.2%	68.8%
Gambia	0.08	0.09	0.029	100.0%	0.0%
Ghana	8.84	8.80	1.200	10.7%	89.3%
Guinea	0.74	0.79	0.195	73.3%	26.7%
Guinea-Bissau	0.05	0.06	0.011	100.0%	0.0%
Liberia	0.44	0.47	0.330	100.0%	0.0%
Mali	0.45	0.48	0.114	56.1%	43.9%
Mauritania	0.15	0.16	0.105	41.9%	58.1%
Niger	0.33	0.24	0.063	100.0%	0.0%
Nigeria	14.56	15.67	5.888	67.1%	32.9%
Senegal	1.41	1.52	0.235	100.0%	0.0%
Sierra Leone	0.23	0.25	0.128	96.9%	3.1%
Togo	0.61	0.10	0.034	88.2%	11.8%
Regional Total	**31.78**	**33.82**	**9.446**	**58.8%**	**41.2%**

A tender for the privatization of Benin's water and electricity utility, Societe Beninoise d'Electricite et de l'Eau (SBEE)is expected before the end of 2003. Plans call for SBEE to separate its water and electricity operations into two separate companies. As part of the privatization, Benin plans to create an agency for rural electrification that would bring power to areas where it was not economically feasible to extend SBEE's grid.

Hydroelectricity is the primary source of Ghana's power. Ghana's current hydroelectric capacity of 1.072 GW is located at Akosombo (912 MW) and Kpong (160 MW). The Ghanaian government is considering additional hydroelectric projects to be built on a Build Operate Transfer (BOT) financing scheme. The $700-million, Bui hydroelectric project would be located on the Black Volta. The Bui project would have a generation capacity of 400 MW. In addition to increasing the domestic electricity supply, power generated from Bui could be exported to Burkina Faso, Mali and Cote d'Ivoire. A second BOT facility, located on the Pra River, would have a total generating capacity of 125 MW. Recent low rainfall has forced power cuts from Ghana's hydroelectric facilities, similar to the power shortages experienced in 1997/1998.

Ghana plans to reduce its reliance on hydroelectric power by increasing and expanding thermal generating capacity. Current thermal facilities are located at Tema and Takoradi. Additional capacity is planned at Tano (gas-fired barges) and at Tema. VRA and GNPC have constructed transmission lines and substations at Essiama and Elubo in the Western Region to feed the power generated at Tano into the national grid. The VRA is currently in the process of installing 110 MW at the Strategic Reserve Plant (SRP), near the Electricity Company of Ghana (ECG) main substation in Tema. In April 2003, U.S.-based CMS Energy announced that it is planning a $100 million expansion of its thermal power plant at Takoradi. The upgrade would convert the plant from burning crude oil to natural gas which it would receive from Nigeria through the WAGP. CMS Energy has a 90 per cent stake in the Takoradi facility, and the VRA holds the remaining 10 per cent. ECG is responsible for electricity distribution to the Ashanti, Western, Central, Eastern, Greater Accra and Volta regions. VRA is responsible for generation and for the distribution of electricity in the Brong Ahafo, Northern, Upper East and Upper West regions. When the WAGP is completed, VRA plans to convert oil-fired facilities at Takoradi and Tema to natural gas.

A large portion of the Liberia Electricity Corporation's (LEC) generation and distribution infrastructure was damaged or destroyed during the civil war. LEC estimates that it will cost more than $107 million and take over five years to repair the entire electricity generation and distribution system

Sierra Leone's Bumbuna hydroelectric project was nearly complete (85 per cent) when civil war disrupted the construction. The government hopes to restart construction in early 2003, but funding is an issue. The cost of completing the project is estimated at $40.2 million, and the African Development Bank (AfDB) has pledged to provide $8.7 million. Sources for the remaining $31.5 million have yet to be identified.

Guinea is the source of several major West African rivers (including the Gambia and Niger Rivers) and has a hydroelectric potential (technically feasible) estimated at 19,400 Gigawatthours per year (Gwh/yr). Only about 1 per cent of Guinea's technically feasible potential has so far been developed. The 75-MW Garafiri hydroelectric facility, on the Konkoure River, was commissioned in 1999; and an 80-MW project is planned 60 miles (100 km) downstream at Kaleta.

Senegal's Societe Nationale d'Electricite (SENELEC) handles the generation, transmission and distribution of the majority of the country's electricity. The government's ownership in SENELEC was reduced to a 41 per cent share in March 1999, when a consortium comprised of the utilities Hydro-Quebec of Canada and Elyo of France acquired a 34 per cent interest in SENELEC. In 2000, a series of power cuts prompted the government to take back the stake sold to the Franco-Canadian venture.

Privatization negotiations with France's Vivendi fell through in February 2001, and government negotiations with U.S.-utility AES Corp ended unsuccessfully in July 2002.

Gambia's National Water and Electricity Company (NAWEC) signed an agreement with South Africa's Eskom in June 2000. Eskom was to acquire a 50 per cent stake in NAWEC and participate in a $75-million investment program to overhaul NAWEC generating plants over the next five years. The agreement, which was to include financing from the World Bank, fell through three months later. In December 2000 the AfDB awarded a $3.8-million loan to finance a rural electrification project in The Gambia. The project consists of construction of six power stations (combined capacity of 6.2 MW), the installation of 141 miles (227 km) of transmission and distribution lines to supply power to 46 towns and villages. The total cost of the rural electrification project is estimated at $19 million, with the additional financing coming from international donors.

Niger is seeking international and multilateral donors to help in the construction of the Kandadji hydroelectric project. Kandadji, first conceived in the mid-1970's, would be located on the Niger River approximately 120 miles (200 km) upstream of Niamey. The 165-MW facility (originally proposed to be 230 MW) is expected to cost $270 million. The government of Niger hopes to begin construction in 2004 and complete it by 2012. Smaller dams on the Niger River at Gambou (122 MW) and Dyodyonga (26 MW) have been considered.

Regional Projects

In October 2000, 14 ECOWAS members signed an agreement to launch a project to boost power supply in the region. The West African Power Pool (WAPP) agreement reaffirmed the decision to develop energy production facilities and interconnect their respective electricity grids. According to the agreement, the WAPP will be accomplished in two phases but is planned to be fully implemented by 2005. The first phase involves countries that are already interconnected, including Nigeria, Benin, Burkina Faso, Cote d'Ivoire, Ghana, Niger and Togo. The second phase involves countries which are yet to have interconnection facilities, which include Guinea, Guinea-Bissau, Liberia, Mali, Senegal, Gambia and Cape Verde. Under the agreement, WAPP is expected to harmonize the regulatory framework that governs the electricity sector in each member country. Nigeria and the AfDB signed a $15.6 million loan agreement in December 2002 for the interconnection of NEPA (Nigerian Electric Power Authority) and Compagnie Electrique du Benin (CEB) networks. CEB is the electricity transmission company for Benin and Togo.

The Organization for the Development of the Senegal River (OMVS), which consists of Mali, Mauritania and Senegal, has constructed two dams. The Diama dam, located in Senegal, was completed in 1986 and its primary function is to stop the upstream encroachment of seawater from the Atlantic Ocean. The Manatali dam, built by the OMVS on the Bafing River, the main tributary of the Senegal River, in Mali, was completed in 1987. The Manatali project was also to include a 200-MW power station and an 800-mile (1,300-km) network of transmission lines to the capitals of Mali (Bamako), Mauritania (Nouakchott) and Senegal (Dakar). Cost overruns, coupled with political and military tensions between Mauritania and Senegal, initially canceled the construction of the power facilities. In March 2000, AfDB approved a $33.5 million loan for the Manatali energy project. The Manatali's generating facilities came online in December 2001, supplying power to Mali's grid. Senegal connected its power grid to Manatali in July 2002 and Nouakchott was connected in November 2002. The OMVS signed a new charter in May 2002 to allocate water resources and hydro-electric power, and approved the restructuring of the Manatali Water Management Company (SOGEM). SOGEM will maintain ownership of infrastructure and equipment at Manatali, but Eskom will handle marketing and distribution of power generated at Manatali.

The AfDB has also issued a $2.13 million grant to finance the study on electricity production and transmission to the member states of the Organization for the Development of the River Gambia (OMVG), a regional organization whose members are the Gambia, Guinea, Guinea Bissau and Senegal. The study aims to strengthen regional integration of OMVG member states in the energy sector, particularly in the electricity sub-sector. A feasibility study of the Sambangalou hydroelectric project on the Gambia River, and the interconnection network project linking the facility to the distribution networks of the four countries will be conducted.

Growing demands for power have prompted Burkina Faso to seek import electricity from neighboring Cote d'Ivoire. A 225-kilovolt (KV) power line, connecting the city of Ferkessedougou in northern Cote d'Ivoire with the Burkinabe capital, Ouagadougou, is expected to begin operations in 2005. Burkina Faso employs diesel generators to produce electricity, but high production costs attributed to fluctuating oil prices prompted the government to begin interconnecting Burkina Faso's grid with that of neighboring countries like Ghana and Cote d'Ivoire to import additional electricity requirements.

Work began in early 2003 on a project to connect portions of Niger to Nigeria's electricity grid. The project involves the construction of three separate networks at an estimated cost of $16 million. The imported power will be much cheaper than the domestically oil-generated electricity currently consumed, and the project will help eliminate several diesel-powered generators.

32

Ecuador

Ecuador is important in world energy markets, as one of Latin America's largest crude oil exporters. The country currently is building a second oil export pipeline, which will more than double Ecuador's oil export capacity.

General Background

Faced with a plummeting currency (the sucre) and a sharp economic decline (-7 per cent real economic growth in 1999), Ecuador strictly tied the sucre to the U.S. dollar in September 2000, fixing the exchange rate at 25,000 sucres per dollar. Ecuador's economic and political situations, which had been unstable for years, improved following this decision to "dollarize." In addition, strong oil prices, large-scale International Monetary Fund (IMF) assistance, and foreign investment associated with construction of a new oil export pipeline all are helping to boost Ecuador's fortunes, at least in the short run.

For 2002, Ecuador's real gross domestic product (GDP) is estimated to have grown by 2.9 per cent, down from the rapid, 5.4 per cent growth experienced in 2001, but still solid. For 2003, Ecuador's economy is expected to grow by 3.8 per cent, one of the best growth rates in Latin America. Ecuador's inflation rate fell sharply in 2002, to 5.6 per cent, down from 37 per cent in 2001 and 96 per cent in 2000. For 2003, inflation is forecast to fall below 4 per cent. Unemployment remains relatively high, however, at around 15 per cent, and around 70 per cent of the country's population is considered poor. In addition, the country has a very high debt-to-GDP ratio of around 78 per cent.

Ecuador's oil sector accounts for around one-fifth of the country's economy, and is the country's most important source of foreign exchange, ahead of coffee, fish, and bananas. This reliance on oil exports makes Ecuador's economy vulnerable to sharp fluctuations in oil prices. Ecuador also is vulnerable to any economic downturn in the United States, since remittances from Ecuadorian workers living in the United States are Ecuador's second largest source of foreign exchange earnings (around

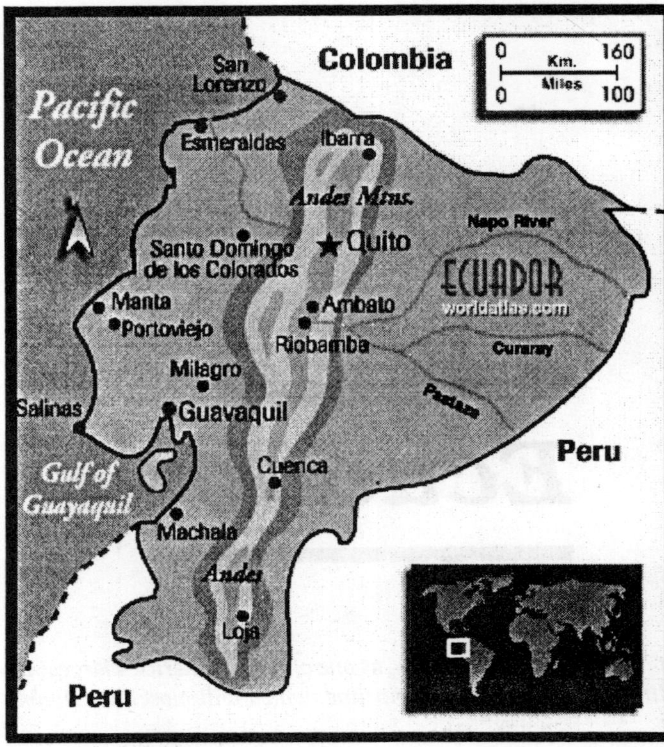

$1.4 billion in 2001). Meanwhile, revenues from construction of a $1.1 billion heavy oil pipeline are helping to fund the country's fiscal stabilization fund and to help pay off the country's huge debt.

In December 2001, the IMF disbursed the final $95 million of a $304 million loan package it had agreed to in April 2000. The IMF deal with Ecuador had been conditioned on Ecuador's pursuit of structural reforms (*i.e.*, reduction of domestic fuel subsidies, loosened restrictions on foreign investment). In line with IMF lending guidelines, Ecuador continues to stress structural reform, increased foreign investment, privatization, and fiscal stabilization. Progress in these areas has been slowed by political opposition from unions, indigenous groups, nationalists, and others. Over 90 per cent of foreign investment goes to the oil sector.

In August 2001, Ecuador's Constitutional Court declared that an increase in the country's Value Added Tax (VAT), from 12 per cent to 15 per cent, was unconstitutional. The VAT increase had been strongly supported by the IMF, and by President Noboa, as a means to place Ecuador's finances on a sounder footing. Meanwhile, around 13 foreign oil companies operating in Ecuador have filed lawsuits claiming that they are entitled to a refund of the VAT, and that the Ecuadorian government owes them at least $150 million. Ecuador disputes this, saying that the companies' oil exploration contracts already adequately compensate them for the VAT. In September 2002, the United States extended the Andean Trade Preferences Act to Bolivia, Colombia, and Peru, but not to Ecuador, possibly in reaction to the VAT issue. In December 2002, Ecuador agreed to international arbitration on the VAT dispute.

Ecuador held national elections in October 2002, with a runoff on November 24, and elected leftist former army colonel Lucio Gutierrez as president. Among other things, Gutierrez–Ecuador's fifth president since 1997–has promised to implement "war economy" measures: expanding social programs, creating jobs, redistributing wealth, ending political corruption, ending many free-market policies, and increasing the state's role in the economy. This platform is in stark contrast to former President Noboa's policies, as well as to IMF guidelines. Since his election, however, Gutierrez has stressed the need for urgent fiscal reform measures, has stated that he will maintain "dollarization," has given assurances that the government will not default on Ecuador's $2.7 billion foreign debt payments due in 2003, and has vowed to cooperate with the IMF. This could mean tough austerity measures and continuation of free-market reforms.

Oil

Since oil was discovered in Ecuador in the 1970s, it has become an increasingly important part of the Ecuadorian economy. The country has 4.6 billion barrels of proven oil reserves, with crude production of around 390,000 barrels per day (bbl/d) during the first 10 months of 2002, down from 422,000 bbl/d in 2001. Of this production, Petroecuador accounts for about 55 per cent of Ecuador's total output, with private companies accounting for the remaining 45 per cent or so. In 2002, Petroecuador reported that its oil output was at a 10-year low point, due mainly to underinvestment and ageing fields. The company hopes to increase production in 2003, but as of September 2002 it reportedly planned to reduce its exploration and production investment budget by more than 40 per cent. Ecuador consumes around 137,000 bbl/d domestically, with the remaining 263,000 bbl/d being exported.

Most of Ecuador's oil reserves are located in the eastern Amazon region, known as the Oriente, with significant new oil deposits suspected to exist in the southeastern Oriente. Major oil fields include: Shushufindi, Sacha, Libertador, Cononaco, Cuyabeno, Lago Agrio, and Auca. Smaller fields include Anaconda, Culebra, and Yulebra. Another field that Ecuador hopes to develop is Palo Azul, located in Orellana province, with potential output of 10,000 bbl/d. Also, in October 2002, Sipetrol, the international unit of Chile's state oil company Enap, signed a contract with Petroecuador to develop four fields–Paraiso, Biguno, Huachito, and Mauro Davalos Cordero–at a cost of $80 million over four years. The fields have estimated reserves of 51 million barrels and possible output of 18,000 bbl/d. Yet another project, at the Eden-Yuturi fields in Block 15, is expected to add 45,000 bbl/d to Ecuador's oil output. Eden-Yuturi is to be developed by Occidental Petroleum, which signed a contract in September 2001. In November 2001, Occidental received approval to build a 100,000-bbl/d pipeline to link Eden-Yuturi to the new OCP heavy oil pipeline.

Petroecuador is attempting to attract foreign investment in the country's largest oil fields, and to boost its production from 230,000 bbl/d (around 60 per cent of national production) today to 600,000 bbl/d by 2005. In early 2001, however, Ecuador's Congress rejected legislation to allow joint ventures, and in December 2000, Ecuador's Constitutional Tribunal rejected a government reform plan which would have allowed private companies to take operational control (not ownership) of Petroecuador's top five oil fields. Meanwhile, Petroecuador is trying to streamline its own operations, is cutting its workforce, and is seeking greater independence over its fiscal and administrative affairs. The company also is planning to step up its marketing efforts internationally, in anticipation of increased exports beginning in 2003. Ecuador's main export market currently is the U.S. West Coast.

In August 2001, Petroecuador released details of a major new oil project aimed at developing the Ishpingo, Tambococha and Tiputini (ITT) oilfields. Combined, the ITT fields contain an estimated 2 billion barrels in heavy oil reserves. Potential output could be as high as 200,000 bbl/d. The project also could involve construction of a plant to upgrade the 14°-16° API gravity crude oil. In May 2002, Petroecuador reported positive results from its Ishpingo 3 test well at ITT. Petroecuador reportedly would like a private company to invest $3.5 billion in the ITT fields, and reportedly would like to award development rights on the field to a foreign government, which would in turn award the operating contract to a private company. In December 2002, Petroecuador postponed awarding a multi-billion contract for ITT development until after President-elect Gutierrez took office on January 15, 2003.

Ecuador has experienced recurring problems between oil producers and the government on one side, and indigenous peoples on the other. Native tribes inhabit most of the oil-rich Oriente area of Ecuador, and have staged numerous peaceful demonstrations against the Ecuadorian government and oil production policies. In May 2001, U.S. District Judge Jed Rakoff dismissed a $1.5 billion lawsuit filed by Ecuadorian Indians against Texaco for alleged environmental and cultural damages, on the grounds that the cases should be tried in Ecuadorian, not U.S., courts.

On December 18, 2002, Petroecuador announced that it would start accepting offers for the four blocks included in its ninth bidding round, which closes in April 2003. The four blocks offered include Blocks 4 and 5, located onshore in southwestern Ecuador, plus 39 and 40, located offshore in the Gulf of Guayaquil. Total investment of $1 billion is expected on these blocks. Other blocks scheduled to be awarded in 2003 include: Atacapi-Parahuacu, Conanaco-Armadillo, Libertador, Mariann-Cuyabeno, Sasha Norte, and Vista Sur.

In July 2002, Petroecuador reportedly decided that it would stop the practice of awarding oil exploration contracts directly to oil companies without a prior tender process. This move followed accusations by Petroecuador workers of corruption and "irregular actions" by the company. Meanwhile, President Gutierrez has accused the previous government of pre-selling Petroecuador's 2003 exports in order to generate emergency funds during its final months in office. In November 2002, Ecuador's Congress rejected a bill that would have amended the national Hydrocarbons Law. If the bill had passed, it would have allowed Petroecuador to sign contracts with other state oil companies without holding a bidding process.

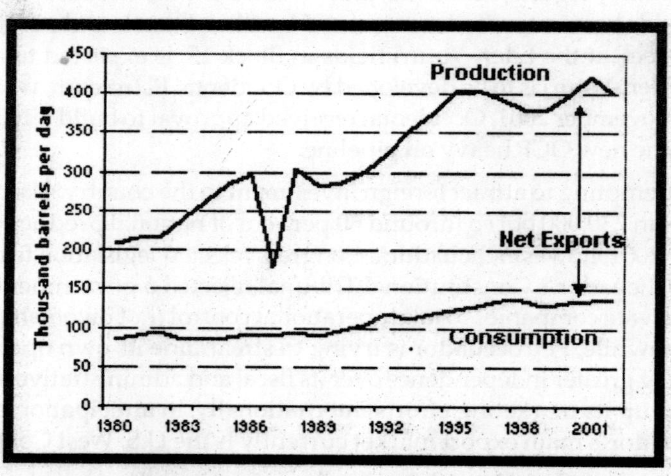

ECUADOR'S OIL PRODUCTION AND CONSUMPTION, 1980-2001

In early October 2002, a British oil worker was kidnapped near the village of Sardinas, 60 miles southeast of Quito. In late December, Ecuador's prime minister indicated that the man most likely was murdered. Argentine oil company Techint, which is building an oil pipeline in the area, had offered a $40,000 reward on December 15. In October 2000, Ecuadorian and Colombian police arrested 57 alleged members of a gang that had kidnapped 10 foreign oil workers in the same region. Also, in December 2002, indigenous groups in Pastaza province of the Amazon region took hostage nine oil workers from Argentina's CGC. The groups are demanding that CGC halt work in the area, specifically on Block 23.

Oil Transportation

Oil pipeline infrastructure in Ecuador is the major constraint on increasing Ecuador's oil production. The major pipeline, the Trans-Ecuadorian (SOTE), extends about 300 miles from the Lago Agrio area in the Oriente to the Balao terminal near the port city of Esmeraldas. The pipeline was built in the early 1970s and has been expanded several times since, with capacity reaching 390,000 bbl/d following work completed in June 2000. The SOTE pipeline transports a variety of crude gravities, reflecting the Oriente's wide range of crude production. In early January 2003, Ecuador's government ordered private oil companies to reduce the amount of low-quality, relatively heavy crude oil they move through SOTE by 5 per cent -18 per cent in order to free up room for higher-grade, lighter crude used by the 46,000-bbl/d capacity La Libertad refinery.

In October 2001, villagers cut a valve on the SOTE and threatened to interrupt oil production if their demands (mainly jobs and economic assistance) were not met. In June 2002, SOTE was shut down after villagers in the Amazon province of Napo briefly occupied the El Salado pumping station and shut valves, stopping the flow of oil for a few days. In November 2002, Petroecuador announced that it would reroute the Papallacta-Lago Agrio section of SOTE due to the danger of landslides following eruption of the Reventador volcano, while another section of the line would be buried.

Construction on a new, 450,000-bbl/d (maximum capacity: 518,000 bbl/d), $1.3 billion, heavy oil pipeline (Oleoducto de Crudos Pesados–OCP) began in June 2001 after the Ecuadorian government granted permission following months of delays and legal objections. The contract for building and maintaining the pipeline, signed in February 2001, went to a five-company consortium headed by Canada's EnCana (formerly Alberta Energy, with a 31.4 per cent share), and also including Spain's Repsol-YPF (25.7 per cent), Argentina's Perez Companc (15.0 per cent) and Techint (4.1 per cent), U.S.-based Occidental (12.3 per cent) and Kerr-McGee (4.0 per cent), and Italy's Agip (7.5 per cent). Westdeutsche Landesbank, Citibank and JP Morgan Chase also have significant financial roles in the project. Construction on the OCP pipeline is expected to be completed by late 2003, at which point Ecuador's oil transport capacity could increase to as much as 850,000 bbl/d, allowing for significant increases in oil production and export capacity. However, Petroecuador will need to increase oil output in order to fill both OCP and SOTE. Currently, it appears that at least initially, only 250,000 bbl/d of OCP's 450,000-bbl/d capacity will be utilized. Of this amount, an estimated 160,000 bbl/d will represent oil currently transported by SOTE, with perhaps 90,000 bbl/d of new production as well.

The OCP pipeline is to run from oil fields in the eastern Ecuadorian Amazon rainforest region, across the Andes Mountains, to a port on the Pacific Ocean. For most of its length, OCP runs parallel to SOTE, except for a 100-mile stretch near Quito, where OCP takes a different, "northern route." Environmental and indigenous opposition is based on claims that pipeline construction or an accident could do serious harm to one of the world's premier bird sanctuaries, the Mindo Nambillo Cloudforest Reserve, which also is a major destination of ecotourists. The Reserve contains an estimated 450 species of birds, including 12 classified as threatened or near-threatened, as well as other rare species like the spectacled bear. In September 2002, an independent report written by Dr. Robert Goodland, former head of the World Bank's Environmental Department, found that the OCP was in "substantial non-compliance with.World Bank Group Social and Environmental Safeguard Policies," such as ISO 14000, an international legal requirement for granting environmental licenses. OCP construction has been delayed somewhat by protests by local and international environmental groups, landowners, and others who want to see the pipeline rerouted around Mindo Nambillo. Also, Ecuador's Environment Ministry temporarily suspended the OCP's license in March 2002, following damage to the forest caused by road construction in the area.

A smaller export pipeline was built in the late 1980s, a 25-mile spur off the Transandino to Colombia (OTA). It carries about 45,000 bbl/d of Ecuadorian crude through Colombia to the port of Tumaco. The OTA is the target of frequent Colombian guerrilla bombings and is susceptible to mudslides and mechanical problems. In June 2001, authorities in Colombia arrested 59 people whom police said were responsible for kidnapping 10 foreign oil workers from a camp near the Colombian-Ecuadoran border.

In December 2002, Petroecuador extended the bidding for construction of a $100 million multi-purpose oil products pipeline connecting Guayaquil, Machal, and Cuenca. The project is aimed at making distribution of gasoline, diesel, and LPG more efficient. Currently, these products are transported by road.

In January 2001, an oil tanker spilled around 240,000 gallons of fuel oil near the world-famous, environmentally unique Galapagos Islands. President Noboa declared a state of emergency and called for international assistance. Fortunately, winds and ocean currents dispersed the oil and pushed it away from the threatened islands. In December 2001, about 40 square miles of ocean surrounding the Galapagos Islands were declared a UNESCO natural heritage site as part of an effort to provide long-term protection for the islands. In July 2002, a smaller diesel oil spill took place off the the Galapagos coast, as a study indicated that mortality among marine iguanas had risen alarmingly, likely as a result of the 2001 spill.

Natural Gas

Ecuador has only small proven natural gas reserves of 345 billion cubic feet (Bcf). Consequently, there is no natural gas market of any significance in Ecuador, while natural gas produced in association with oil is flared. This could change, however, with the development of fields in the Gulf of Guayaquil and the eastern Oriente region. U.S.-based Noble Affiliates (and its subsidiary, Energy Development Corporation Ecuador Ltd.) has signed a 15-year agreement with Petroecuador to tap the estimated 177 billion cubic feet (Bcf) of recoverable gas reserves of the Amistad field in Block 3 of the Gulf of Guayaquil. As of November 2001, Noble reportedly was looking for a partner to help it develop Amistad. In July 2001, the Amistad 7 well tested at 19.4 million cubic feet per day. Ecuador also is attempting to increase recovery of associated natural gas from the Sacha and Shushufindi oilfields.

Electricity

In 2000, Ecuador had an installed electric power capacity of about 3.5 million kilowatts, with power generation of 10.4 billion kilowatthours (bkwh). About 75 per cent of Ecuador's electricity generation in 2000 was hydropower, and the remainder was primarily oil-fired. Roughly 50 per cent -65 per cent of the country's electricity comes from one hydroelectric plant, Paute, making the country vulnerable to disruptions there, particularly during the country's driest time of year (October-

November). In early October 2001, for instance, Ecuador's government declared a state of emergency in the country's electricity sector due to a severe drought which has reduced production at the countries' hydroelectric power plants, including Paute.

Meanwhile, Ecuador's power demand is growing rapidly, around 4 per cent -5 per cent per year, but generation capacity, transmission and distribution infrastructure are having trouble keeping pace. Many people remain without access to electricity, and demand is increasing in areas that are electrified. In addition, there has been insufficient capital for maintenance, and as a result the power grid is in very poor condition, with transmission losses as high as 35 per cent. In November 2001, the World Bank approved a $23 million loan to support Ecuador's efforts at modernizing and expanding the country's electric power and telecommunications services. Overall, Ecuador estimates that it might need more than $4 billion, much of which will need to come from foreign investors, to upgrade its power sector.

New generating capacity is being added, including a $90 million, 130-megawatt (MW), simple cycle, natural-gas-fired plant near Machala which began operation in September 2002. Machala is powered by natural gas piped from the Amistad field in the Gulf of Guayaquil; this gas will allow for expansion to 207 MW by mid-2004 and 312 MW by mid-2005. The plant, Ecuador's first natural gas-fired commercial generating facility, is being run by Machala Power, a subsidiary of Houston-based Energy Development Corporation (EDC), and will supply power to Guayaquil, the largest city in Ecuador. Machala will help Ecuador avoid power outages during dry periods, when hydroelectric power output is low.

In another effort aimed at averting dry-season power shortages, Ecuador has awarded a contract to build a backup reservoir and small hydropower plant south of Paute. The facility, known as Mazar, will cost $500 million and have a power generating capacity of 186 MW. In November 2002, Conelec, the national electric power regulator, suspended a concession that had been awarded to a Spanish consortium led by Union Fenosa, to build the plant, reportedly due to irregularities in the contract.

Privately-financed projects undertaken to augment the electricity infrastructure include constructing a 230-kilovolt line to link Milagro and Machala on the Pacific coast, linking the northern and southern parts of the country, and linking a new hydropower station into the central coastal grid. These projects should improve existing interconnections with Colombia and make possible interconnection projects with Peru.

In September 2001, energy ministers from Ecuador, Colombia, Peru, and Venezuela signed an agreement on integrating their countries' power grids. Along with Bolivia, these countries make up the Comunidad Andina de Naciones (CAN) group. In November 2002, Peru announced that it would begin exporting electricity, via a new $15 million, 120-MW transmission line, to Ecuador beginning in late 2004. Also, in December 2002, Ecuador and Colombia completed work on a 260-MW, 131-mile power line connecting those two countries, although startup has now been delayed until January 30 due to system integration glitches.

Restructuring

In 1961, Ecuador's 22 regional power companies were joined in the *Instituto Ecuatoriano de Electrification* (INECEL), which controlled generation, transmission, and distribution of electricity in Ecuador until 1998. INECEL has now split into six generation companies (HidroPaute, HidroAgoyan, HidroPucara, TermGuayas, TermoPichincha, and TermoEsmeraldas), one transmission company (Transelectric), and 19 distribution companies. These companies are owned by *Fondo de Solidaridad*, the newly created Solidarity Fund.

In May 2002, the sale of Emelec, a state-owned electric distributor in the Guayaquil area, and Ecuador's largest utility, was canceled by Conelec, marking the latest setback in Ecuador's power privatization program. In March 2002, President Noboa had suspended the auction of 10 state electric power distributors in the Amazon and Andean regions due to opposition from local government authorities. Privatization of the distributors has been delayed several times, in part due to opposition from labor unions and Indian rights organizations. Also, in October 2001, the country's Constitutional Tribunal ruled that sale of the 17 power distributors would be illegal (the government rejected the ruling, leaving the situation uncertain).

In preparation for privatization, power subsidies theoretically are slated to end, increasing the consumer electricity price. Ecuadorians pay an estimated less than 30 per cent of the actual cost of their electricity. However, fears of social unrest and mass protests by indigenous organizations have stalled this process as well, with President Noboa announcing in April 2002 that any rate hikes would be postponed until the end of his term in office.

On March 20, 2002, Ecuador's Congress passed a resolution rejecting the privatization of the country's power sector, and called on the government to heed a Constitutional Court ruling that had voided a measure allowing the sale of public assets. In mid-March, President Noboa had withdrawn 10 power companies (out of 17 total) from the auction scheduled for April 12, but opponents of privatization–including indigenous organizations, unions, local governments, and others–continued to demand that the entire plan be scrapped.

Country Overview

Independence: May 24, 1822 (from Spain)

Population (2002E): 13.1 million

Location/Size: Northwestern South America, 105,037 sq. mi., slightly smaller than Nevada

Major Cities: Quito (capital), Guayaquil, Cuenca, Machala, Portoviejo, Manta, Ambato, Santo Domingo, Esmeraldas

Languages: Spanish (official), Quechua and other Indian languages

Ethnic Groups: Mestizo (mixed Spanish and Indian, 55 per cent), Amerindian (25 per cent), Spanish (10 per cent), Black (10 per cent)

Religion: Roman Catholic (95 per cent)

Economic Overview

Currency: Sucre/U.S. Dollar

Official Exchange Rate (1/27/03): $1 = 25,000 sucres

Gross Domestic Product (GDP, 2002E): $20.2 billion

Real GDP Growth Rate (2002E): 2.9 per cent *(2003F):* 3.8 per cent

Inflation Rate (CPI, 2002E): 5.6 per cent *(2003F):* 3.6 per cent

Unemployment Rate (2002E): around 15 per cent

Merchandise Exports (2002E): $4.9 billion

Merchandise Imports (2002E): $5.3 billion

Current Account Balance (2002E): -$1.4 billion (-6.9 per cent of GDP)

Major Trading Partners: United States, Colombia, Peru, Venezuela, Chile, Italy, Brazil

Major Export Products (2000E): Crude oil (44 per cent), bananas (27 per cent), shrimp (6 per cent)

Major Import Products (2000E): Raw materials (45 per cent), capital goods (25 per cent), consumer goods (22 per cent)

Total External Debt (2002E): $18.9 billion

Energy Overview

Proven Oil Reserves (1/1/03E): 4.6 billion barrels

Oil Production (2002E): 400,000 barrels per day (bbl/d), of which 390,000 bbl/d was crude oil

Oil Consumption (2002E): 137,000 bbl/d

Net Oil Exports (2002E): 263,000 bbl/d

Crude Oil Refining Capacity (1/1/03E): 176,000 bbl/d

Natural Gas Reserves (1/1/03E): 345 billion cubic feet (Bcf)

Natural Gas Production (2000E): 5 Billion cubic feet (Bcf)

Natural Gas Consumption (2000E): 5 Bcf

Recoverable Coal Reserves (1996E): 26 million short tons (MMST)

Coal Production and Consumption (2000E): None

Electric Generation Capacity (2000E): 3.5 million kilowatts

Electricity Generation (2000E): 10.4 billion kilowatthours (about 75 per cent hydroelectric)

Environmental Overview

Total Energy Consumption (2000E): 0.35 quadrillion Btu (0.1 per cent of world total energy consumption)

Energy-Related Carbon Emissions (2000E): 5.5 million metric tons of carbon (0.1 per cent of world carbon emissions)

Per Capita Energy Consumption (2000E): 27.9 million Btu (vs. U.S. value of 351.0 million Btu)

Per Capita Carbon Emissions (2000E): 0.4 metric tons of carbon (vs. U.S. value of 5.6 metric tons of carbon)

Energy Intensity (2000E): 19,575 Btu/ $1995 (vs. U.S. value of 10,918 Btu/ $1995)**

Carbon Intensity (2000E): 0.30 metric tons of carbon/thousand $1995 (vs. U.S. value of 0.17 metric tons/thousand $1995)

Sectoral Share of Energy Consumption (1998E): Transportation (32.8 per cent), Industrial (28.4 per cent), Residential (27.2 per cent), Commercial (11.6 per cent)

Sectoral Share of Carbon Emissions (1998E): Transportation (44.8 per cent), Industrial (26.8 per cent), Residential (18.3 per cent), Commercial (10.0 per cent)

Fuel Share of Energy Consumption (2000E): Oil (76.3 per cent), Hydroelectricity (14.7 per cent), Natural Gas (1.3 per cent), Coal (0.0 per cent)

Fuel Share of Carbon Emissions (2000E): Oil (89.5 per cent), Natural Gas (10.2 per cent), Coal (0.0 per cent)

Renewable Energy Consumption (1998E): 115 trillion Btu* (2 per cent decrease from 1997)

Number of People per Motor Vehicle (1998): 22.2 (vs. U.S. value of 1.3)

Status in Climate Change Negotiations: Non-Annex I country under the United Nations Framework Convention on Climate Change (ratified February 23rd, 1993). Ratified the Kyoto Protocol on January 13th, 2000).

Major Environmental Issues: Deforestation; soil erosion; desertification; water pollution; pollution from oil production wastes.

Major International Environmental Agreements: A party to the Antarctic-Environmental Protocol, Antarctic Treaty, Biodiversity, Climate Change, Desertification, Endangered Species, Hazardous Wastes, Nuclear Test Ban, Ozone Layer Protection, Ship Pollution, Tropical Timber 83, Tropical Timber 94, Wetlands and Whaling.

The total energy consumption statistic includes petroleum, dry natural gas, coal, net hydro, nuclear, geothermal, solar, wind, wood and waste electric power. The renewable energy consumption statistic is based on International Energy Agency (IEA) data and includes hydropower, solar, wind, tide, geothermal, solid biomass and animal products, biomass gas and liquids, industrial and municipal wastes. Sectoral shares of energy consumption and carbon emissions are also based on IEA data.

Oil and Gas Industries

Organization: Petroecuador, formerly CEPE, serves as the holding company for all state-owned petroleum operations.

Foreign Oil Company Involvement: Agip, Alberta Energy Co., Enap, Kerr-McGee, Occidental Petroleum, Perez Companc, Repsol-YPF, Techint.

Major Oil Fields: Shushufindi, Sacha, Libertador, Cononaco, Cuyabeno, Lago Agrio, and Auca. Smaller fields include Anaconda, Culebra, and Yulebra.

Major Refineries (capacity, bbl/d, 1/1/01E): Esmeraldas (110,000), La Libertad (46,000), Shushufindi (20,000)

Major Pipelines (capacity, bbl/d, 2001E): SOTE Transecuadorean (390,000), Transandean to Colombia (45,000)

Major Terminals: Guayaquil, Esmeraldas/Balao, La Libertad, Tumaco (Colombia–for export of 40,000 bbl/d of Lago Agrio crude oil via Trans-Andean pipeline)

33

Egypt

Egypt is a significant oil producer and a rapidly growing natural gas producer. Projects currently under construction will likely make Egypt an exporter of liquefied natural gas (LNG) by early 2005. The Suez Canal and Sumed Pipeline are strategic routes for Persian Gulf oil shipments, making Egypt an important transit corridor.

General Background

After several years of strong growth in the late 1990s, Egypt's economic growth has slowed markedly over the last three years. Real growth in the country's gross domestic product (GDP) was only 1.6 per cent in 2002, after a figure of 3.3 per cent in 2001, and having been as high as 6.0 per cent as recently as 1999. GDP growth for 2003 is projected at 2.9 per cent for 2003.

Several of Egypt's key sources of hard currency revenue have been negatively impacted as a result of regional tensions and fears of war and terrorism. The tourism sector has been hit hardest, and revenues from the Suez Canal also have declined. In a normal year, tourism revenues account for about 5 per cent of Egypt's GDP, and are among the country's five main sources of hard currency inflows (the others being remittances from Egyptian workers abroad, oil exports, Suez Canal tolls, and foreign aid). Over the longer term, Egypt's macroeconomic prospects may be more favorable, provided progress is made on such structural issues as privatization, trade liberalization, and deregulation. Egypt's main challenge is matching employment growth to the nearly 800,000 new job seekers coming into the labor market each year. Unofficial estimates put Egypt's unemployment rate in the 15 per cent -25 per cent range, roughly twice the official figure. To lower unemployment, Egypt needs to maintain a high rate of GDP growth and to bring in more foreign investment.

Egypt's government plans to accelerate its program for the privatization of state-owned enterprises (SOEs), though to date the privatization program has moved slowly due to the large debts of SOE's and severe overstaffing (layoffs are still difficult due to labor regulations). In recent years, the private

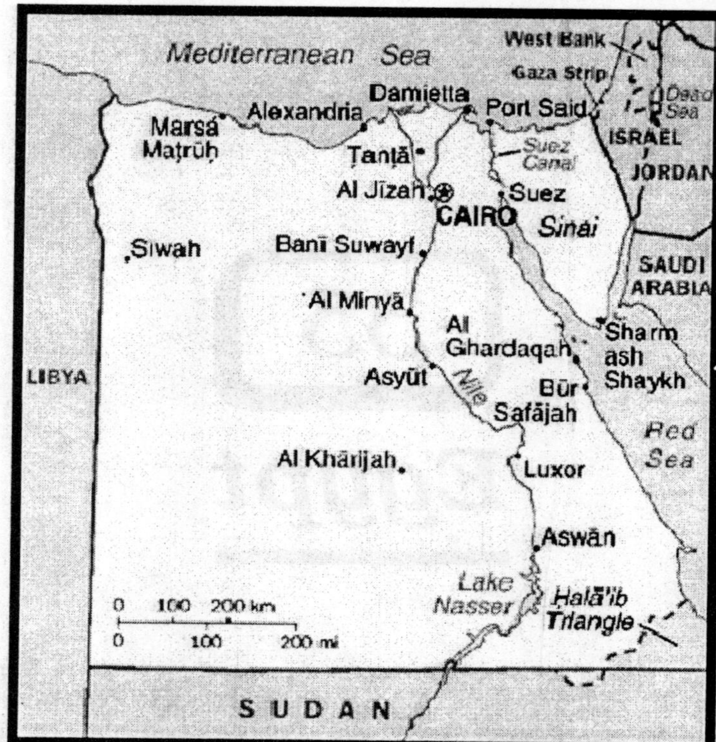

sector percentage of overall Egyptian GDP has been growing by around 1.5 per cent per year, with 110 SOEs having been privatized since 1994. In the future, the government plans to target "strategic" areas for privatization, including telecommunications and other utilities, including the Egyptian Electricity Authority, although the Egyptian General Petroleum Corporation (EGPC) and the new natural gas entity, Egypt Gas (EGAS), remain off limits.

Energy will continue to play an important role in Egypt's economy in coming years. While oil exports have been declining as production has fallen at mature oilfields and domestic consumption has risen, natural gas exports are expected to become a major source of hard currency revenues over the next decade.

Oil

Egypt produced an average of about 631,616 barrels per day (bbl/d) of crude oil in 2002, down sharply from 748,000 bbl/d in 2000, but only slightly below the 639,260 bbl/d produced in 2001. Egyptian crude oil production had peaked at 922,000 bbl/d in 1996. Demand for petroleum products has declined slightly since 1998, after rapid growth during the previous five-year period. This is due in part to the weakness of the economy, but also to reductions in subsides for petroleum products consumption and the increased use of compressed natural gas (CNG) as a fuel for motor vehicles. Egypt is hoping that exploration activity, particularly in new areas, will discover sufficient oil in coming years slow the decline in output. Egyptian oil production comes from 4 main areas: the Gulf of Suez (about 60 per cent), the Western Desert, the Eastern Desert, and the Sinai Peninsula.

Oil from the Gulf of Suez basin is produced mainly by Gupco (Gulf of Suez Petroleum Company), a joint venture between BP and the Egyptian General Petroleum Corporation. Production in the Gupco fields, with most wells in operation since the 1960s and 1970s, has fallen in recent years. Gupco is attempting to slow the natural decline in its fields through significant investments in enhanced oil recovery as well as increased exploration. BP is undertaking a program to invest $450 million over six years (starting in 1999) in technology to prolong the productive life of Gulf of Suez fields. Egypt's second largest oil producer is Petrobel, which is a joint venture between EGPC and Agip of Italy. Petrobel operates the Belayim fields near the Gulf of Suez, and also is undertaking an upgrade program to stem declining production. Other major companies in the Egyptian oil industry include Badr el-Din Petroleum Company (EGPC and Shell); Suez Oil Company (EGPC and Deminex); and El Zaafarana Oil Company (EGPC and British Gas–BG). A new oil find was reported in October 2001 in the Gulf of Suez. Canada's Cabre Exploration reported a drilling success in the offshore East Zeit block which tested at around 8,000 bbl/d, with reserves estimated at 60 million barrels. A new concession was awarded in August 2002 to Dover Petroleum of Canada for the East Wadi Araba area in the Gulf of Suez. Dover announced in January 2003 that its initial well had shown oil, though assessment of the well is still underway.

Egypt's overall oil production has declined more slowly than in the Gulf of Suez fields, due to new output from independent producers like Apache and Seagull Energy at smaller fields, especially in the Western Desert and Upper Egypt. Crude oil production in the Qarun block in southern Egypt reached around 60,000 bbl/d by early 2000, but has since fallen to 36,000 bbl/d. Apache and Seagull have developed the Beni Suef IX field in the East Beni Suef concession in Upper Egypt, which produces over 5,000 bbl/d. The field is said to contain around 100 million barrels of crude oil. A joint venture between EGPC and Agip also is producing about 50,000 bbl/d from an area in the Qattara Depression in the Western Desert, in the Meleiha and West Razzaq blocks.

Offshore oil production possibilities in the Mediterranean are beginning to be explored. The largest concession awarded went to Shell, in February 1999, for a large deepwater area off Egypt's Mediterranean coast. BP Amoco and TotalFinaElf also were awarded a large offshore block from the same bidding round. A smaller offshore concession was awarded to Italy's ENI-Agip. While most discoveries offshore from the Nile Delta have been natural gas, it is believed that there may also be large quantities of oil in the area. Shell reportedly is optimistic about the prospects for its North East Mediterranean Deepwater concession, and its first two test wells have been successful.

Suez Canal/Sumed Pipeline

In addition to its role as an oil exporter, Egypt has strategic importance because of its operation of the Suez Canal and Sumed (Suez-Mediterranean) Pipeline, two routes for export of Persian Gulf oil. Tanker traffic and revenues have declined over the last decade as a result of competition from oil pipelines and the alternate route around the Cape of Good Hope in South Africa. The decline seems to have stopped recently, with revenues rising slightly in 1999, in part due to new pricing offered by the Suez Canal Authority. The SCA offers a 35 per cent discount to liquefied natural gas (LNG) tankers, with even deeper discounts for the largest LNG tankers, as well as other discounts for oil tankers.

The SCA is continuing enhancement and enlargement projects on the canal. The canal has been deepened so that it can accept the world's largest bulk carriers, but it will need to be deepened further to 68 or 70 feet, from the current 58 feet, to accommodate fully laden very large crude carriers (VLCCs). The SCA has attempted to reach an agreement with its main competition for northbound crude traffic, the Sumed pipeline. Such an agreement could bar any tanker small enough to traverse the canal from transporting oil through the pipeline. The SCA offers incentives for tankers to off-load a portion of its cargo through the Sumed, allowing for passage through the canal, and reloading at the other end of the pipeline.

The Sumed pipeline is an alternative to the Suez Canal for transporting oil from the Persian Gulf region to the Mediterranean. The 200-mile pipeline runs from Ain Sukhna on the Gulf of Suez to Sidi Kerir on the Mediterranean. The Sumed's original capacity was 1.6 million bbl/d, but with completion of additional pumping stations, capacity has increased to 2.5 million bbl/d. The pipeline is owned by the Arab Petroleum Pipeline Company (APP), a joint venture between Egypt (50 per cent), Saudi Arabia (15 per cent), Kuwait (15 per cent), the U.A.E. (15 per cent), and Qatar (5 per cent). The APP also has been increasing storage capacity at the Ain Sukhna and Sidi Kerir terminals.

Refining

Egypt's nine refineries are able to process 726,250 bbl/d of crude, with the largest refinery being the 146,300-bbl/d El-Nasr refinery at Suez. The government has plans to increase production of lighter products, petrochemicals, and higher octane gasoline by expanding and upgrading existing facilities.

The new 100,000-bbl/d MIDOR (Middle East Oil Refinery Ltd.) refinery in Alexandria commenced operation in April 2001. While it had originally been planned as a primarily export-oriented project, most of its products are now sold locally. The Israeli company Merhav, which had been the largest Israeli investor in Egypt, sold its 20 per cent stake in the refinery to the National Bank of Egypt in June 2001. Negotiations are reportedly underway with several potential Arab investors, including the Libyan government, for the sale of the Bank's ownership stake.

Natural Gas

Due to major recent discoveries, natural gas is likely to be the primary growth engine of Egypt's energy sector for the foreseeable fututre. Foreign oil companies began more active exploration for natural gas in Egypt beginning in the early 1990s, and very quickly found a series of significant natural gas deposits–in the Nile Delta, offshore from the Nile Delta, and in the Western Desert. Today, Egypt's natural gas sector is expanding rapidly, with production having nearly doubled between 1999 and 2002. Natural gas production in Egypt stood at about 3.0 billion cubic feet per day (bcf/d) in late 2002, and is expected to rise to around 5.0 bcf/d by 2007, with much of the increased volume being exported as LNG. Major foreign companies involved in natural gas exploration and production in Egypt include BG, BP, ENI-Agip, and Shell. Apache also produces gas from its concessions in the Western Desert. The Egyptian government formed a new state-owned entity in 2001 to manage the natural gas sector, Egyptian Natural Gas Holding Company (EGAS), separating those assets out from EGPC.

Egypt's government released a revised estimate of proven natural gas reserves in October 2002, which put the figure at 58.5 trillion cubic feet (Tcf), based on several new finds. Probable reserves are believed to be around 120 Tcf. Most of this increase has come about as a result of new natural gas discoveries offshore from the Nile Delta, and some finds in the Western Desert. In the Nile Delta, which has emerged as a world-class natural gas basin, recent offshore field developments include Port Fuad, South Temsah, and Wakah. In the Western Desert, the Obeiyed Field is an important natural gas area currently under development.

The International Egyptian Oil Company (IEOC), a subsidiary of Italy's ENI-Agip group, is Egypt's leading natural gas producer, operating in the Gulf of Suez, the Nile Delta, and the Western Desert regions. In cooperation with BP Amoco, IEOC has been concentrating its natural gas exploration and development efforts in the Nile Delta region. On November 4, 1997, BP (along with its partners EGPC and IEOC) announced plans to develop the giant Ha'py gas field in the Ras el-Barr concession of the Nile Delta region at an estimated cost of $248 million. The field came onstream in February 2000, and has reached an output of 280 million cubic feet per day (Mmcf/d). In September 1997, IEOC tested the Temsah gas field (located offshore from the Nile Delta) at 11.6 Mmcf/d. In October 1998, BP (25 per cent owner) and ENI-Agip signed a natural gas sales agreement with EGPC (50 per cent owner) and IEOC (25 per cent owner) for Temsah. Temsah's gas reserves are estimated at 3.9 Tcf, and the gas sales agreement was for 35 Mmcf/d initially in 2000, increasing to 480 Mmcf/d by 2003. ENI reported another find in its East Delta Deep Marine concession, which may hold as much as 1 Tcf of additional reserves. Canadian independent Centurion Energy reported a new discovery in the El Manzala concession, onshore in the Nile Delta, in August 2001. Centurion signed a contract in October 2002 with the Egyptian government to begin natural gas deliveries from El Manzala of 35 Mmcf/d in mid-2003.

Two areas in the Western Desert–Obeiyed and Khalda–have shown great potential for increasing Egypt's natural gas production in the near future. Obeiyed is producing 300 Mmcf/d, after the completion of a pipeline linking it to Alexandria. Production in the Khalda concession is currently around 200 Mmcf/d. Apache reported another gas discovery in Khalda in August 2001. Output from Obeiyed and Khalda is transported to Alexandria by a 180-mile pipeline.

Much of the growth in natural gas production in the next decade, however, will be from areas offshore from the Nile Delta. Several major new natural gas finds currently are under development. In May 1999, the Italian firm Edison and the BG Group made a large find ("Scarab/Saffron") in their West Delta Deep Marine concession, which tested at 45 Mmcf/d, followed by another ("Simian") which tested at 44 Mmcf/d in October 1999. The two companies announced in July 2000 that their second and third wells at the field also had tested successfully at a similar flow rate, which was contrained by the capacity of the equipment. Another successful test well drilled on another structure within the same concession also was announced in September 2000. The Scarab/Saffron finds are currently under development, with commercial production of 600 MMcf/d to begin in 2003. Bids were solicited in November 2002 for the development of the Simian field, which is to link into the same pipeline to the Egyptian coast as the Scarab/Saffron fields. BP Amoco and Shell also have concessions offshore from the Nile Delta, and initial seismic survey work and exploratory drilling has indicated significant probable reserves. Shell has announced that probable reserves in its Northeast Mediterranean (NEMED) concession are 15 Tcf. ExxonMobil also holds a 25 per cent stake in this concession. BP and the IEOC also are preparing to bring several fields off the Nile Delta coast into production. BP reported a new find estimated at 500 Bcf in the offshore North Alexandria Concession Area in July 2001.

Natural gas demand has grown rapidly in Egypt as thermal power plants, which account for about 65 per cent of Egypt's total gas consumption, have switched from oil to gas. Domestic natural gas consumers are to be served by several private distributors, franchises which were awarded in late 1998. One of the franchises, awarded to a team headed by BG and including the Egyptian construction firm Orascom and Edison of Italy, is developing distribution infrastructure in Upper Egypt as far south as Asyut, where no piped natural gas had been available. After the initial phase, valued at $220 million, a possible later phase may extend the natural gas grid south to Aswan.

The rapid rise in natural gas reserves has led to a search for export options, which has become particularly important to Egypt's future international balance of payments due to the decline in oil exports. In late 1999, the Egyptian government stated that natural gas reserves were more than sufficient for domestic needs, and that foreign firms producing gas in Egypt should seek export customers. In early 2000, the government announced a moratorium on new purchase agreements by EGPC for domestic consumption, as previously signed agreements will meet projected demand over the next several years. It also announced in September 2000 a new pricing policy which includes ceiling and floor prices, designed to protect both consumers and producers from the risks of prices indexed to oil.

The idea of exporting natural gas to Israel had been under discussion for several years, but appeared by mid-2001 to have been sidelined for the time being by the deterioration in Egyptian-Israeli relations as a result of renewed violence between the Palestinians and Israel. The most ambitious version of the scheme would have involved the construction of an offshore pipeline from El-Arish in Sinai up the coast of Israel, with a possible extension onward to Turkey. The East Mediterranean Gas Company (a consortium of EGPC, Merhav of Israel, and Egyptian businessman Hussein Salem) had been set up to pursue the project. ENI completed a pipeline up Egypt's Mediterranean coast to El-Arish, which could have served as a starting point for the export pipeline. All these plans are now on hold.

A smaller export pipeline to Jordan is currently under construction, and set for completion in late 2003. Egypt is building the section from the existing pipeline terminus at El-Arish to Aqaba in Jordan, with a subsea section in the Gulf of Aqaba bypassing Israeli waters. Egypt, Jordan, and Syria agreed in principle in early 2001 to extend the pipeline into Syria, with eventual natural gas exports to Turkey, Lebanon, and possibly Cyprus. The feasibility of this option is questionable, though, as Turkish demand probably would not support another source of piped gas.

Egypt's other option for exports is LNG. Two LNG projects are currently underway. The Spanish firm Union Fenosa is building a two-train liquefaction facility at Damietta, which is scheduled to begin commercial production in late 2004. Unlike most previous LNG projects, this one is not tied in directly with upstream natural gas production. Union Fenosa has contracted with EGAS for the supply of natural gas from its distribution grid, and will take all of the LNG output itself for use at the company's power plants and distribution to other users in Spain and elsewhere in Europe. ENI also has become involved in the project recently, having purchased a 50 per cent stake in Union Fenosa's natural gas business in December 2002. The second LNG export project, at Idku, is to be built by BG in partnership with Edison of Italy. The project is tied in to natural gas reserves from BG's Simian/ Sienna offshore fields, and is scheduled to begin production in 2005. Gaz de France is to be the main offtaker for the Idku LNG project, having signed a contract in October 2002 for 127 Bcf per year beginning in 2005.

Electricity

Egypt has installed generating capacity of 13.3 gigawatts (GW), with plans to add 9.3 additional GW (mainly gas-fired) by 2010. Around 79 per cent of Egypt's electric generating capacity is thermal (natural gas), with the remaining 21 per cent hydroelectric, mostly from the Aswan High Dam. All oil-fired plants have been converted to run on natural gas. With electricity demand growing, though at a slower rate than in the late 1990s due to the country's sluggish economy, Egypt is building several power plants and is considering limited privatization of the electric power sector. Egypt's power sector is currently comprised of seven regional state-owned power production and distribution companies, which were held by the Egyptian Electricity Authority (EEA). In July 2000, the EEA was converted into a holding company, though still owned by the state. Previous privatization plans have stalled, and the future direction of government policy in the electric utilities sector is unclear.

Egypt has several privately-owned power plants currently under construction which were financed under Build, Own, Operate, and Transfer (BOOT) financing schemes. BOOT projects are used to fund large-scale public infrastructure without affecting the country's debt profile. Private developers are allowed to recover their costs of construction through ownership and operation of the plant for a fixed period before handing it over to the state. The first BOOT project was a gas-fired steam power plant with two 325-megawatt (MW) generating units, located at Sidi Kerir on the Gulf of Suez, which began commercial operation in late 2001. The plant cost $450 million. Electricity from the plant is priced at 2.54 cents per kilowatthour. This competitive price stems largely from the availability of cheap natural gas–to be supplied by Egypt's Gasco–as a feedstock. U.S.-based InterGen (a joint venture of Bechtel Enterprises and Shell Generating Ltd.), along with local partners Kato Investment and First Arabian Development and Investment, have the 20-year BOOT contract for Sidi Kerir. The second BOOT power project award went to Electricite de France (EDF), for two gas-fired plants to be located near the cities of Suez and Port Said. Each plant will have an installed capacity of 650 MW, and the project cost will total around $900 million. The price for power from the EDF plants will be 2.4 cents per Kilowatt hour (Kwh), the lowest price yet offered for a BOOT plant. The project reached financial close in April 2001. The future of BOOT financing in Egypt is unclear, however, and recent government statements indicate that no new BOOT projects are likely in the near future.

Planned future projects are two 750-MW plants planned for Nubariya in the western Nile Delta near Alexandria, a 750-MW addition to the Cairo North power complex, and smaller hydroelectric projects at Nag Hammadi and Asyut. The addition to Cairo North finally is moving forward after several years of delays, now that funding has been secured from multilateral donors. The new generating units will be owned by EEHC. The 64-MW Nag Hammadi hydropower project also is slated to begin construction this year, with European Investment Bank financing.

Egypt also is planning to build a part-solar power plant at Kureimat as a BOOT project, which will have 30 MW of solar capacity out of a total planned capacity of 150 MW. The World Bank will provide a financing package from its Global Environmental Facility which will offset the cost difference between the solar capacity and thermal capacity. A Netherlands-funded project is building 60 MW of wind power units in the Suez Canal area. Egypt also has a 22-MW nuclear research reactor at Inshas in the Nile Delta, built by INVAP S.A. of Argentina, which began operation in 1997.

Work has recently been completed on the interconnection of Egypt's electric transmission grid with other countries in the region. The Five-Country interconnection of Egypt's system with those of Jordan, Syria, and Turkey was completed by 2002. Egypt also activated a link to Libya's electric grid in December 1999.

Environment

In a country that is predominantly desert, the Nile River provides the lifeblood for Egypt's population. With 96 per cent of Egyptians living astride the river, environmental issues are a central component of Egyptian life. Population growth, modernization, and increased economic development have brought environmental problems to the forefront, especially air pollution. In Cairo, emissions from vehicles and lead smelters, together with sand blowing in from the adjacent Western Desert, have created high levels of particulate matter in the air–a deadly combination for public health in the densely-populated capital.

Country Overview

Independence: February 28, 1922 (from the United Kingdom)

Population (7/02E): 70.7 million

Location/Size: Northern Africa/1,001,450 sq. km (386,662 sq. miles), about the size of Texas and New Mexico

Major Cities: Cairo (capital), Alexandria, Aswan, Asyut, Giza, Ismailiya, Port Said, Suez, Tanta

Languages: Arabic (official), English, French

Ethnic Groups: Egyptian, Bedouin, and Berber compose 99 per cent of the population

Religions: Sunni Muslim (94 per cent), Coptic Christian (6 per cent)

Defense (8/98): Army (320,000), Air Defense Command (80,000), Air Force (30,000), Navy (20,000), Reserves (254,000)

Economic Overview

Currency: Egyptian Pound (LE)

Market Exchange Rate (1/21/03): LE 4.65 = $1 U.S.

Nominal Gross Domestic Product (GDP) (2002E): $81.4 billion

Real GDP Growth Rate (2002E): 1.6 per cent

Inflation Rate (2002E): 3.3 per cent

Current Account Balance (2002E): -$0.4 billion

Major Trading Partners (2002): United States, Italy, Germany, Japan, South Korea

Merchandise Exports (2002E): $5.3 billion

Merchandise Imports (2002E): $15.2 billion

Merchandise Trade Balance (2002E): -$9.9 billion

Major Export Products: Crude oil and petroleum products; cotton yarn and textiles; engineering and metallurgical goods; agricultural goods and raw cotton

Major Import Products: Machinery and transport equipment; livestock; food and beverages

Total External Debt (2002E): $28.2 billion

Energy Overview

Proven Oil Reserves (1/1/03E): 3.7 billion barrels

Oil Production (2002): 757,213 barrels per day (bbl/d), of which 631,616 bbl/d is crude oil

Oil Consumption (2002E): 538,000 bbl/d

Net Oil Exports (2002E): 219,213 bbl/d

Crude Refining Capacity (1/1/03E): 726,250 bbl/d

Natural Gas Reserves (1/1/03): 58.5 trillion cubic feet (Tcf)

Natural Gas Production (2000E): 646 Bcf

Natural Gas Consumption (2000E): 646 Bcf

Recoverable Coal Reserves (12/31/99E): 24 million short tons (Mmst)

Coal Production (2000E): 0.4 Mmst

Coal Consumption (2000E): 2.2 Mmst

Electric Generation Capacity (1/1/00E): 13.3 gigawatts (79 per cent thermal, 21 per cent hydroelectric)

Electricity Generation (2000E): 64.7 billion kilowatthours

Environmental Overview

Total Energy Consumption (2000E): 2.0 quadrillion Btu (0.51 per cent of world total energy consumption)

Energy-Related Carbon Emissions (2000E): 33.2 million metric tons of carbon (0.51 per cent of world carbon emissions)

Per Capita Energy Consumption (2000E): 31.8 million Btu (vs. U.S. value of 351.0 million Btu)

Per Capita Carbon Emissions (2000E): 0.5 metric tons of carbon (vs. U.S. value of 5.6 metric tons of carbon)

Energy Intensity (2000E): 25,531 Btu/ $1995 (vs. U.S. value of 10,918 Btu/ $1995)**

Carbon Intensity (2000E): 0.42 metric tons of carbon/thousand $1995 (vs. U.S. value of 0.17 metric tons/thousand $1995)

Sectoral Share of Energy Consumption (1998E): Residential (21.3 per cent), Industrial (53.3 per cent), Transportation (20.2 per cent), Commercial (5.2 per cent)

Sectoral Share of Carbon Emissions (1998E): Transportation (19.3 per cent), Industrial (54.0 per cent), Commercial (4.5 per cent), Residential (22.2 per cent)

Fuel Share of Energy Consumption (2000E): Oil (56.2 per cent), Natural Gas (33.2 per cent), Coal (2.4 per cent)

Fuel Share of Carbon Emissions (2000E): Oil (65.9 per cent), Natural Gas (30.6 per cent), Coal (3.4 per cent)

Renewable Energy Consumption (1998E): 179 trillion Btu* (1 per cent increase from 1997)

Number of People per Motor Vehicle (1998): 33.3 (vs. U.S. value of 1.3)

Status in Climate Change Negotiations: Non-Annex I country under the United Nations Framework Convention on Climate Change (ratified December 5th, 1994). Signatory to the Kyoto Protocol (signed March 3, 1999- not yet ratified).

Major Environmental Issues: Agricultural land being lost to urbanization and windblown sands; increasing soil salinization below Aswan High Dam; desertification; oil pollution threatening coral reefs, beaches, and marine habitats; other water pollution from agricultural pesticides, raw sewage, and industrial effluents; very limited natural fresh water resources away from the Nile which is the only perennial water source; rapid growth in population overstraining natural resources.

Major International Environmental Agreements: A party to Conventions on Biodiversity, Climate Change, Desertification, Endangered Species, Environmental Modification, Hazardous Wastes, Law of the Sea, Marine Dumping, Nuclear Test Ban, Ozone Layer Protection, Ship Pollution, Tropical Timber 83, Tropical Timber 94, Wetlands and Whaling.

The total energy consumption statistic includes petroleum, dry natural gas, coal, net hydro, nuclear, geothermal, solar, wind, wood and waste electric power. The renewable energy consumption statistic is based on International Energy Agency (IEA) data and includes hydropower, solar, wind, tide, geothermal, solid biomass and animal products, biomass gas and liquids, industrial and municipal wastes. Sectoral shares of energy consumption and carbon emissions are also based on IEA data.

Oil and Gas Industries

State Oil Company: Egyptian General Petroleum Corporation (EGPC) plus 11 smaller state oil companies

State Pipeline Companies: Sumed-Arab Petroleum Pipeline Company (APP), Domestic pipelines-Petroleum Pipelines Company (PPC), Export gas pipelines-Egypt Trans-Gas Company (EGTC)

Major Foreign Oil Company Involvement: Apache, British Gas, BP-Amoco, Deminex, TotalFina-Elf, ENI-Agip, Exxon-Mobil, Marathon, Norsk Hydro, Novus, Repsol, Royal Dutch Shell, Samsung, Texaco

Major Ports: Alexandria, Port Said, Sidi Kerir, Ras Shukheir, Suez, Ain Sukhna

Major Oil Fields: Belayim Marine, October, Morgan, Belayim, Badri, Ras Budran

Major Gas Fields: Abu Madi, Abu Qir/North Abu Qir, Shukheir, Badreddin

Major Pipelines (capacity): Sumed pipeline (2.5 million bbl/d)

Major Oil Refineries (crude oil capacity): Cairo Petroleum Refining Company–Mostorod (145,000 bbl/d), Tanta (35,000 bbl/d); El-Nasr Petroleum Company–Suez (146,300 bbl/d), Wadi Feran (8,550 bbl/d); Alexandria Petroleum Company–El Mex (100,000 bbl/d); Ameriya Petroleum Refining Co. (78,000 bbl/d); Suez Oil Processing Company–Suez (66,400 bbl/d); Assiut Petroleum Refining Co. (47,000 bbl/d); Middle East Oil Refinery (MIDOR) (100,000 bbl/d).

34

Equatorial Guinea

Recent offshore oil discoveries and the prospects for additional finds make Equatorial Guinea one of the leading areas for oil exploration in sub-Saharan Africa.

General Background

Equatorial Guinea is located in the western part of sub-Saharan Africa and is comprised of a mainland area (Rio Muni province) and a series of small islands, the largest of which, Bioko, houses the country's capital, Malabo. Equatorial Guinea is notable as the only Spanish-speaking country in Africa, having gained independence from Spanish rule in 1968. The current president, Teodoro Obiang Nguema, came to power in 1979 through a self-initiated coup that overthrew Francisco Macias Nguema, Obiang's uncle and the country's first president. In 1992, the government adopted legislation establishing a multiparty democracy. Since then, Obiang has been re-elected twice, most recently at the end of 2002, but both times amid his opponents' allegations of election fraud.

Since independence, Equatorial Guinea's government has been routinely charged with engaging excessively in corruption, human rights abuse and political repression, factors responsible for the closing of the US embassy in Malabo in 1995. With the rapid growth of US private investment in Equatorial Guinea since 1996, the increasing US interest in the region prompted a July 2002 decision to reopen the US embassy. The reintroduction of US diplomatic engagement in Equatorial Guinea is also aimed at fostering improvements in the country's governance, particularly in light of continued allegations of corrupt government behavior.

For many years, agriculture (primarily cocoa, coffee and timber) formed the basis of the Equatoguinean economy. Since 1995, significant offshore oil discoveries in the Gulf of Guinea have caused oil to displace cocoa as the country's most important export commodity, and has made oil production the major driver of Equatorial Guinea's growth. Real gross domestic product (GDP) growth was 45.5 per cent in 2001, up from the 16.9 per cent growth in 2000, and narrowly missing the record

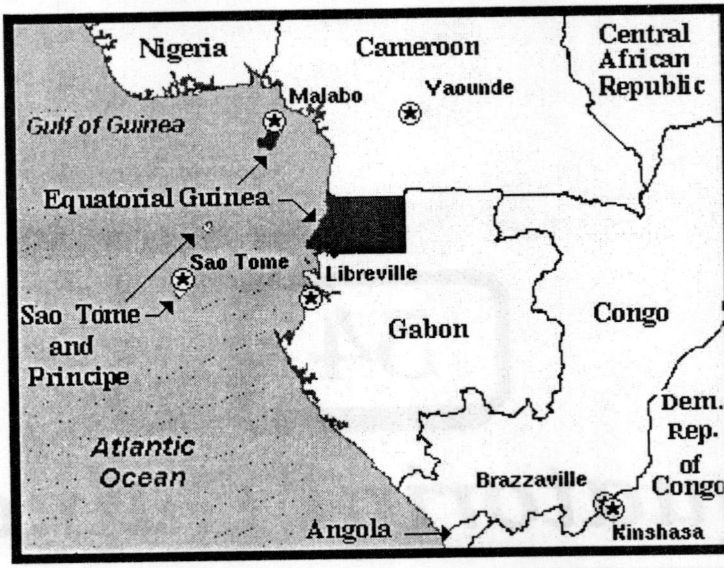

71.2 per cent GDP growth experienced in 1997. Economic growth is expected to remain strong with estimates of real GDP growth of 24 per cent in 2002 and 15 per cent in 2003. U.S. investment (almost exclusively in the oil/energy sectors) has increased dramatically since 1996 and Equatorial Guinea is now the fourth largest destination for American investments in sub-Saharan Africa, after South Africa, Nigeria, and Angola.

Despite rapid growth in real GDP, there is strong evidence of government misappropriation of oil revenues, in particular, for lavish personal expenditures. Furthermore, the failure of the government to inject oil revenues into the country's economy, especially to fund much-needed improvements in the country's infrastructure, has meant little improvement in the economic and social welfare of most Equatoguineans. While real per capita GDP has doubled in the last five years, there has been little positive change in social indicators. Meanwhile, the rapidly expanding economy has begun to create inflationary pressure, translating into an average consumer price index (CPI) of 6 per cent for the past few years. In 2002, the government responded to civil protests about rising food costs by fixing prices on several food staples.

Foreign aid historically has been an important part of the Equatorial Guinea's economy, but during the early 1990's, several donors suspended aid programs in Equatorial Guinea due to non-performance. The World Bank (WB) originally severed its relationship with Equatorial Guinea in 1993, but resumed lending to the country in April 2002 with a program aimed at the development of Equatorial Guinea's transport sector. An IMF-negotiated enhanced structural adjustment facility (ESAF) with Equatorial Guinea expired in 1996. In Article IV consultations, which were concluded in August 2001, the IMF stated that renewed assistance from the fund would hinge on the government's demonstration of improvements in macroeconomic management and governance, as well as the timeliness and transparency of statistical information required to monitor performance. The EU also has suspended its involvement in Equatorial Guinea, but stated that a gradual resumption of aid would occur in step with democratic reforms. Equatorial Guinea has retained normal relations with the African development Bank (AfDB), which has financed 24 loans and grants to the country.

Oil

Oil is Equatorial Guinea's most valuable asset. Since the discovery of the Zafiro field in 1995, oil production has increased more than tenfold, and oil has quickly become the country's most important export commodity, accounting for nearly 90 per cent of the value of total exports in 2001. Furthermore, the location of Equatorial Guinea's reserves in the hydrocarbon-rich Gulf of Guinea (estimates of probable reserves as high as 10 per cent of world total), has prompted large amounts of foreign investment (primarily by US companies) in further exploration and development of the country's oil sector.

The government's involvement in the oil industry has expanded slowly since 1998, when the government introduced more liberal regulatory and profit-sharing arrangements for hydrocarbon exploration and production activities, including revised and updated Production Sharing Contracts (PSC). As a result, government oil revenues increased from 13 per cent to 20 per cent of receipts earned from oil exports. Although significant, the government's share is still relatively small by international standards. In 2001, GE Petrol was formed as the new state-run oil company. Although GE Petrol was originally proposed to focus on downstream activities, its activities have included managing the government's interest (currently, about 5 per cent) in PSCs. Plans to increase the government's stake in new and existing contracts have been announced, but not formally pursued. In 2002, the government announced an expansion of GE Petrol's powers to include the creation of agencies and subsidiaries abroad as well as investment in other national or foreign companies. GE Petrol has since formed a joint-venture with Swiss-based Glencore for exploration of Block O, between Bioko and Cameroon.

Reserves/Exploration

Equatorial Guinea's total proven reserves are estimated at 1.1 billion barrels of oil. Presently, the majority of the country's oil reserves are found in the Zafiro field, which is located offshore, northwest of Bioko and south of Nigeria's Edop, Etin and Ofun fields. The Zafiro field was discovered in 1995 by ExxonMobil and Ocean Energy (merged with Devon Energy in 2003) and constitutes the northern portion of Block B, for which the two companies hold an exploration license. The Zafiro field's probable reserves are estimated to be as much as 1 billion barrels.

ExxonMobil has focused on increasing production from the Zafiro field in recent years and therefore, additional drilling has been carried out to accomodate this plan. Other discoveries that have been made on Block B include the Opalo East, Topacio, Amatista, Rubi and Serpentina fields. ExxonMobil and Devon have also pursued joint exploration of Block C, which lies between Block B and Bioko's western coast. During 2003, Exxon has planned to drill 5 additional wells in Block B.

Equatorial Guinea's second major producing oil field is located just offshore of Rio Muni in exploration Block G, and is estimated to contain 300 million barrels of oil. The Ceiba field, which was discovered by Amerada Hess (formerly Triton Energy) in 1999, began production just 14 months after its discovery, a record that was facilitated by the government's strong interest in accomodating oil sector development. Amerada Hess has an 85 per cent interest in Blocks G and F (directly north of G), for which it shares licenses with South Africa's Energy Africa (15 per cent interest). These two companies have experienced continued successful exploration of Block G, resulting in additional discoveries in 2001 (Okume and Oveng fields) and 2002 (Akom, Elon and Abang fields). Initial exploration of Block F was unsuccessful until the early 2002 discovery of the Ebano Field in the southern part of Block F, although subsequent drilling in other wells in Block F has also come up dry. Following the discovery of the Abang field in June 2002, Amerada Hess and Energy Africa planned to submit a plan to the Equatoguinean government for the development of their six recent discoveries in

Blocks F and G. The companies hope to achieve production of up to 100,000 barrels per day (bbl/d) from the new fields by early 2004.

Following the discovery of the Ceiba field, the first major discovery of oil offshore of Rio Muni, interest in petroleum exploration of surrounding areas has been significant, with many new companies acquiring licenses in exploration blocks further offshore in the Rio Muni basin. International companies with interests in one or more exploration blocks include Chevron (US), Vanco Energy (US), Atlas Petroleum International (US), Devon Energy (US), Roc Oil (Australia), Petronas (Malaysia), Sasol Petroleum (South Africa), and Glencore (Switzerland). As many as five wells are planned to be drilled in the Rio Muni Basin area in 2003.

To the east of the Zafiro field, and just north of Bioko, is the Alba field, discovered in 1991. Original estimates of reserves in the Alba field were around 68 million barrels of oil equivalent (BOE), but recent exploration has increased new estimates significantly. Unlike the Zafiro or Ceiba fields, exploration and production in the Alba field has focused on natural gas, including condensates, which is described in further detail below.

The expansion of offshore oil exploration by Equatorial Guinea and its neighbors has increased the importance of maritime borders used to delineate each country's rights to offshore oil deposits. In March 1999, President Obiang signed a decree unilaterally adopting an equidistant median line to define territorial boundaries as stipulated under the UN Convention on the Law of the Sea. Cameroon, Sao Tome & Principe, and Nigeria subsequently accepted the decision as an improvement over oft-disputed traditional boundaries. Relations remained strained, however, because of Nigeria's ongoing border dispute with Cameroon, which continues following Nigeria's refusal to accept the October 2002 decision of the International Court of Justice (ICJ) to award sovereignty over the disputed area to Cameroon.

Equatorial Guinea currently is involved in a similar dispute with Gabon, over the ownership of the island of Mbagne in the Gulf of Guinea. In February 2002, Equatorial Guinea's nominal control of Mbagne was threatened when the government of Gabon seized control of the island. Since then, the governments of both countries have met to discuss the issue, including the possibility of joint-exploration of the waters surrounding Mbagne, but have not achieved any resolution to their dispute.

On a positive note, tensions have been dispelled between the Nigerian and Equatoguinean governments over the latter's claim of sole ownership of the Zafiro field. At issue was whether the Zafiro was a separate field, or part of an oil structure that straddled the territorial waters of both countries. Tensions began when TotalFinaElf (Total, formerly Elf) (licensee of the Nigerian Block OPL 102, 2 miles north of Zafiro) and the Nigerian government reported that seismic data of the Zafiro field confirmed that it extended into Nigerian territory. In 1998, Total drilled two wells on OPL 102 and announced a discovery called Ekanga, which was subsequently challenged by Equatorial Guinea's claims that the Ekanga wells were drilled in its territorial waters. In September 2000, Equatorial Guinea and Nigeria signed a pact delineating their maritime boundaries. This was followed by an April 2002 treaty in which the countries agreed to joint exploration of crude oil in the Zafiro-Ekanga field.

Production

Oil production from Equatorial Guinea is expanding rapidly, averaging 210,000 bbl/d in 2002, a tremendous increase from the 1996 level of 17,000 bbl/d. During 2003, production improvements and expansions have pushed petroleum output even higher to near 350,000 bbl/d. Meanwhile, the government has denied rumors that it would cap production levels to extend the life of the country's petroleum reserves.

The Zafiro field is Equatorial Guinea's largest oil producer. Production at Zafiro has risen from its initial level of 7,000 bbl/d in August 1996 to approximately 180,000 bbl/d at the end of 2002. Since 1998, ExxonMobil has invested over $1 billion in expanding production from the Zafiro field. Important developments have included the completion of the Jade Platform in June 2000, which added 12,000 bbl/d to production. More significantly, in July 2003, the producer began operation of a new Floating Production Storage and Offloading (FPSO) vessel in the Southern Expansion Area (SEA) of the Zafiro field. This expansion project was implemented under Exxon Mobil's Early Production System, which succeeded in bringing the project online in 19 months, increasing production at the Zafiro field to 300,000 bbl/d.

Production at the Ceiba field rose dramatically during 2002, following production improvements. The Ceiba field originally began production in December 2000 with a target of 50,000 bbl/d; however, in Februrary 2001, Triton Energy reported that technical difficulties at the Ceiba-1 well were limiting production to 45,000 bbl/d. Following Amerada Hess' acquisition of Triton in 2001, the company announced a plan to replace the existing Sendje Berge FPSO in the Ceiba field with a larger FPSO. In February 2002, Amerada Hess completed installation of the Sendje Ceiba and production in the Ceiba field increased to above 50,000 bbl/d. By the end of 2002, production at the Ceiba field was scheduled to increase to 90,000 bbl/d.

Downstream

Equatorial Guinea's domestic petroleum consumption is estimated at 2,000 bbl/d, primarily in the form of motor fuel. The company Getotal, jointly owned by TFE and the government of Equatorial Guinea, has a monopoly on the distribution of petroleum products, all of which are imported.

Natural Gas

Equatorial Guinea' natural gas reserves are located offshore of Bioko, primarily in the Alba and Zafiro oil and gas fields. At present, Alba, the country's largest natural gas field, contains 1.3 trillion cubic feet (Tcf) of proven reserves, but probable reserves are estimated to be at least 4 Tcf. The Alba field was discovered in 1991 by the Spanish firm, Cespa, at which time it was also estimated to contain 68 million barrels of condensate. Throughout the 1990's, Alba was used primarily for condensate production, averaging 6,700 bbl/d of condensate and liquified petroleum gas (LPG) production in 1999, while much of the natural gas from the field was flared off. With a 63.3 per cent interest, Marathon Oil (acquired CMS Energy's assets in late 2001) is the largest stakeholder in the Alba field, with Noble Energy holding a 36.7 per cent interest. Recently, Marathon has expanded exploration of natural gas in the Alba field in conjunction with its other expansion plans for production and processing of natural gas. In July 2003, Marathon reported a new discovery in Block D, named the Bococo field, located 6 miles west of the Alba field.

Natural gas and condensate production in Equatorial Guinea has expanded rapidly in the last five years in response to new investments by major stakeholders in the Alba nautral gas field. In 2001, the Atlantic Methanol Production Company (AMPCO) completed construction of a $450 million methanol plant on Bioko, which now produces 20,000 bbl/d of methanol using 120 million cubic feet (MMcf/d) of natural gas from the Alba field. Nobel Affiliates and CMS Energy formed AMPCO in 1999 as a joint venture, equally sharing a 90 per cent interest in the plant with the remaining 10 per cent held by the Equatoguinean government. The U.S. Overseas Private Investment Corporation (OPIC) provided funding for the project in the form of a $173 milllion loan guarantee, OPIC's largest-ever African loan.

At the end of 2001, production in the Alba gas field was up to 17,000 bbl/d of natural gas condensate, 2,700 bbl/d of LPG and 240 MMcf/d of natural gas. Earlier this year, Marathon was planning the sale of $400 million of its non-core assets to enable it to focus on further development of its natural gas assets in Equatorial Guinea. In late 2002, Marathon Oil received approval from the government to expand condensate production to 46,000 bbl/d and LPG production, to 16,000 bbl/d by October 2004. The first part of this expansion is expected to utilize 800 MMcf/d of wet natural gas for condensate production, and is scheduled for completion by the end of 2003.

More recently, in May 2003, the government approved Marathon's proposal to construct a liquified natural gas (LNG) facility on Bioko that will supply 3.4 million tons of LNG. Marathon has signed a Memorandum of Understanding (MOU) with British Gas to export the LNG to its U.S facilities for a 17-year period beginning in 2007, the facility's expected completion date. While a financing scheme for the plant has not yet been reported, the government of Equatorial Guinea has stated that it is considering a 25 per cent interest in the project. The LNG facility is expected to increase the importance of Equatorial Guinea in the region's natural gas market with the possibility of it serving as a point of completion for future natural gas production from neighboring countries.

Natural gas consumption in Equatorial Guinea has slowly increased in recent years, as increased production has made natural gas conveniently accessible. Natural gas consumption in 2001 was approximately 1 Bcf.

Electricity

Estimates of Equatorial Guinea's generating capacity vary, with 15.4 megawatts (MW) of certain installed capacity, and 5-30 MW of estimated additional capacity. About 5.0 MW are located on the mainland, including 4 MW of oil-fired thermal capacity and 1 MW of hydroelectric capacity. Bioko is known to receive electricity from at least two thermal plants and one hydroelectric plant. The expansion of natural gas production at the Alba field in recent years has provided a convenient fuel source for new generation. The 10.4-MW, natural gas-fired Punta Europa plant began operation in 1999 and supplies electricity to Bioko while another natural gas-fired, 4-6 MW of generation is currently under construction at the AMPCO complex on the island. Equatorial Guinea is estimated to have 2,600 MW of hydropower potential.

Equatorial Guinea's electricity sector is owned and operated by the state-run monopoly, SEGESA. In general, the power supply is somewhat unreliable, due to aging equipment and poor management, as demonstrated by prolonged blackouts in Malabo in May 2002 and again, in April 2003. As a result, small, diesel generators are widely-used as a back-up source of power supply.

Country Overview

Independence: October 12, 1968 (from Spain)

Population (July 2003E): 510,500

Location/Size: West Central Africa–the mainland region of Rio Muni borders the Atlantic Ocean just north of the Equator between Cameroon and Gabon; the islands of Bioko, Annobon, Elobey Grande, Elobey Chico and Corisco lie in the Atlantic Ocean / 28,050 square kilometers (10,831 square miles), about the size of Maryland

Major Cities: Malabo (capital) on island of Bioko, Bata on mainland enclave of Rio Muni

Languages: Spanish and French (official languages), Fang, Pidgin English, Bubi

Ethnic Groups: Fang (constitute majority of mainland population), Bubi (on Bioko), Annobonese, Ndowe, Kombe, and Bujebas

Religion: Christian (predominantly Roman Catholic), Traditional Beliefs

Economic Overview

Currency: Communauté Financière Africaine (CFA) franc; CFA franc fixed to the Euro since Jan. 1, 1999

Market Exchange Rate (09/12/03): US$1 = 580.4 CFA Fr

Gross Domestic Product (GDP) (2002E): $2.4 billion; *(2001E):* $3.6 billion

Real GDP Growth Rate (2002E): 24 per cent ; *(2003E):* 15 per cent

Consumer Price Inflation Rate (2002E): 7.2 per cent ; *(2003E):* 6.8 per cent

Current Account Balance (2000): -$300 million; *(2001):* -$400 million

Major Trading Partners: United States, Spain, France, China, Cameroon, United Kingdom, Japan

Merchandise Exports (2002E): $2.2 billion; *(2003E):* $3.0 billion

Merchandise Imports (2002E): $400 million; *(2003E):* $400 million

Merchandise Trade Surplus (2002E): $1.8 billion; *(2003E):* $2.5 billion

Major Export Products: Oil, timber, cocoa

Major Import Products: Food, clothing, petroleum products, automobiles, machinery, iron and steel

Total External Debt (2001): $260 million

Energy Overview

Secretary of State for Energy: Miguel Ekua Ondo

Proven Oil Reserves (1/1/03E): 1.1 billion barrels

Oil Production (2001): 181,000 barrels per day (bbl/d), all of which is crude oil

Oil Consumption (2001): 2,000 bbl/d

Net Oil Exports (2001E): 179,000 bbl/d

Natural Gas Reserves (1/1/03E): 1.3 trillion cubic feet (Tcf)

Electric Generation Capacity (1/1/03): 15.4 megawatts (of which 80 per cent is thermal an 20 per cent hydroelectric)

Electricity Generation (2001): 24 million kilowatthours

Environmental Overview

Total Energy Consumption (2001): 0.005 quadrillion Btu (<0.001 per cent of world total energy consumption)

Energy-Related Carbon Emissions (2001): 0.57 million metric tons of carbon (<0.01 per cent of world carbon emissions)

Per Capita Energy Consumption (2001): 10.5 million Btu (vs. U.S. value of 341.8 million Btu)

Per Capita Carbon Emissions (2001): 1.2 metric tons of carbon (vs. U.S. value of 5.5 metric tons of carbon)

Fuel Share of Energy Consumption (2001): Oil (84.7 per cent), Natural Gas (15 per cent), Hydropower (0.3 per cent)

Fuel Share of Carbon Emissions (2001): Oil (10.9 per cent), Natural Gas (89.1 per cent), Coal (0.0 per cent)

Status in Climate Change Negotiations: Equatorial Guinea signed and ratified both the United Nations Framework Convention on Climate Change and the Kyoto Protocol on August 16, 2000.

Major Environmental Issues: Deforestation, wildlife destruction, non-potable tap water and desertification.

Major International Environmental Agreements: A party to the Conventions on Biodiversity, Desertification, Endangered Species, Law of the Sea and Ship Pollution.

Oil and Gas Industries

Organization: Petroguinea state company in upstream activities (not yet active); downstream activities–Getotal (owned by TotalFinaElf-80 per cent and Equatorial Guinea-20 per cent)

Major Oil Fields (2002 production): Zafiro (180,000 bbl/d), Ceiba (50,000 bbl/d), Alba (6,500 bbl/d)

Major Foreign Oil Company Involvement: ExxonMobil, Amerada Hess (formerly Triton), ChevronTexaco, Energy Africa, Marathon Oil, Noble Affiliates, Devon Energy, Petronas

Refineries: None

Pipelines: None

Terminals: Proposed LNG terminal on Bioko (expected completion date of 2007)

35

France

France is one of the world's largest nuclear power producers, but has limited fossil fuel resources. The 2000 merger of its top two oil companies formed the fourth largest oil publicly-traded company in the world.

General Background

With an estimated GDP of $1.4 trillion in 2002, France is the fifth-largest industrialized economy. France is a founding member of the European Union (EU) and a member of the Group of Seven (G-7) industrialized nations, the General Agreement on Tariffs and Trade (GATT)/World Trade Organization (WTO), the International Energy Agency (IEA), and the International Atomic Energy Agency (IAEA). On January 1, 1999, France adopted the European single currency, the euro. In recent years, France's economy has experienced stronger growth in comparison to its neighbors. However, this growth is slowing down. France's economy grew 1.8 per cent in 2001 and is estimated to have grown only 1.2 per cent in 2002, its lowest level since 1996. Recently, the French government has indicated that its public deficit was likely to have already breached the 3 per cent ceiling allowed by the EU's Stability and Growth Pact. Such a breach would trigger Brussels to issue a formal "excessive deficit procedure" against Paris, which could result in fines.

Traditionally, the role of the state has been stronger in France than in other Western European countries. France is one of the most economically centralized countries in Europe, with a strong history of state ownership in the aviation, telecommunications, and energy industries. However, the role of the government is now changing. Important economic and political developments in France include privatization and increasingly frequent mergers and acquisitions (M&As) and hostile corporate takeovers, once virtually unheard of in the country.

Drivers of this movement towards decentralization include increased pressures of globalization, compliance to EU privatization directives, and political initiatives from the newly formed center-right government. Previously, the moderate socialist Prime Minister, Lionel Jospin, and the Gaullist President,

Jacques Chirac headed the French government under the French system of governmental "cohabitation." The divided government moved slowly toward privatizing France's energy industry, despite EU directives, calling for member states to relinquish control of their energy companies. This gridlock has left France trailing its EU neighbors in opening its natural gas and power markets to competition. It has also sparked hostility, particularly in Spain and Italy, to France's state-owned utility company, Electricité de France (EdF), which acquired shares in some of their utilities.

President Jacques Chirac and newly appointed prime minister, Jean-Pierre Raffarin, have reemphasized their desire to liberalize France's natural gas and electricity markets. In August 2002, they announced plans for the partial privatization of EdF and state-owned Gaz de France (GdF). This came as part of a program to allow the private sector to invest in state-owned companies, as well as to raise money to reduce the government's budget deficit. In December 2002, the French parliament adopted the European Directive on natural gas deregulation into law, an act that the EU required in August 2000.

Energy

French energy policy has been relatively consistent in recent decades, with the main objectives including: securing energy supply, achieving international competitiveness, and protecting the environment. The focus on energy security has led France to become one of the world's top producers and consumers of nuclear power. However, the French government has recently organized a national energy policy debate, which will focus on energy sources for the next thirty years, particularly the status of nuclear power and the future of renewables.

Oil

Oil's role in France's energy sector has decreased substantially since the 1970s. Overall, the contribution of oil to primary energy consumption fell from 68 per cent in 1973 to 38 per cent in 2001. Nuclear power has largely replaced oil as a primary source for electricity generation.

Exploration and Production

France is highly dependent on oil imports. In 2002, France imported about 1.85 million barrels per day (bbl/d) of its approximate 1.96 million bbl/d oil consumption. France has oil reserves totaling only 148 million barrels as of January 2003. France's domestic crude oil production comes from numerous small-sized wells, with total crude production of 26,900 bbl/d.

Despite France's limited domestic oil reserves and production, the French oil industry is an important actor in world energy markets. Major oil assets of French oil companies are located in the North Sea, Africa, and Latin America. French imports come primarily from Saudi Arabia and Norway, followed by the United Kingdom, Iraq, Iran, Nigeria, and Russia.

In early 1999, French oil company Total merged with Belgian oil company Petrofina to create TotalFina, the world's sixth-largest oil company and the third-largest oil company in Europe. In 2000, TotalFina merged with Elf Aquitaine to create TotalFinaElf. After the deal was completed, TotalFinaElf became the fourth-largest publicly listed oil company in the world, after ExxonMobil, Royal Dutch/ Shell, and BP. The company has proven reserves of about 10.8 billion barrels of oil equivalent (boe) and production of about 2.1 million boe/d. TotalFinaElf aims to increase hydrocarbon production by nearly 40 per cent by 2005. The company owns more than 50 per cent of the refinery capacity in France, and is the seventh-largest refiner in the world.

During the 1990s, Total and Elf Aquitaine reportedly negotiated with Iraq on development rights for the Majnoon and Nahr Umr oil fields. Majnoon is the largest of Iraq's oilfields slated for post-sanctions development, with reserves of 12-30 billion barrels. In July 2001, angered by France's perceived support for the U.S. "smart sanctions" plan, Iraq announced that it would no longer give French companies priority in awarding oil contracts, and would reconsider existing contracts as well.

In 2002, TotalFinaElf joined nine other companies in supporting the construction of the Baku-Tbilisi-Ceyhan oil pipeline project. The 1,100-mile route will carry oil from a terminal near Baku to the port of Ceyhan on the Mediterranean coast. The pipeline will be capable of carrying over one million barrels of oil per day, and should be completed by 2005.

Downstream

France's crude oil refining capacity is 1.9 million bbl/d. The country's largest refinery is TotalFinaElf's refinery at Gonfreville l'Orcher, with a capacity of 323,643 bbl/d. Increasingly strict EU environmental regulations for refineries are in large measure behind recent upgrades in the French refining sector. The regulations will become stricter in 2005, and substantial investment in the refining sector will be necessary to meet these new mandatory targets. ExxonMobil has begun adapting its Port Jerome refinery to 2005 EU specifications.

Because oil security has been such a concern for French energy policy-makers, there is a French law allowing the French government to refuse to close a refinery if it believes its supply or price security is at risk. Essentially, this gives the French government veto power over EU legislation regarding refineries. This could become an important issue as the EU's environmental standards are strengthened.

Natural Gas

Natural gas has increased its share of France's total energy consumption from 7.2 per cent in 1973 to 15.3 per cent in 2002. France has limited natural gas resources (506 billion cubic feet (BcF) as of January 2003), and therefore imports almost all of the natural gas it consumes. Natural gas consumption was estimated at 1.48 trillion cubic feet (TcF) in 2001, an increase of 4.5 per cent over 2000.

Sector Organization

The French natural gas industry is run by Gaz de France (GdF), the state-held company with a monopoly on importation and distribution of natural gas in France. By 2003, GdF aims to possess sufficient reserves to produce at least 15 per cent of the natural gas it sells. The company's annual production capacity currently stands at more than 70 Bcf. GdF also has the largest underground storage capacity in Western Europe, with 318 Bcf, about 3 months supply.

During the last few years, GdF has sought to increase its holdings, particularly in the North Sea. In 2001, GdF acquired interests in Norway's Snøhvit (12 per cent) and Njord (20 per cent) fields. Snøhvit natural gas reserves are estimated at 7.3 Tcf. In the same year, GdF acquired from Texaco about 21 per cent equity interest in twelve exploration licenses in the UK North Sea. GdF has also signed an agreement with Sonatrach to explore the Ahnet Basin in Algeria, as well as with two Egyptian utilities, EGPC and EGAS, to increase these companies' capacity to supply natural gas to Europe. Last April GdF paid a reported 213 million euros for CalEnergy Gas (UK), which included four gas-producing fields in the southern region of the U.K. North sea–Johnston, Schooner, Windermere and Anglia–as well as three exploration permits and four fields under study for gas development. Most recently, GdF acquired Preussag Energie of Germany. This purchase increased GdF's natural gas reserves to 2.8 Tcf. TotalFinaElf has also expanded its natural gas portfolio. In November 2002, it finalized an agreement to buy 35 Bcf per year from Sonatrach. TotalFinaElf is also involved in developing the South Pars natural gas field in Iran.

Norway is France's top natural gas import source, followed by Russia and Algeria. Natural gas imports from Russia have been declining in recent years, while imports from Algeria have been rising. Imports will increase from Algeria when the Medgaz natural gas pipeline between Algeria and Algeria, Spain is completed. The pipeline will have a capacity 564.8 Bcf per year. TotalFinaElf and GdF both have 12 per cent shares in the pipeline and are expected to take their gas into France. GdF also imports liquefied natural gas (LNG) to its two terminals. In addition to long-term contracts, GdF buys natural gas on the spot market or with short-term contracts from the UK's North Sea.

GdF is establishing France as a hub for Western European natural gas. In October 1998, France for the first time became linked via pipeline to a foreign production field. The NorFra pipeline linked Norway's Troll gas field in the North Sea to the French natural gas grid. The pipeline is 521 miles long, and is the longest undersea natural gas pipeline in the world. By 2005, the Norwegian pipeline is expected to supply one-third of France's total natural gas consumption. GdF is near completing the construction of the Les Marches du Nord-Est pipeline. It has already completed the first 124-mile part, which went operational in October 2001. The second 186-mile part was expected to be operational in late 2002. GdF signed a 25-year contract with Italy's Snam for delivery of 212 Bcf of Norwegian natural gas through the pipeline. GdF plans to spend $2.5 billion 2001-2003 on developing its pipeline network and installations in France.

Liquefied Natural Gas

France is the largest importer of LNG in Europe. It receives nearly 90 per cent of its LNG from Algeria. GdF has the only two LNG terminals: the 159-Bcf-per-year-capacity facility at Fos-sur-Mer near Marseilles and the 353-Bcf-per-year-capacity facility at Montoir-de-Bretagne, near Nantes. GdF announced in July 2002 that it plans to build a second terminal at Fos-sur-Mer. The new terminal will have a capacity of 288 Bcf per year and will come into operation in 2006.

In its recent report on France's natural gas market, the French Electricity Regulation Commission (Commission de Régulation de l'Énergie) was critical of GdF's two LNG terminals, stating that the company should allow access at these terminals for other market entrants. CRE insists that GdF must reserve part of its new terminal in Fos-sur-Mer for new market entrants.

Increasing France's importance as a transit center, GdF receives Nigerian LNG at its Montoir-de-Bretagne terminal that is swapped out to Italy's Enel. The terminal receives 141 Bcf annually, 124 Bcf under the Italian contract and 18 Bcf under a contract signed by GdF. In October 2002, GdF announced that it has agreed to purchase the entire output of "Train 1" of the Egyptian LNG project. The project is located at Idku, about 31 miles east of Alexandria.

Natural Gas Market Privatization

The French government has been slow to liberalize its natural gas market according to the European Directive (98/30/EC). In December 2002, the French parliament, almost 3 years past the EU deadline, made the Directive into national law. The Directive requires EU members to make 20 per cent of the natural gas market open to competition. GdF had, nonetheless, already opened its grid to third-party access in August 2000, allowing France's largest industrial consumers to choose their suppliers.

The French Electricity Regulation Commission (CRE) explained in its recent report that the market limits competition because it is inundated with long-term contracts with foreign groups. This allows GdF to dominate supply. CRE plans to ensure competitive conditions, which will include introducing more flexibility into contractual terms, particularly when the commission takes complete regulatory control of the natural gas market in 2004.

Movement towards changing the status of GdF from a state-owned enterprise to a joint stock company has been slow. France has been one of the slower countries in the EU to pave the way for competition. Although France adopted draft legislation in May 2000, the full national parliament did not pass a law to open the market until recently. The newly formed government has indicated that they plan to begin this process. Prime Minister Raffarin pledged in August 2002 partly to privatize GdF by early 2004. In 2002, the French government already allowed the privatization of the country's natural gas transport network, which GdF and TotalFinaElf purchased.

Coal

France has relatively small coal reserves of only 40 million short tons. France neither produces nor consumes significant amounts of coal. Since 1973, coal production decreased by 91 per cent and consumption by 57 per cent. Coal-fired electricity has been mostly replaced by nuclear power. Coal imports come from Australia, the United States, Poland, Germany and South Africa.

The French government has supported the coal industry since the 1994 National Coal Pact between Charbonnages de France (CdF), the state coal company, and French coal miners unions. According to the agreement, the industry would receive state support as it was gradually phased out all together. All French coal mines are slated to be shut down by 2005. In May 2001, the EC authorized France to pay 991 million euros in state aid to the coal industry.

Electricity

France is the second-largest electricity market, consumer and generator in Europe behind Germany. It is the largest exporter of electricity in Europe. France generates roughly 77 per cent if its electricity from nuclear power plants.

Sector Organization

France's electricity sector is dominated by the wholly state-owned utility company, Electricité de France (EdF), which produces, transports, and distributes over 95 per cent of electricity in France. EdF is the last major state-run electricity monopolist in the EU, as most of France's neighbors have privatized their electricity companies. There has been, however, partial liberalization of some aspects of France's electricity sector.

There are currently only two companies of any size in France that may be able to compete with EdF on a limited basis. The first is CNR, Compagnie Nationale du Rhône. France's second-largest electricity group, which produces about 3 per cent of France's electricity. In August 2001, CNR and Electrabel of Belgium created a company called Energie du Rhône, which markets electricity produced by both of these companies. The second producer is SNET, Société Nationale d'Electricité et de Thermique, a subsidiary of French coal utility Charbonnages de France (CdF). SNET's shareholders include CdF (51 per cent), EdF (18.7 per cent), and Spain's Endesa (30 per cent). Endesa has recently announced interest in increasing its share in SNET.

Electricity Market Privatization

The 1996 EU Directive on the electricity market required that at least 26 per cent of electricity sales in EU member countries be opened to competition, beginning in February 1999. This requirement increased to 28 per cent in February 2000 and to 33 per cent in 2003. In February 2000, a full year after

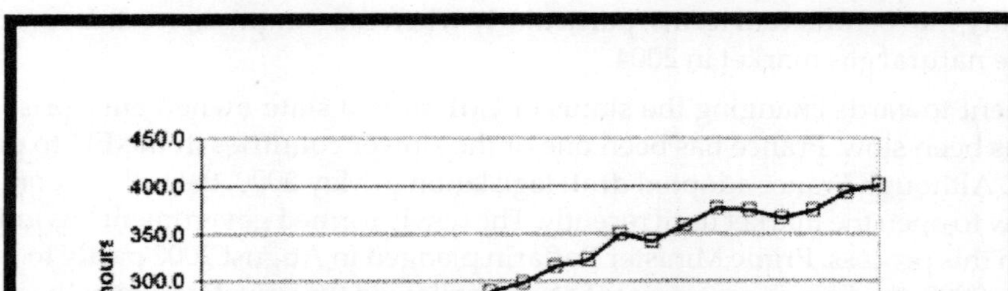

FRANCE'S ELECTRICITY GENERATION BY FUEL, 1980-2001

the first EU deadline, France passed legislation that began the electricity sector's liberalization. Since that time, about 1,800 large industrial and commercial consumers (those using more than 16 million kilowatthours per year), comprising about 30 per cent of the market, have been able to choose their electricity supplier. As of February 2003, 37 per cent of the French market is open to competition. All facilities with annual electricity consumption above 7 GWh are able to choose their supplier.

Another step toward liberalization was the French government's creation of the Electricity Transmission Network (Reseau de Transport d'Electricité, RTE). RTE, which became official on July 1, 2000, operates France's high-tension transmission network and is independent of EdF. RTE's mission is to assure all clients fair access to the network. It also oversees natural gas deregulation.

In late November 2001, the Powernext electricity trading market was launched in France. Powernext auctions standard hourly contracts for physical delivery of electricity to business customers under responsibility of the RTE and guaranteed by Clearnet, a subsidiary of the Euronext stock exchange. Powernext aims to trade 10 per cent of the French market by 2003-2004, and also to act as a price reference for the electricity market. Powernext plans to launch French electricity futures trading, which will include hedging products for all power-related risk (gas, electricity futures contracts, CO_2, weather derivatives).

In recent years, EdF has acquired a number of large foreign companies, such as London Electricity and German utility EnBW. EdF acquired UK gas and electricity utility Seeboard in June 2002, and is attempting to take over AEP Energy Services in Norway. EdF's acquisitions, however, have not been without criticism, particularly from other EU member states and companies. Many claim that EdF has made aggressive acquisitions in Europe's deregulated electricity market while France has been slow to open its doors to competition in the utility sector, which has protected EdF's market share in France. Another charge is that EdF's status as a state-owned monopoly has made it easier for it to purchase

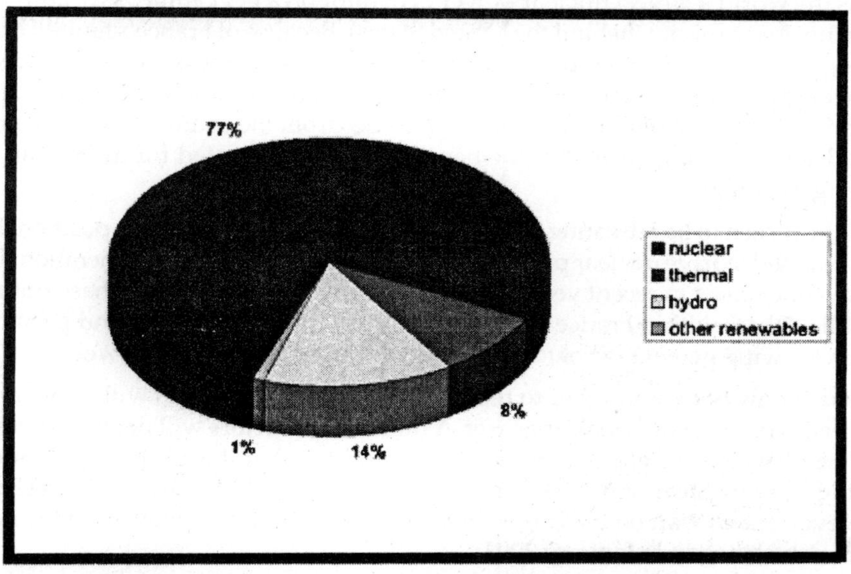

FRANCE'S ELECTRICITY GENERATION BY SOURCE IN 2001

and outbid competitors abroad. Of particular controversy is the company's status as an Etablissment Publique et Industrielle (EPIC) which grants unlimited state guarantees against insolvency and bankruptcy. The EU Competition Commission recently announced in March that it will investigate these grants.

Spain and Italy have also attempted to block the state-owned EdF from entering their privatized electricity markets. Both countries adopted regulations, which limited EdF's voting rights in the companies in which it acquired shares. Spain, however, compromised in October 2002 by allowing the takeover of part of Hidrocantabrico in return for France increasing the interconnection between the two countries from 1,000 MW to 4,000 MW by 2006.

In the case of Italy, EdF and Fiat took control of Edison in 2001. Edison is 76 per cent controlled by Italenergia, an energy consortium in which EdF has an 18 per cent stake. In response, the Italian government passed a decree to reduce EdF's voting rights in Italenergia to 2 per cent in the hope of blocking the French state-owned company from having partial control of the company.

EdF and the Italian government have been trying to reach a compromise so that the voting restrictions can be lifted. Italy would like to see the Italian electricity group Enel be allowed to buy 20-30 per cent stakes in four French nuclear power plants. A final agreement is still pending.

The French government plan to part-privatize EdF by 2004 has faced some difficulties recently. EdF and GdF renegotiated retirement plans with four labor unions. However, the Confederation Generale du Travail (CGT) rejected the plan. Under current law, the companies bear the direct cost of retirement, but are backed by state guarantees. The new plan would take the charges off the companies' books in order make them attractive to private investment. It will remain difficult for the French government to begin the privatization process until this matter is settled.

Nuclear

France is the world's largest nuclear power generator on a per capita basis, and ranks second in total installed nuclear capacity (behind the United States). Because of France's limited domestic energy sources, energy supply security and reliance on imports are major issues in France. Government policy has strongly promoted increases in nuclear power generation over the past three decades. Currently, about 77 per cent of France's electricity comes from the country's 58 nuclear reactors. This represents a dramatic change from 1973, when fossil fuels accounted for more than 80 per cent of French power generation.

France is now seen to be retreating slowly from its staunchly pro-nuclear position. Previously, the government planned to have nuclear power reach 100 per cent of electricity generation. Environmental objections have increased in recent years, however. Germany's decision to phase out nuclear power started a public debate within France about the future of its own industry, and public opinion polls showed that a growing percentage of the public favors an end to nuclear power.

France must now decide whether to replace obsolete nuclear plants with more modern nuclear plants, or to begin phasing out nuclear power. A number of reactors will need to be replaced around 2015-2020. The new center-right government has indicated that it is preparing to authorize EdF to begin planning a new nuclear unit, based on the Franco-German EPR (European Pressurized Reactor) design. However, it will wait on making a final decision until the conclusion of the national energy policy debate, which began in January 2003.

In September 2001, the French government restructured its nuclear sector into a single government holding company, Areva. The Areva Group is a combination of Cogema, Framatome, CEA Industrie, and the Commissariat à l'Energie Atomique (CEA), the French Atomic Energy Agency, which is the major shareholder of the Areva Group with nearly 80 per cent. The group presides over the country's major nuclear enterprises, including mining, fuels, treatment, recycling, decontamination and engineering. The French government plans in the long term to offer Areva shares to outside investors, although the CEA will retain control.

France is one the few countries in the world with a nuclear reprocessing plant. Cogema's La Hague facility received authorization from the French nuclear energy regulator to start operations of two new facilities, hull and end-pieces compacting and plutonium purification and conditioning, in January 2002.

Environment

France has been victim to several major oil spills in recent years that resulted in severe environmental damage to France's coastline and caused serious economic harm to France's tourism and fishing industries. In response, the French government is taking a proactive approach to preventing marine pollution by establishing an extended ecological zone into the Mediterranean Sea, as well as attempting to impose more stringent conditions on the oil tanker transportation.

Air pollution, especially in Paris, is still a problem, despite the adoption of measures to mitigate the effects of increased transportation and growing energy consumption from France's transportation sector. However, efforts by successive French governments to improve energy efficiency and to promote conservation have reduced the country's level of energy intensity. Likewise, France's level of carbon intensity is lower than that of its peers in Western Europe, largely due to its reliance on nuclear energy. France's commitment to the use of nuclear power has allowed the country to cut its carbon emissions over the past 20 years, although the growth in France's carbon emissions in the mid-1990s may make it difficult for the country to live up to its commitments under the Kyoto Protocol.

By European standards, France's development and use of renewable energy resources has been fairly limited. Market barriers thus far has stifled the use of renewables for electricity and heat production in France. Nevertheless, despite (or perhaps because of) France's lack of fossil fuel resources, most energy-related environmental trends in France appear to be headed for greater efficiency and less environmental impact. France's plan for an "environment charter" to be added to its constitution is geared to make the country a world leader in promoting environmental concerns, making the outlook for France's environment quite favorable if the French government can follow through with that legislation and continue to improve its energy efficiency efforts.

Country Overview

Independence: 486 (unified by Clovis)

Population (2002E): 59.4 million

Location/Size: Western Europe, bordering the Bay of Biscay and English Channel, between Belgium and Spain southeast of the UK; bordering the Mediterranean Sea, between Italy and Spain/547,030 sq km (slightly less than twice the size of Colorado)

Language: French 100 per cent, rapidly declining regional dialects and languages (Provencal, Breton, Alsatian, Corsican, Catalan, Basque, Flemish)

Ethnic groups: Celtic and Latin with Teutonic, Slavic, North African, Indochinese, Basque minorities

Religions: Roman Catholic 83-88 per cent, Protestant 2 per cent, Jewish 1 per cent, Muslim 5-10 per cent, unaffiliated 4 per cent

Defense (8/98): Army 203,200; Air Force 78,100; Navy 63,300

Economic Overview

Currency: Euro (EUR)

Exchange Rate (3/21/03): 1 U.S. Dollar = EUR 0.942

Gross Domestic Product (GDP, 2002E): $1.42 trillion

Real GDP Growth Rate (2002E): 1.2 per cent *(2003F):* 1.4 per cent

Inflation Rate (consumer prices, 2002E): 1.9 per cent *(2003F):* 1.7 per cent

Unemployment Rate (2002E): 8.8 per cent *(2003F):* 9.0 per cent

Exports of Goods and Services (2002E): $309.1 billion

Imports of Goods and Services (2002E): $299.1 billion

Major Trading Partners: Germany, Italy, Belgium, the United Kingdom, the United States

Major Export Products: Machinery and transport equipment, agricultural products, chemical products

Major Import Products: Machinery and transport equipment, agricultural products, chemical products, and energy

Energy Overview

Proven Oil Reserves (1/1/03E): 148 million barrels

Oil Production (2002E): 76,917 barrels per day (bbl/d), of which 26,900 bbl/d is crude oil

Oil Consumption (2002E): 1.96 million bbl/d

Net Oil Imports (2002E): 1.85 million bbl/d

Crude Oil Refining Capacity (1/1/03E): 1.9 million bbl/d

Natural Gas Reserves (1/1/03E): 506 billion cubic feet (Bcf)

Natural Gas Production (2001E): 0.07 trillion cubic feet (Tcf)

Natural Gas Consumption (2001E): 1.48 Tcf

Net Natural Gas Imports (2001E): 1.41 Tcf

Coal Reserves (2001E): 39 million short tons (Mmst)

Coal Production (2001E): 2.53 Mmst

Coal Consumption (2001E): 20.89 Mmst

Electricity Generation (2001E): 520 billion kilowatthours (bkwh), nuclear (77 per cent), hydro (14 per cent), thermal (8.2 per cent), less than 1 per cent other renewables

Electricity Consumption (2001E): 415 bkwh

Net Electricity Exports (2001E): 72.6 bkwh

Environmental Overview

Total Energy Consumption (2001E): 10.5 quadrillion Btu* (2.6 per cent of world total energy consumption)

Energy-Related Carbon Emissions (2001E): 108.1 million metric tons of carbon (1.6 per cent of world carbon emissions)

Per Capita Energy Consumption (2001E): 177.8 million Btu (vs. U.S. value of 341.8 million Btu)

Per Capita Carbon Emissions (2001E): 1.83 metric tons of carbon (vs. U.S. value of 5.51 metric tons of carbon)

Energy Intensity (2001E): 5,805 Btu/ $1995 (vs U.S. value of 10,736 Btu/ $1995)**

Carbon Intensity (2001E): 0.06 metric tons of carbon/thousand $1995 (vs U.S. value of 0.17 metric tons/thousand $1995)

Fuel Share of Energy Consumption (2001E): Oil (40 per cent), Natural Gas (15.3 per cent), Coal (4.6 per cent)

Fuel Share of Carbon Emissions (2001E): Oil (67 per cent), Natural Gas (21.5 per cent), Coal (11 per cent)

Status in Climate Change Negotiations: Annex I country under the United Nations Framework Convention on Climate Change (ratified March 25th, 1994). Signatory to the Kyoto Protocol (April 29th, 1998) and ratified the Treaty on May 31, 2002.

Major Environmental Issues: Some forest damage from acid rain; air pollution from industrial and vehicle emissions; water pollution from urban wastes and agricultural runoff.

Major International Environmental Agreements: A party to Conventions on Air Pollution, Air Pollution-Nitrogen Oxides, Air Pollution-Sulphur 85, Air Pollution-Sulphur 94, Air Pollution-Volatile Organic Compounds, Antarctic-Environmental Protocol, Antarctic Treaty, Biodiversity, Climate Change, Desertification, Endangered Species, Hazardous Wastes, Law of the Sea, Marine Dumping, Marine Life Conservation, Ozone Layer Protection, Ship Pollution, Tropical Timber 83, Tropical Timber 94, Wetlands and Whaling. Has signed, but not ratified: Air Pollution-Persistent Organic Pollutants, Climate Change-Kyoto Protocol.

36

Gabon

Gabon is sub-Saharan Africa's third largest producer and exporter of crude oil, the country's main export commodity. Recent declines in oil production have led the government to encourage further exploration and exploitation of oil reserves, while concerns over a long-term decline in proven oil reserves has brought about further development of the non-oil sector.

General Background

Since its independence from France on August 17, 1960, Gabon has remained relatively peaceful and stable. Gabon's President Omar Bongo, re-elected on December 6, 1998, is Africa's second longest serving head of state, in power since 1967. In December 2001, the Parti democratique gabonais (PDG), the President's party, won a sweeping victory in the parliamentary elections.

Gabon's economy continues to rely heavily on exports of crude oil, which account for nearly 65 per cent of the government's budget, 80 per cent of total export revenues, and 43 per cent of gross domestic product (GDP). Gabon's current account balance worsened significantly in 2001 due to a combination of slightly lower world oil prices, weaker forestry exports, and a high increase in imports. This followed a dramatic improvement in Gabon's current account balance in 2000 spurred by higher world oil prices. Non-oil sector GDP increased by about 4 per cent, mainly in the commerce, agriculture, and wood processing industries. Timber products typically have been Gabon's second largest export, comprising around 12 per cent of total exports and about 5 per cent of GDP. Exports of manganese also contribute significantly to Gabon's GDP, and are expected to grow by 600,000 metric tonnes per year now that the new Industrial Complex of Moanda has been built.

In part as a result of the lower-than-expected oil export revenues, Gabon's real GDP grew by 1.7 per cent in 2001, falling short of projections. Consumer prices rose by 2.5 per cent in 2001. In 2002, Gabon's real GDP is expected to grow by 1.5 per cent while inflation is projected to grow by 0.7 per cent. Gabon is still recovering from an economic crisis in 1999, which prompted the national government

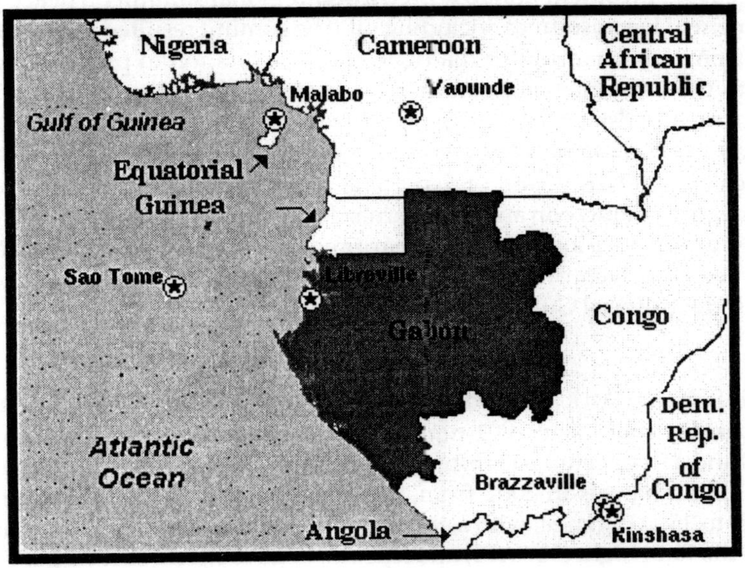

to take steps to increase the country's economic growth by diversifying the structure of its exports, restructuring and reducing its civil service, and privatizing several parastatal industries. In October of 2001, following a ten-year debate, legislation was enacted to create a duty free zone in a 5.8 square mile area surrounding Port-Gentil. Companies operating within the zone will be exempt from minimum investments, customs duties on imports and exports, price controls, import/export quotas, and licenses.

In April 2002, the International Monetary Fund (IMF) allowed an agreement providing for an 18-month stand-by credit of $119 million to lapse. The IMF cited Gabon's failure to meet necessary criteria for keeping it in place. The IMF credit line, conditional upon the attainment of several quantitative and structural benchmarks, had been designed to support Gabon's economic reform program in the event that oil prices fell in 2001. The agreement ended a break in relations with the IMF which occurred after an earlier three-year structural adjustment plan expired in March 1999.

Gabon continues to have one of the highest per capita incomes (around $3,700) in sub-Saharan Africa, but its income distribution is extremely uneven and almost half of the population lives below the poverty line. Since the 1994 devaluation of the CFA franc by 50 per cent, Gabon's informal sector has grown dramatically due to higher domestic prices, relatively lower salaries, and a scarcity of official or formal–particularly public sector–employment opportunities (unemployment was estimated at 25 per cent in 1997).

Gabon is aiming for greater regional economic integration. In January 1999, the Central Africa Economic and Monetary Community (CEMAC), which includes Gabon and five other countries, agreed to take steps towards convergence of their macroeconomic policies, stabilization of their common currency, creation of a common market, and harmonization of their sectoral policies. In December 1998, Gabon's currency, the CFA franc, was officially pegged to the Euro, the common currency adopted by 11 European members of the European Community.

In November 1999, President Bongo hosted a sub-regional summit on peace prospects, stability, and the management of key resources, including oil. The summit resulted in the creation of the Gulf of Guinea Commission, a mechanism for dialogue and consultation to prevent, manage, and resolve conflicts linked to the economic and commercial exploitation of the natural resources within the territorial limits of member states. The meeting was called by Nigeria's President Olusegun Obasanjo and was attended by Presidents Denis Sassou Nguesso of Congo, Teodoro Obiang Nguema of Equatorial Guinea, Miguel Trovoada of Sao Tome and Principe, and Foreign Ministers Augustin Kontchou Kouomegni of Cameroon and Joao Bernardo de Miranda of Angola. Since the creation of the Gulf of Guinea Commission, a long-standing maritime border dispute between Gabon and Sao Tome and Principe has been resolved, and the Commission is likely to debate the issue of maritime borders between Gabon and Equitorial Guinea.

OIL

Gabon is sub-Saharan Africa's third largest oil producer. The country's total oil production of 302,000 barrels per day (bbl/d) in 2001 represented a 7.4 per cent decline from the 326,000 bbl/d produced in 2000 and a 9 per cent drop from 1999 production levels. Gabon's proven oil reserves have increased to 2.5 billion barrels in 2002 from 1.3 billion barrels in 1996. Most of this crude oil has gravities in the 30° to 35° API range, with a small amount of heavier, 25° API output. Gabon's primary crude exports are the Rabi Light (34 ° API) and Mandji Blend (30 ° API and 1.1 per cent sulfur) streams. Gabon exported about 140,000 bbl/d of crude oil to the United States in 2001, accounting for over 46 per cent of Gabon's crude oil production. Western Europe is also a key destination for Gabonese crude, with occasional spot shipments to the Far East. Gabon officially left OPEC in 1996, citing the organization's high annual dues of $1.9 million.

Although Gabon's proven oil reserves have nearly doubled since 1996, the government is concerned about a longer-term trend of diminishing oil reserves. Production from Gabon's largest oil

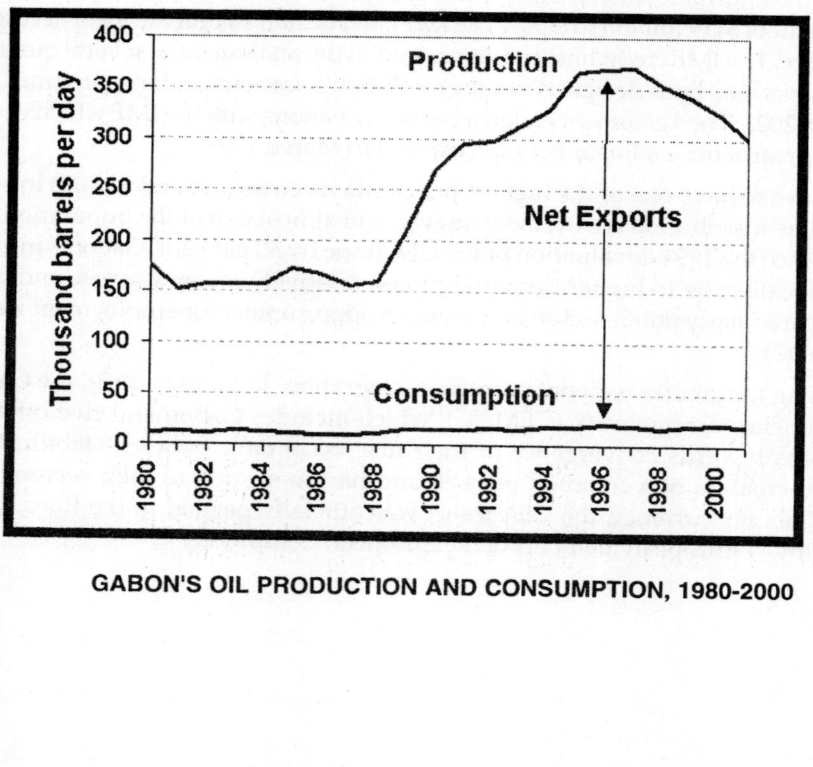

GABON'S OIL PRODUCTION AND CONSUMPTION, 1980-2000

field, Rabi-Kounga, fell to 100,000 bbl/d in 2000, down from the 1997 peak of 217,000 bbl/d. Only one quarter of the Shell -operated Rabi-Kounga field's estimated recoverable reserves of 850 million barrels remains available for further production. Former Shell Gabon chairman John Barry estimates that Rabi-Kounga will remain Gabon's biggest producing field over the next five years, even as output declines.

To help boost reserves and production, Gabon's oil ministry has revised its production-sharing contracts to attract new investors and has increased the number of exploration permits issued. In its ninth oil licensing round, which opened in April 2000 and closed in January 2001, Gabon offered 27 blocks, of which 13 were onshore, five were located in shallow water and nine were found in deep water. The government of Gabon described the response from international petroleum companies as disappointing and pledged to continue to promote the blocks.

Despite the possibility of declining reserves, and the disappointing response to the new exploration permits, several foreign oil companies have invested in Gabon's oil sector in recent years. Most recently, in April 2002, Energy Africa signed a production sharing agreement that allows the company to operate in the area covered by the Akoum permit. At present, Energy Africa is planning at least one exploration well, and holds a 100 per cent interest, although Gabon's government has an option to buy a 10 per cent share at first production. In December 2000, Shell Gabon signed a production sharing agreement for the Douka Marin and Panga Marin blocks.

Meanwhile, Pioneer Natural Resources Co. has a production-sharing contract covering the Olowi block that lies three miles south of Gabon's Gamba oil field. The company drilled a new well named Awena Marin -1 in April 2002, with 34° oil flowing at non-commercial rates. Overall, reservoir quality was lower than in the previous Olowi Marin discovery.

In 2001, Elf Gabon's (a subsidiary of TotalFinaElf) net profit from operations in Gabon declined by 31 per cent (to $124.5 million) due to lower oil prices and a 4.6 per cent decrease in production. Faced with declining onshore and shallow-water production, Elf Gabon is turning to deep-water exploration. In April 2001, the company signed a concessional agreement for Olonga Marin which covers 4,000 square miles. Oil in the concession is believed to lie at a depth between 9,900 and 11,500 feet below the Atlantic floor. In August 2000, Elf Gabon also began a seismic survey in the Akori block, 36 miles off Port Gentil.

In October 2001, Vaalco Energy executed a five-year extension of its Production Sharing Contract for the 1,864 square mile Etame Block. Vaalco Energy formed a consortium in 1998 to evaluate recent discoveries on the block including those of Etame and Tchibala. This consortium is comprised of Vaalco Energy (30.4 per cent), Pan African Energy Corporation Ltd (32.5 per cent), Sasol Petroleum International (30 per cent), PetroEnergy Resources Corporation (4.5 per cent), and Nissho Iwai Corporation (2.6 per cent). In June 2002, Vaalco Energy announced that the first of three wells to be drilled on the Etame Development Project had been completed successfully. Known as the ET-3H well, it flowed at a rate of 7,630 bbl/d. One month later the second well (ET-4H) was completed, flowing at a rate in excess of 9,800 bbl/d. The third well, known as ET-1VA, was completed in September 2002 and flowed at a rate of 4,130 bbl/day. The Etame Project is expected to increase Gabon's oil production by around 15,000 bbl/d when the field is brought fully onstream.

Gabon's oil industry received a modest boost in March 2001 when the Atora field came on stream, producing 20,000 bbl/d. Atora is estimated to contain between 50 million and 100 million barrels. The field is operated by Elf Gabon, with Amerada Hess Gabon controlling a 40 per cent share and Shell Gabon managing a 20 per cent share. The production from Atora made Elf Gabon the largest producer of oil in the country. Another modest boost is expected from Shell Gabon's 2002 discovery of a new oil field on the Ozigo concession, which is expected to increase Gabonese oil production by a further 20,000 bbl/d when fully developed.

In February 2001, PanAfrican Energy signed a concession agreement to reactivate the Remboue field, where initial production was due to begin in July 2001. Output was expected to reach 1,800 bbl/d by mid-2002 as additional wells are brought on stream. In addition, PanAfrican has built an onshore, $2.2-million pipeline from its 1,000-bbl/d Obangue field to the nearby Avocette field, operated by TotalFina Elf, to replace a river barge that has carried oil to market since Obangue was brought on stream in 1998.

In December 1999, Eni, the Italian oil and natural gas company, through its subsidiary Agip Gabon, signed a farm-out agreement with Petronas Carigali (Overseas), a subsidiary of the Malaysian national oil company Petronas, for oil exploration in Gabon's southern deep waters. Under the agreement Eni was required to sell a 25 per cent stake in each of three exploration blocks offshore Gabon to Petronas Carigali. The three blocks, Mpolo, Chaillu, and Meboun, cover a total area of 22,508 square km (9,000 square miles) in the Gabon Basin. Petronas, jointly with Energy Africa, also has equity participation in four other blocks in Gabon. Eni has been active in Gabon since 1981 and currently produces oil from the Limande offshore field.

Refining

The Sogara refinery at Port Gentil is Gabon's only refinery. Opened in 1967, Sogara is jointly owned by the Gabonese government (25 per cent), private investors (9.2 per cent), and a consortium of international oil firms led by TotalFina Elf. The refinery is currently operating at 82 per cent (17,300 bbl/d) of its 21,000 bbl/d nameplate capacity.

Natural Gas

Gabon's natural gas reserves total 1.2 trillion cubic feet. The majority of the natural gas produced in Gabon is used in the generation of electricity or as a refinery fuel. Carbide Holdings is proposing to establish an iron works that would utilize gas currently flared, as well as gas that is re-injected into the Rabi-Kounga field.

Electricity

Hydroelectric stations account for 71 per cent of Gabon's electricity production. The primary sites are at Tchimbele (69 MW) and Kinguele (57.6 MW) on the M'Bei River, and Poubara on the Ogooue River. There is an estimated total hydropower potential of 6,000 MW if all possible sites were to be developed.

In March 1997, Gabon announced that a 20-year concession to run the state-owned electricity and water utility, the Société d'Electricité et d'Eaux du Gabon (SEEG), was awarded to the French firm VIVENDI (formerly known as Compagnie Générale des Eaux). This is the first privatization of a sub-Saharan water and electric utility that entails full commitment for future investment by the private operator to upgrade and modernize the systems. VIVENDI received a 51 per cent interest in the SEEG, 5 per cent was offered to SEEG's employees, and the remaining 44 per cent was offered to the general public.

Country Overview

Independence: August 17, 1960 (from France)

Population (2001E): 1.3 million

Location/Size: West Central Africa, bordering the Atlantic Ocean at the Equator between Cameroon, the Congo and Equatorial Guinea/267,670 square kilometers (103,000 square miles), about the size of Colorado

Major Cities: Libreville (capital); Port-Gentil; Franceville

Languages: French (official), Fang; Myene; Bateke; Obamba; Bapounou/Eschira; Bandjabi

Ethnic Groups: About 40 Bantu tribes, including four major tribal groupings (Fang, Eschira, Bapounou, Bateke); about 100,000 expatriate Africans and Europeans (27,000 French)

Religion: Christian (55 per cent -75 per cent); Muslim (less than 1 per cent); traditional beliefs

Defense (8/98): Army (3,200); Navy (500); Air Force (1,000); Paramilitary Forces 4,800 (including a Gendarmerie of 2,000)

Economic Overview

Currency: Communauté Financière Africaine (CFA) franc

Market Exchange Rate (9/06/02): US$1 = 662 CFA

Gross Domestic Product (GDP) (2001E): $4.7 billion

Real GDP Growth Rate (2001E): 1.7 per cent *(2002E)*: 1.5 per cent

Inflation Rate (2001E): 2.5 per cent *(2002E)*: 0.7 per cent

Current Account Balance (2001E): $200.3 million

Major Trading Partners: France, United States, Japan, China, Côte d'Ivoire

Merchandise Exports (2001E): $2.5 billion

Merchandise Imports (2001E): $0.9 billion

Major Export Products: Oil, timber, manganese

Major Import Products: Machinery and transport equipment, other manufactures, food

Total External Debt (2002E): $4.0 billion

Energy Overview

Proven Oil Reserves (1/1/02E): 2.5 billion barrels

Oil Production (2001E): 302,000 barrels per day (bbl/d), of which almost all is crude oil *(2002E):* average of 299,000 bbl/d (data for first six months of 2002).

Oil Consumption (2001E): 19,000 bbi/d

Net Oil Exports (2001E): 283,000 bbl/d

Major Crude Oil Customers: United States (nearly 50 per cent), Western Europe

Crude Oil Refining Capacity (1/1/02E): 17,300 bbl/d

Natural Gas Reserves (1/1/02E) : 1.2 trillion cubic feet

Natural Gas Production (2000E): 3.5 billion cubic feet (Bcf)

Natural Gas Consumption (2000E): 3.5 Bcf

Electric Generation Capacity (1/1/2000E): 0.3 gigawatts

Electricity Production (2000E): 0.85 billion kilowatthours (of which thermal electricity was 29 per cent, hydroelectric was 71 per cent)

Environmental Overview

Minister of Forest Economy, Water, Fishing, and in charge of Environment: Emile Doumba

Total Energy Consumption (2000E): 0.05 quadrillion Btu* (<0.1 per cent of world total energy consumption)

Energy-Related Carbon Emissions (2000E): 1.68 million metric tons of carbon (<0.1 per cent of world carbon emissions)

Per Capita Energy Consumption (2000E): 41.7 million Btu (vs U.S. value of 351.0 million Btu)

Per Capita Carbon Emissions (2000E): 1.4 metric tons of carbon (vs U.S. value of 5.5 metric tons of carbon

Energy Intensity (2000E): 9,262 Btu/ $1995 (vs U.S. value of 10,918 Btu/ $1995)**

Carbon Intensity (2000E): 0.31 metric tons of carbon/thousand $1995 (vs U.S. value of 0.17 metric tons/thousand $1995)

Sectoral Share of Energy Consumption (1998E): Industrial (37.6 per cent), Transportation (17.8 per cent), Residential (45.0 per cent), Commercial (2.5 per cent)

Sectoral Share of Carbon Emissions (1998E): Transportation (48.6 per cent), Industrial (37.3 per cent), Residential (11.3 per cent), Commercial (2.6 per cent)

Fuel Share of Energy Consumption (2000E): Oil (80.3 per cent), Natural Gas (7.3 per cent), Coal (0.0 per cent)

Fuel Share of Carbon Emissions (2000E): Natural Gas (56.0 per cent), Oil (44.0 per cent), Coal (0.0 per cent)

Renewable Energy Consumption (1998E): 44 trillion Btu* (0 per cent increase from 1997)

Number of People per Motor Vehicle (1998): 34.5 (vs U.S. value of 1.3)

Status in Climate Change Negotiations: Non-Annex I country under the United Nations Framework Convention on Climate Change (ratified January 21st, 1998). Not a signatory to the Kyoto Protocol.

Major Environmental Issues: Deforestation and poaching.

Major International Environmental Agreements: A party to Conventions on Biodiversity, Climate Change, Desertification, Endangered Species, Law of the Sea, Marine Dumping, Nuclear Test Ban, Ozone Layer Protection, Ship Pollution, Tropical Timber 83, Tropical Timber 94 and Wetlands.

The total energy consumption statistic includes petroleum, dry natural gas, coal, net hydro, nuclear, geothermal, solar, wind, wood and waste electric power. The renewable energy consumption statistic is based on International Energy Agency (IEA) data and includes hydropower, solar, wind, tide, geothermal, solid biomass and animal products, biomass gas and liquids, industrial and municipal wastes.

Oil and Gas Industries

State Oil Company: Société Nationale Petrolière Gabonaise

Major Refineries (capacity–bbl/d): Port Gentil (17,300)

Oil Terminals: Onshore: Cap Lopez, Oguendjo, Gamba, Port Gentil *Offshore:* Lucina, M'Bya

Major Foreign Oil Company Involvement: Amerada Hess, Broken Hill Petroleum, Devon Energy, Energy Africa, Eni, Marathon, PanAfrican Energy, Perenco, Petrofields, Petronas, Pioneer Natural Resources, Shell, Sasol Petroleum International, TotalFina Elf (Elf Gabon), Vaalco, Vanco

37

Germany

Germany is one of the world's largest energy consumers. Because the country has limited indigenous energy resources (except for coal), Germany imports most of its energy.

General Background

Germany is one of the largest economies in the world, a founding member of the European Union (EU), a North Atlantic Treaty Alliance (NATO) member, and a member of the Group of Seven (G-7) industrialized nations. It joined the common European currency, the euro, on January 1, 1999, and Frankfurt is the seat of the European Central Bank. Germany is the largest consumer of oil and second largest of natural gas in Western Europe.

Germany's economy has struggled over the last two years, with real gross domestic product (GDP) growing by only 0.2 per cent, to $1.99 trillion, in 2002. After the economy contracted in the first two quarters of 2003, Germany fell into a technical recession. Since the country's economy is heavily reliant on exports, continued sluggish global demand and a stronger euro have been particularly damaging.

Economic malaise has also weakened German equity markets and consumer confidence, as the German stock exchange, the DAX, lost half of its value over the course of 2002, while consumer spending contracted throughout the year. The resulting declines in tax revenues caused Germany's public deficit to exceed 3 per cent of its GDP in 2002, the maximum allowed under the European Union's 1997 Stability and Growth Pact, prompting an official warning from Brussels. If Germany breaches the Pact in 2004, the third consecutive year in row, Ecofin, the council of EU finance ministers, would have the right to impose binding fiscal sanctions against Germany.

In March 2003, Chancellor Schröder outlined a number of proposals to stimulate the economy under the program "Agenda 2020." The main focus of the program was to increase labor market

flexibility by introducing initiatives which would reduce the high cost of labor and create incentives for people to look for work. In 2002, unemployment in Germany stood at 9.8 per cent. In June 2003, the Chancellor decided to bring forward to 2004 tax cuts originally planned for 2005 in the hope of providing an additional stimulus to the economy.

In September 2002, Gerhard Schröder won his second term as German Chancellor, leading a coalition between Germany's Social Democrats (SPD), and the environmentalist Green Party. Gerhard Schröder has served as the German Chancellor since 1998. President Johannes Rau has served as the German head of state since 1999.

Energy

Germany has limited domestic fossil-fuel resources and therefore relies on imports to meet its energy needs. In 2001, gross imports accounted for 64 per cent of Germany's total primary energy consumption of 14.35 quadrillion British thermal units (Btus). Oil accounted for 41 per cent of domestic German energy consumption in 2001, followed by natural gas (23 per cent), coal (23 per cent), and nuclear power (11 per cent).

Germany is a world leader in developing renewable energy. In 2000, the government set a goal to double the proportion of renewable energy sources by 2010. By 2050, half of Germany's entire energy demand should be met by solar, wind, biomass, hydro, and geothermic sources, according the German government. The government has set these goals in order to help meet the country's greenhouse gases reduction quotas: reduction of CO2 emissions by 25 per cent of 1990 levels by 2005 and reduction of the six greenhouse gases cited in the Kyoto Protocol by 21 per cent between 2008 and 2012, within the context of the EU burden-sharing program. Germany will most likely play a significant role in the EU's new international emissions trading scheme, which the European Parliament approved on July 2, 2003. The European Union-wide market will allow companies to buy and sell the difference between the pollution they produce and their limit set by member countries. The market is set to begin in 2005.

European Union Marker Liberalization Requirements

Germany's energy policy is greatly influenced EU regulations. In 1996, the EU began the process of creating an internal market at the European level for energy. An eventual goal of this initiative is to harmonize existing rules and to identify complementary measures in order to avoid 15 separate markets. In 1998, the European Commission introduced two directives, requiring the EU member states to liberalize their electricity and natural gas markets. These directives seek to abolish exclusive rights, requiring member states to unbundle network activities from generation and supply activities, as well as to promote transparency and non-discrimination in services.

In Germany, the energy market, which is in theory completely open to competition, are dominated by a handful of large power companies. Electricity and natural gas prices remain some of the highest in Europe and third-party access to national grids limited. Germany also remains the only EU member yet to have established a regulatory agency for natural gas and electricity markets, a requirement of the EU. However, the German government announced April 2003 that it would create an agency in accordance with EU requirements by mid-2004.

Following reunification of the country in 1990, the major task of German energy policy was to merge successfully the radically different energy sectors of the East and West. West Germany had a diversified and mainly privately-owned system of energy supply with a high standard of energy efficiency and a commitment to environmental protection. In contrast, East Germany's energy sector was highly centralized, predominantly state-owned, and mainly dependent upon lignite, or "brown coal" as its primary fuel. Lignite, because of its characteristic chemical impurities, results in higher emissions of toxins into the atmosphere than other fuels when burned. To date, a great deal of progress has been made in conforming the former East Germany's energy sector to the standards of the West in the areas of privatization and environmental regulation.

Oil

In 2002, Germany consumed about 2.7 million barrels per day (bbl/d) of oil. Nearly all of Germany's oil is imported, making the country the third-largest oil importer in the world. Since 1991, oil imports have accounted for approximately 94 per cent of total petroleum consumption in Germany. Over the past decade, imports from Non-OPEC suppliers, particularly Russia, have increased. German imports oil from Russia, Norway, the United Kingdom, Libya, Kazakhstan, and Syria.

Production In 2002, Germany produced around 69,000 bbl/d of crude oil, of which approximately 18,000 bbl/d came from the country's offshore oil production, the *Mittelplate* field. The field is located in the Wattenmeer tidal flatlands off the western coast of German state Schleswig-Holstein. Mittelplate's crude oil, which is extracted and processed on the Mittelplate production island, is then transported ashore by barge and routed via pipeline to refineries in Schleswig-Holstein. Another 16,000 bbl/d are extracted from the Mittelplate offshore hydrocarbon structure through onshore extended-reach horizontal wells located at the Dieksand Land Station on the western coast of Schleswig-Holstein. Combined, the Mittelplate offshore production island and the Dieksand Land Station produced roughly

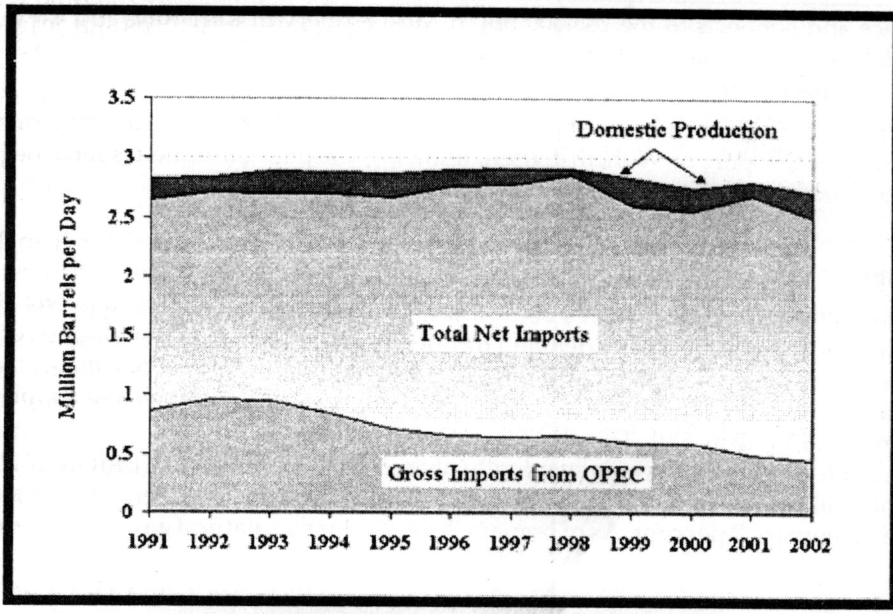

OIL CONSUMPTION IN GERMANY, 1991-2002

33,000 bbl/d in 2001, representing half of Germany's domestic crude oil production. The Mittelplate project is a 50/50 joint venture between two German companies, RWE Dea and Wintershall. Germany's other oil fields are located primarily in the north and northeast of the country.

Consumption Forecast

Oil has accounted for roughly 41 per cent of Germany's total energy consumption every year since 1991. Two different studies on Germany's petroleum demand were released in the last year, one by the German firm PROGNOS, and another from the German oil industry group *Mineralölwirtschaftsverband*. Both organizations predicted a modest increase in petroleum demand in the short-run (through 2005), but their forecasts differed in the long-term. While PROGNOS predicted stable oil demand from 2005 to 2020, *Mineralölwirtschaftsverband* anticipated a 14 per cent decline in oil demand during this period. EIA expects that German energy consumption could reach 3.1 million barrels per day by 2005, with modest growth through 2020. Factors contributing to the future of German oil consumption are: the declining role of coal and nuclear power; Germany's rapidly increasing energy efficiency; and the country's Eco-Tax, which was introduced in 1999 and levies a tax on fossil fuels in an effort to discourage the use of polluting fuels. Motor gasoline and diesel fuels are heavily taxed in Germany, with Germans paying about twice as much for motor gasoline than Americans.

Downstream

As of January 2003, Germany had a crude oil refinery capacity of 2,267,100 bbl/d, the seventh highest in the world. Germany receives a large portion of its *crude oil imports through pipelines*. For example, Germany's three major southern refineries receive Mediterranean sea borne cargoes from Syria, Libya, southern Russia, and Kazakhstan through two pipelines: the *SPSE* line which starts in

Lavera, France and connects to the 285,000 bbl/d Miro refinery in Karlsruhe and the *Trans-Alpine* pipeline which begins in Trieste, Italy, and connects to the 262,300 bbl/d Bayernoil and the 70,000 bbl/d OMV refinery in Burghausen.

The German government has also shown interest in a plan that would extend the *Odessa-Brody oil pipeline* to the German port of Wilhemshaffen, allowing Caspian oil to be refined in Europe and transported to another continent. However, it currently remains unclear whether this project will materialize.

Natural Gas

Germany is the European Union's second largest consumer of natural gas after the United Kingdom. Germany meets most of its demand through imports, as the country produces insufficient natural gas to satisfy domestic consumption. In 2002, the country produced 0.79 trillion cubic feet (Tcf) of natural gas, while consuming 3.4 Tcf. Natural gas accounted for about 23 per cent of total energy consumption in Germany in 2001. This share is expected to rise over the decade, especially for electric power generation as nuclear power is eventually phased out. In 2001, Russia provided 42 per cent of Germany's natural gas imports, the Netherlands 26 per cent, Norway 25 per cent, the United Kingdom 4 per cent and Denmark 3 per cent. As of January 2003, Germany's natural gas proven reserves stood at 11.3 Tcf.

Production

Natural gas production in Germany has remained roughly stable in recent years, while consumption has grown markedly. Most of Germany's natural gas resources are located onshore, in the North of the country, between the Wesser and Elbe rivers. Germany's only offshore natural gas

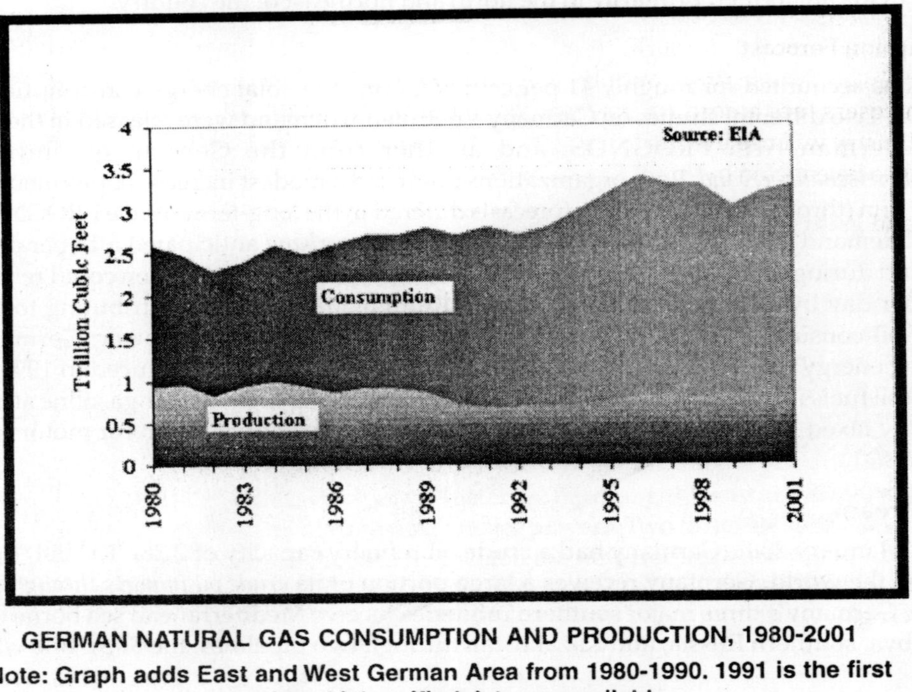

GERMAN NATURAL GAS CONSUMPTION AND PRODUCTION, 1980-2001

Note: Graph adds East and West German Area from 1980-1990. 1991 is the first year for which unified data are available

field is located in the German North Sea, about 190 miles offshore, where production began in September 2000. Known as the *Deutsches Nordseekonsortium* (German North Sea Consortium), the field is operated by a consortium of Wintershall (operator), BEB Erdgas und Erdoel, BASF, and RWE-DEA. New pipelines will transmit the anticipated 116.5 million cubic feet per day of production. The field is expected to produce for 16 years.

Sector Organization

Energy legislation passed in 1998 liberalized the German natural gas sector in accordance with the *EU's Gas Directive 98/30/EC*. Although the German natural gas market is open to competition, many industry players have petitioned the government for more reform, citing that a few power companies dominate the market. Currently, Germany has a system where the natural gas network operators regulate themselves according to the framework agreement known as the *Verbändevereinbarung*, which was originally signed in July 2000.

Regulatory Agency for the German Natural Gas Market

The German government announced in April 2003 its intentions to create a regulatory agency. According to the EU Gas Directive, EU member states are required to improve third-party access to their natural gas grids and to create state-controlled regulatory agency by July, 2004. Germany is the only EU member country which has not created a regulatory agency. Many German gas firms and industry players, however, remain opposed to the introduction of a market regulator.

Monitoring Report

In addition, the German Federal Ministry of Economics and Labor released on September 1, 2003 a status report on the natural gas liberalization process in Germany. The report suggested plans to simplify the current self-regulating system in order to make the market more competitive and to improve third-party access to the national grid. One suggestion entails a plan to change the current point-to-point grid access model in favor of an entry-exit system based on a number of trading zones. Companies wishing to transport natural gas from one zone to another will be charged one flat fee for entering and exiting a zone, irrespective of how many different networks are used. The current system charges pipeline users fees according distance that natural gas traveled and the number of networks used. This Ministry will use this report as basis to develop a new bill to improve the overall grid access. The law has to be enacted by July 1, 2004 according to EU law.

E. On and Ruhrgas Merger

In January 2002, the German Cartel Office rejected the merger between E.On and Ruhrgas, but the German federal government overruled the ruling, citing national interest in creating a "national champion." The decision prompted responses from nine smaller German natural gas firms, as well as from the Energy Ministry of the United Kingdom. The U.K. argued that creating a dominant energy company prior to market liberalization ran contrary to increasing competition across Europe. The nine firms dropped their objections after E.On/Rurhgas made concessions in an out-of-court agreement to make access to Ruhrgas' grid more flexible and simpler.

In September 2002, the German government imposed the following conditions (divestures) in order for E. On to take over Ruhrgas: their combined 42.1 per cent stake in Verbundnetz Gas (VNG), of which 26.8 per cent must goto a strategic shareholder which will build VNG up as a national challenger to E. On and RWE; stakes in EWE, a regional natural gas and electricity distributor; a 44 per cent stake in Bayerngas, which municipal utilities of Münich, Augsburg, Regenburg, and Ingolstadt and the city of Landshut purchased in July 2003; stakes in utility Stadtwerke Bremen; and stakes in water company Gelsenwasser. Ruhrgas was also required to auction off 200 billion kilowatts of natural gas, around 671 Bcf, from its long-term supply contracts in six annual auctions. The first auction, known as the *Gas Release Program*, took place on July 30, 2003.

Pipelines

Germany is both a major destination point and transit center for Europe's natural gas pipelines. Germany has six major pipelines on land, three from the North Sea, and several in the construction and planning stages. Pipelines from the Czech Republic transport Russian natural gas. The existing pipelines include: 1) The MEGAL pipeline from the Czech Republic to France through Germany, with annual capacity of 777 billion cubic feet (Bcf); 2) the TENP pipeline from the Netherlands to Germany and onward to Switzerland and Italy, with an annual capacity of 247 Bcf; 3) the STEGAL pipeline from the Czech Republic to Germany, with an annual capacity of 283 Bcf; 4) the NETRA pipeline from Etzel/Wilhelmshaven to Steinitz/Bernau, with an annual capacity of 706 Bcf; 5) the MIDAL pipeline from the port of Emden to Ludwigshafen with an annual capacity of 459 Bcf; 6) the JAGAL pipeline which feeds Russian natural gas into the Wingas network from the Mallnow distribution station, located on Germany's eastern border.

The pipelines that bring Norwegian natural gas ashore are Norpipe, which lands at Emden, and Europipe I & II, which land at Dornum. From the Dornum receiving station, the natural gas is linked either to the NETRA pipeline or to the metering station at Emden, where the MIDAL pipeline begins. The TENP pipeline can also bring in UK gas by way of the Netherlands. Wingas, which already owns the MIDAL and STEGAL pipelines, is planning to construct a pipeline with a capacity of 353-424 Bcf per year from Burghausen in Southwest Germany to the states of Badem-Württemburg and Bavaria in Southeastern Germany. Ruhrgas is the largest shareholder in the MEGAL, TENP, and NETRA pipelines, though it has a majority stake only in the TENP pipeline.

North Trans-Gas Pipeline

Gazprom, in conjunction with Germany's Wintershall, announced plans in June 2003 to build the North Trans-Gas Pipeline to run from St. Petersburg to Germany onto the United Kingdom. The pipeline project, which has received support from the European Commission, would have a 706 Bcf to 1.1 Tcf per year capacity. However, it still remains unclear whether the project will proceed.

Emden Natural Gas Hub

The large volumes of natural gas entering Germany, particularly on the Northwest coast around Emden, have given rise to efforts to establish Europe's third major natural gas hub at Bunde near the Dutch border. This is the point where the pipeline system of Gasunie of the Netherlands links up to the German networks of Ruhrgas, Wingas, and BEB. Spot trading by about a dozen companies is already occurring in this area, although volumes are small so far. Its location is in close proximity to where very large volumes of natural gas come into Germany, combined with European Commission proposals to unbundle integrated gas companies in the EU make the formation of an important hub likely. A conference was held in May 2001, on the subject of fostering natural gas trade at Bunde, which was attended by 40 companies, including all of Germany's major players. Two firms, EuroHub and HubCo, emerged in 2002 as the front-runners in the race to develop the new trading hub.

Coal

Germany is by far Europe's largest coal producer and consumer. Both production and consumption fell sharply between 1991 and 2001 as the country was restructured during integration, and many of the former East Germany's inefficiently run lignite mines were abandoned. Also, the unified Germany has made efforts to diversify its fuel mix. Coal's role in the German energy sector has stabilized in recent years and has accounted for about 23 per cent of total energy consumption in Germany since 1997, falling from 30 per cent in 1991. Germany consumed 265 million short tons (Mmst) in 2001 and produced 225.6 Mmst. Over 75 per cent of German coal production is used for electricity generation.

Production

Germany produces two grades of coal: anthracite, also known as "hard coal", and lignite, or "brown coal." Hard coal production, which is located primarily in the Ruhr and Saar regions, is expensive in Germany relative to other major coal producers because German coal is located deep underground and therefore is more expensive to extract. Hard coal production has remained a viable industry only through heavy subsidization, which is being reduced, but not ended.

Sector Organization

Energy legislation passed in 1998 liberalized the German natural gas sector in accordance with the *EU's Gas Directive 98/30/EC*. Although the German natural gas market is open to competition, many industry players have petitioned the government for more reform, citing that a few power companies dominate the market. Currently, Germany has a system where the natural gas network operators regulate themselves according to the framework agreement known as the *Verbändevereinbarung*, which was originally signed in July 2000.

Regulatory Agency for the German Natural Gas Market

The German government announced in April 2003 its intentions to create a regulatory agency. According to the EU Gas Directive, EU member states are required to improve third-party access to their natural gas grids and to create state-controlled regulatory agency by July, 2004. Germany is the only EU member country which has not created a regulatory agency. Many German gas firms and industry players, however, remain opposed to the introduction of a market regulator.

Monitoring Report

In addition, the German Federal Ministry of Economics and Labor released on September 1, 2003 a status report on the natural gas liberalization process in Germany. The report suggested plans to simplify the current self-regulating system in order to make the market more competitive and to improve third-party access to the national grid. One suggestion entails a plan to change the current point-to-point grid access model in favor of an entry-exit system based on a number of trading zones. Companies wishing to transport natural gas from one zone to another will be charged one flat fee for entering and exiting a zone, irrespective of how many different networks are used. The current system charges pipeline users fees according distance that natural gas traveled and the number of networks used. This Ministry will use this report as basis to develop a new bill to improve the overall grid access. The law has to be enacted by July 1, 2004 according to EU law.

E. On and Ruhrgas Merger

In January 2002, the German Cartel Office rejected the merger between E.On and Ruhrgas, but the German federal government overruled the ruling, citing national interest in creating a "national champion." The decision prompted responses from nine smaller German natural gas firms, as well as from the Energy Ministry of the United Kingdom. The U.K. argued that creating a dominant energy company prior to market liberalization ran contrary to increasing competition across Europe. The nine firms dropped their objections after E.On/Rurhgas made concessions in an out-of-court agreement to make access to Ruhrgas' grid more flexible and simpler.

In September 2002, the German government imposed the following conditions (divestures) in order for E. On to take over Ruhrgas: their combined 42.1 per cent stake in Verbundnetz Gas (VNG), of which 26.8 per cent must goto a strategic shareholder which will build VNG up as a national challenger to E. On and RWE; stakes in EWE, a regional natural gas and electricity distributor; a 44 per cent stake in Bayerngas, which municipal utilities of Münich, Augsburg, Regenburg, and Ingolstadt and the city of Landshut purchased in July 2003; stakes in utility Stadtwerke Bremen; and stakes in water company Gelsenwasser. Ruhrgas was also required to auction off 200 billion kilowatts of natural gas, around 671 Bcf, from its long-term supply contracts in six annual auctions. The first auction, known as the *Gas Release Program*, took place on July 30, 2003.

Pipelines

Germany is both a major destination point and transit center for Europe's natural gas pipelines. Germany has six major pipelines on land, three from the North Sea, and several in the construction and planning stages. Pipelines from the Czech Republic transport Russian natural gas. The existing pipelines include: 1) The MEGAL pipeline from the Czech Republic to France through Germany, with annual capacity of 777 billion cubic feet (Bcf); 2) the TENP pipeline from the Netherlands to Germany and onward to Switzerland and Italy, with an annual capacity of 247 Bcf; 3) the STEGAL pipeline from the Czech Republic to Germany, with an annual capacity of 283 Bcf; 4) the NETRA pipeline from Etzel/Wilhelmshaven to Steinitz/Bernau, with an annual capacity of 706 Bcf; 5) the MIDAL pipeline from the port of Emden to Ludwigshafen with an annual capacity of 459 Bcf; 6) the JAGAL pipeline which feeds Russian natural gas into the Wingas network from the Mallnow distribution station, located on Germany's eastern border.

The pipelines that bring Norwegian natural gas ashore are Norpipe, which lands at Emden, and Europipe I & II, which land at Dornum. From the Dornum receiving station, the natural gas is linked either to the NETRA pipeline or to the metering station at Emden, where the MIDAL pipeline begins. The TENP pipeline can also bring in UK gas by way of the Netherlands. Wingas, which already owns the MIDAL and STEGAL pipelines, is planning to construct a pipeline with a capacity of 353-424 Bcf per year from Burghausen in Southwest Germany to the states of Badem-Württemburg and Bavaria in Southeastern Germany. Ruhrgas is the largest shareholder in the MEGAL, TENP, and NETRA pipelines, though it has a majority stake only in the TENP pipeline.

North Trans-Gas Pipeline

Gazprom, in conjunction with Germany's Wintershall, announced plans in June 2003 to build the North Trans-Gas Pipeline to run from St. Petersburg to Germany onto the United Kingdom. The pipeline project, which has received support from the European Commission, would have a 706 Bcf to 1.1 Tcf per year capacity. However, it still remains unclear whether the project will proceed.

Emden Natural Gas Hub

The large volumes of natural gas entering Germany, particularly on the Northwest coast around Emden, have given rise to efforts to establish Europe's third major natural gas hub at Bunde near the Dutch border. This is the point where the pipeline system of Gasunie of the Netherlands links up to the German networks of Ruhrgas, Wingas, and BEB. Spot trading by about a dozen companies is already occurring in this area, although volumes are small so far. Its location is in close proximity to where very large volumes of natural gas come into Germany, combined with European Commission proposals to unbundle integrated gas companies in the EU make the formation of an important hub likely. A conference was held in May 2001, on the subject of fostering natural gas trade at Bunde, which was attended by 40 companies, including all of Germany's major players. Two firms, EuroHub and HubCo, emerged in 2002 as the front-runners in the race to develop the new trading hub.

Coal

Germany is by far Europe's largest coal producer and consumer. Both production and consumption fell sharply between 1991 and 2001 as the country was restructured during integration, and many of the former East Germany's inefficiently run lignite mines were abandoned. Also, the unified Germany has made efforts to diversify its fuel mix. Coal's role in the German energy sector has stabilized in recent years and has accounted for about 23 per cent of total energy consumption in Germany since 1997, falling from 30 per cent in 1991. Germany consumed 265 million short tons (Mmst) in 2001 and produced 225.6 Mmst. Over 75 per cent of German coal production is used for electricity generation.

Production

Germany produces two grades of coal: anthracite, also known as "hard coal", and lignite, or "brown coal." Hard coal production, which is located primarily in the Ruhr and Saar regions, is expensive in Germany relative to other major coal producers because German coal is located deep underground and therefore is more expensive to extract. Hard coal production has remained a viable industry only through heavy subsidization, which is being reduced, but not ended.

Lignite, or "brown coal," production, however, is inexpensive in Germany. Germany is the world's largest lignite producer, with about one-fifth of global output. Lignite was the most important fuel in the former East Germany, with East Germany producing about three times as much lignite as West

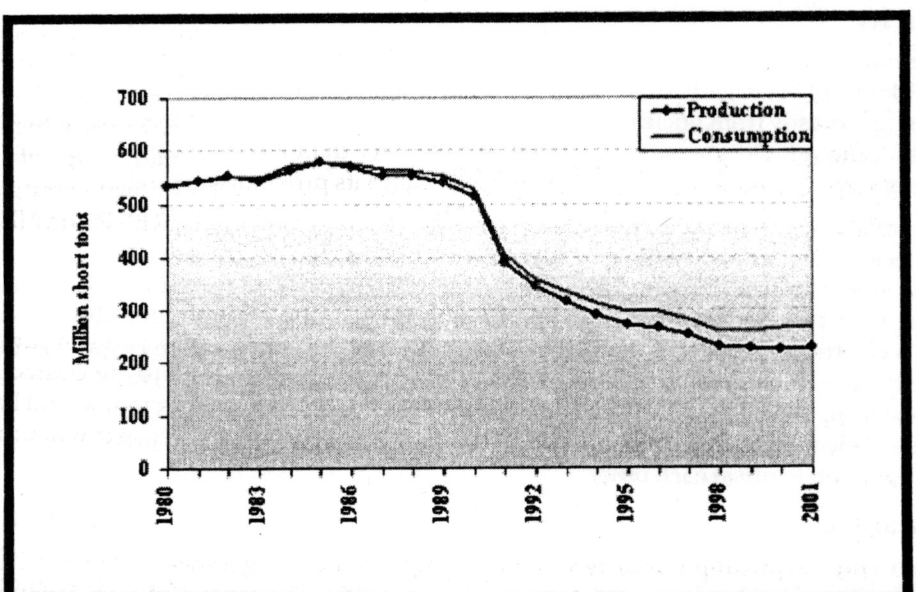

GERMAN COAL CONSUMPTION AND PRODUCTION, 1980-2001

Note: Graph adds East and West German Area from 1980-1990. 1991 is the first year for which unified data are available

Germany in the years prior to reunification. Since reunification, wasteful and environmentally damaging mining methods practiced during Communist rule in the former East Germany have been reformed and lignite output has fallen by about 40 per cent.

Imports

As domestic production declines, Germany is emerging as a significant coal importer. Gross imports of hard coal, coke, and briquettes have more than doubled since unification, and in 2001, for the first time ever, the consumption of imported coal exceeded the consumption of domestic coal. Germany's largest suppliers are Australia, Canada, Poland, South Africa, and Colombia, among others. The Federation of German Coal Importers expects German hard coal imports to increase over the next 20 years, as nuclear power is phased out and domestic production declines.

Future of Coal in Germany

Coal is Germany's only major domestic fuel source and the government considers maintaining hard coal production capacity critical to the country's energy security. For this reason, Germany subsidizes the production of hard coal in order to maintain a cadre of working coal professionals. The coal industry is also a major employer in Germany. Germany's state assistance to coal producers was addressed by the European Union in 2000 and 2001, resulting in temporary agreements authorizing coal subsidies in EU member countries. In July 2002, upon expiry of the 1952 European Coal and Steel Community Treaty, the Energy Ministers of the European Union states agreed to phase out subsidies to domestic coal producers by 2010, and stipulated that beginning in 2003, subsidies will be reduced to below their 2001 levels. The member states have agreed to review this decision in 2007. Other EU member states which provide assistance to their coal industries are France, Spain, and the United Kingdom, but Germany's assistance is by far the largest.

Electricity

Germany has Europe's largest electricity market. In 2001, Germany generated 544.8 billion kilowatt hours (bkwh) of electricity, two-thirds of which came from fossil fuels (mostly coal), with the other third coming mostly from nuclear power along with small amounts of hydropower and other renewable sources. Although Germany produces more electricity than it consumes, the country is a small net electricity importer, because of transmission losses and its proximity to foreign sources of generation.

Germany has an installed electric capacity of 114 million kilowatts, which includes considerable excess generation capacity. Recent German estimates predict 14 per cent growth in electricity consumption between 1999 and 2020. Concurrently, Germany has agreed to phase out nuclear power over the next two decades. Nuclear power, which currently accounts for almost one-third of the county's electricity generation, is likely to be replaced by a combination of gas-fired and coal-fired electricity as well as renewable energy. Electricity imports, which also could rise in coming years, come primarily from France, the Czech Republic, Norway, and Austria. Germany also exports small amounts of electricity to the Netherlands, Switzerland, and Austria, and in 2001 imports and exports almost completely offset each other

Renewables

Germany's consumption of renewable energy (including geothermal, solar, wind, wood and waste electric power) has increased dramatically since 1991. The consumption of renewable electricity, particularly wind power, has more than quadrupled since 1991, to 22.6 billion kilowatt hours in 2001, representing approximately 4 per cent of the country's total electricity consumption. Germany is

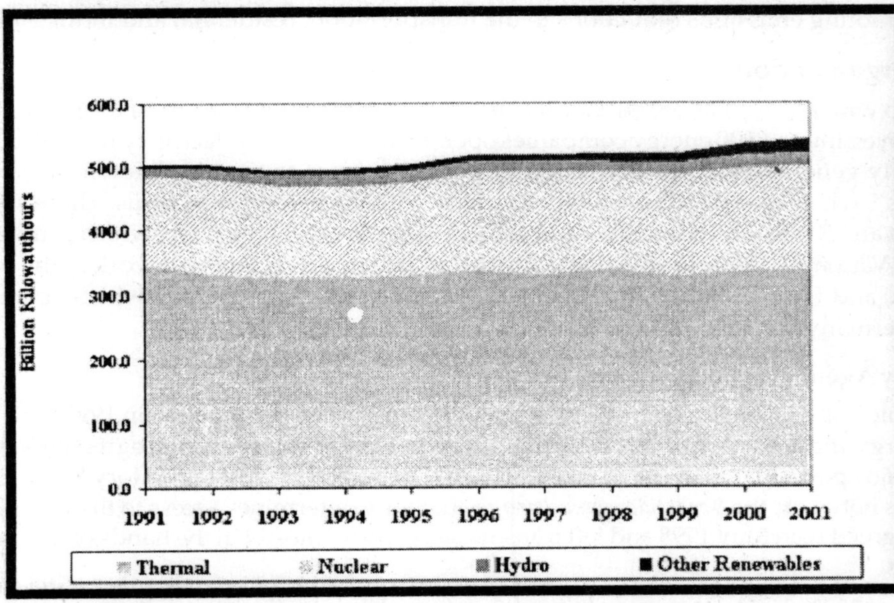

ELECTRICITY GENERATION IN GERMANY, 1991-2001

currently the leader in wind power, with 39 per cent of the world's installed wind capacity. As of the second quarter of 2003, Germany had 14,283 wind turbines. The government also plans to increase offshore wind capacity by 20,000-25,000 MW by 2030.

Promotion of Renewables and Energy Efficiency

The German government has been promoting renewables, as well as energy efficiency initiatives in order to reduce the country's greenhouse gas emissions. In 2000, the government passed the *Renewable Energy Act* (*Das Erneubare Energien Gesetz*) designed to double the amount of electricity generated from renewable sources by 2010. The government's mid-term goal is to increase electricity generated from renewables to 20 per cent by 2020 and to 10 per cent of total energy consumption.

Other Programs

In 1999, the German government began the *100,000 solar panel roof program* (*100,000 Dächer-Solarstrom-Programm*), which provided low interest loans to homeowners who installed rooftop solar panels of at least 1 kilowatt. The program ended in June 2003 after it reached its target of 300 MW. In a move to promote energy efficiency, the German government passed in February 2002 the *Combined Heat and Power Law* (*Kraft-Wärme-Kopplungsgesetz*), which provided 4.5 billion euro of funding.

Through the refurbishment of existing plants and installation of new smaller plants, the government hopes to reduce greenhouse gas emissions by 23 million tons per year by 2010. The government also passed the *Energy Saving Ordinance* (*Energieeinsparverordnung*) in 2002 which requires new buildings to reduce energy consumption by 30 per cent in comparison to current standards. The *Eco-Tax* is another program which seeks to raise prices gradually in all sectors in order to create incentives for the development of new technologies and installation of energy-effienct products. There are other programs which promoting emissions reductions in the transportation, residential and industrial sectors.

Sector Organization

In step with EU legislation, the German power market has become one of the most competitive in Europe. An estimated 900 energy companies operate in the German electricity market, but 80 per cent of electricity generation is still controlled by only four large companies known as "supra-regional companies", which also distribute their electricity with the help of local firms. These supra-regional companies are: RWE, which operates primarily in western Germany; E.On, which produces in central Germany; Vattenfall Europe, which was formed in January 2003 and will work in the New Laender and Berlin; and Energie Baden-Württemberg Aktiengesellschaft (EnBW), which operates in south-western Germany.

Regulatory Agency for the Electricity Market

Complete liberalization of the German electricity sector was completed in 1998 with the adoption of the Energy Industry Act of April 1998. Electricity market reform has progressed via agreements among major participants in the market but is not overseen by any regulatory body. Similar to the natural gas network, the *Verbändervereinbarung* currently determines access to the grid system, which was first agreed to in May 1998 and left transmission control mostly in the hands of the supra-regional companies.

In April 2003, the German government announced its intentions to create a regulatory agency for the electricity markets by mid-2004. The new regulatory agency will be mandated to define conditions of access and use of electricity grids, as well as a designing rules for calculating grid access fees. Over the years, the German system has been criticized for its lack of price transparency and the division of the German market into two distinct trading zones. The EC estimates that about 10 per cent -20 per cent of large electricity customers and less than 5 per cent of smaller customers have switched to competitive suppliers since 1998.

Nuclear Power

Currently, Germany ranks fourth worldwide in installed nuclear capacity, behind the United States, France, and Japan. Germany's 19 nuclear plants comprise about 20 per cent of Germany's electric generation capacity, and about 30 per cent of actual generation. E. On, RWE, HEW, and EnBW own nuclear generation capacity, with E. On holding stakes in 11 of Germany's 19 nuclear power plants.

Nuclear Power Phase Out Nuclear power has become controversial since the September 1998 elections. The Greens, the environmental party that is part of the ruling alliance, are staunchly opposed to the continued use of nuclear power. Chancellor Schröder had decided to close all 19 nuclear reactors in 2005, but he has since amended his position. The government formally signed an agreement with utility companies in June 2001 to gradually phase out nuclear power, and in April 2002, the German Parliament amended the country's atomic energy legislation to reflect this amendment. Each nuclear plant is allowed to produce a finite amount of electricity. Power plants will have a life span of about 32 years. The deal could see the total elimination of nuclear power by 2021, as the newest nuclear plant opened in 1989. Generation volumes are transferable; if an older plant closes before reaching its production ceiling, its remaining allowable production can be transferred to a new plant.

Some observers suggest that there are few economically viable alternatives to replace quickly such a significant portion of the fuel mix, especially in the wake of power-sector liberalization. As

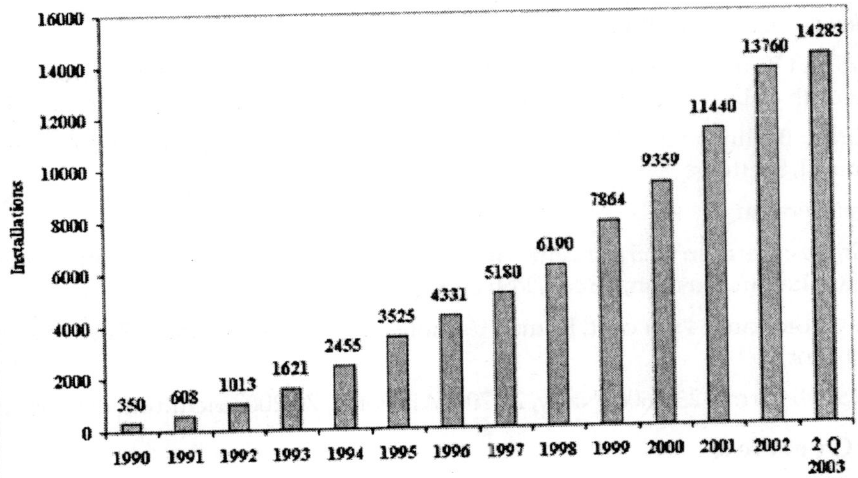

THE NUMBER OF WIND TURBINES IN GERMANY, 1990 TO THE 2ND QUARTER 2003

European markets become more liberalized and more price-sensitive, replacing the mostly amortized plants will prove difficult. Over the longer term, however, high costs (high fixed costs, long depreciation periods and long annual operating times) associated with nuclear generation could work to decrease nuclear generation's role in Germany's power sector. Nuclear installations currently are initiating programs to reduce production costs and waste disposal costs in order to become more price-competitive. Some executives in Germany's nuclear industry have claimed that the June 2001 agreement is not irreversible, and that an electricity shortage and a change in the political climate might lead to a renewal of nuclear energy.

Environment

Germany has a strong commitment to protecting its environment. It has actively promoted the use of renewable energy, both under the Kohl government with the Electricity Feed Law, and now under Schröder's government with eco-taxes. In Germany's eco-tax regime, energy tax revenue is used to fund renewable projects.

In 2001, Germany emitted 223.2 million metric tons of carbon from the consumption of fossil fuels. Germany ranks third in total carbon emissions within the G-7, after the United States and Japan. Germany signed the Framework Convention on Climate Change in Rio de Janeiro in June 1992 and ratified it on December 9, 1993. Signers of the agreement pledged to stabilize per capita CO_2 emissions in the year 2000 and beyond at 1990 levels. In 2000, Germany set a goal of reducing the six greenhouse gases cited in the Kyoto Protocol by 21 per cent between 2008 and 2012, within the context of the EU burden-sharing program. This has been made more achievable given the sharp drop in total German carbon emissions since 1990, due mainly to decreased consumption of energy overall (and in particular lignite) in the former East Germany. Germany ratified the Kyoto Protocol on May 31, 2002.

Country Overview

Independence: January 18, 1871 (reunification of West and East Germany took place on October 3, 1990)

Population (2002E): 83 million

Location/Size: Central Europe, bordering the Baltic Sea and the North Sea, between the Netherlands and Poland, south of Denmark/137,821 square miles (slightly smaller than Montana)

Major Cities: Berlin (national capital since 10/3/90), Hamburg, Munich, Cologne, Frankfurt, Essen, Dortmund, Stuttgart

Language: German

Ethnic Groups: German 91.5 per cent, Turkish 2.4 per cent, other 6.1 per cent (made up largely of Serbo-Croatian, Italian, Russian, Greek, Polish, Spanish)

Religions: Protestant 34 per cent, Roman Catholic 34 per cent, Muslim 3.7 per cent, unaffiliated or other 28.3 per cent

Defense (8/98): Army, 230,600; Navy, 26,700; Air Force, 76,200 (including conscripts)

Economic Overview

Currency: euro

Exchange Rate (09/04/03): 1 US Dollar = 0.922 euro

Gross Domestic Product (GDP, nominal, 2002E): $1.99 trillion *(2003F):* $2.39 trillion

Real GDP Growth Rate (2002E): 0.2 per cent *(2003F):* 0.0 per cent

Inflation Rate (consumer prices, 2002E): 1.3 per cent *(2003F):* 0.9 per cent

Unemployment Rate (2002E) 9.8 per cent *(2003F):* 10.8 per cent

Merchandise Exports (2002E, $U.S.): $617 billion

Merchandise Imports (2002E, $U.S.): $493 billion

Current Account Balance (2002E, $U.S.): $49 billion

Energy Overview

Proven Oil Reserves (1/1/03E): 342 million barrels

Oil Production (2002E): 147,000 barrels per day (bbl/d), of which 69,000 bbl/d was crude oil

Oil Consumption (2002E): 2.72 million bbl/d

Net Oil Imports (2002E): 2.57 million bbl/d

Natural Gas Reserves (1/1/03E): 11.3 trillion cubic feet (Tcf)

Natural Gas Production (2002E): 0.79 Tcf

Natural Gas Consumption (2002E): 3.4 Tcf

Natural Gas Imports (2002E): 2.9 Tcf

Natural Gas Exports (2002E): 0.259 Tcf

Coal Reserves (2001E): 72.8 billion short tons

Coal Production (2001E): 226 million short tons (Mmst)

Coal Consumption (2001E): 265 Mmst

Net Coal Imports (2001E): 39 Mmst

Electric Generation Capacity (2001E): 114 million kilowatts

Electricity Production (2001E): 544.8 billion kilowatt hours

Electricity Consumption (2001E): 506.8 billion kilowatthours

Environmental Overview

Total Energy Consumption (2001E): 14.35 quadrillion Btu* (3.6 per cent of world total energy consumption)

Energy-Related Carbon Emissions (2001E): 223.2 million metric tons of carbon (3.4 per cent of world total carbon emissions)

Per Capita Energy Consumption (2001E): 174.3 million Btu (vs U.S. value of 341.8 million Btu)

Per Capita Carbon Emissions (2001E): 2.7 metric tons of carbon (vs U.S. value of 5.5 metric tons of carbon)

Energy Intensity (2001E): 7,474 Btu/ $1995 (vs U.S. value of 10,870 Btu/ $1995)**

Carbon Intensity (2001E): 0.117 metric tons of carbon/thousand $1995 (vs U.S. value of 0.17 metric tons/thousand $1995)

Status in Climate Change Negotiations: Annex I country under the United Nations Framework Convention on Climate Change (ratified December 9th, 1993). Under the negotiated Kyoto Protocol (signed on April 29th, 1998, and ratified on May 31, 2002), Germany, as a member of the European Union, has agreed to reduce greenhouse gases 8 per cent below 1990 levels by the 2008-2012 commitment period.

Major Environmental Issues: Emissions from coal-burning utilities and industries and lead emissions from vehicle exhausts (the result of continued use of leaded fuels) contribute to air pollution; acid rain, resulting from sulfur dioxide emissions, is damaging forests; heavy pollution in the Baltic Sea from raw sewage and industrial effluents from rivers in eastern Germany; hazardous waste disposal.

Major International Environmiental Agreements: A party to Conventions on Air Pollution, Air Pollution-Nitrogen Oxides, Air Pollution-Sulphur 85, Air Pollution-Sulphur 94, Air Pollution-Volatile Organic Compounds, Antarctic-Environmental Protocol, Antarctic Treaty, Biodiversity, Climate Change, Desertification, Endangered Species, Environmental Modification, Hazardous Wastes, Law of the Sea, Marine Dumping, Nuclear Test Ban, Ozone Layer Protection, Ship Pollution, Tropical Timber 83, Tropical Timber 94, Wetlands, Whaling. Has signed, but not ratified, Air Pollution-Persistent Organic Pollutants.

The total energy consumption statistic includes petroleum, dry natural gas, coal, net hydro, nuclear, geothermal, solar, wind, wood and waste electric power. The renewable energy consumption statistic is based on International Energy Agency (IEA) data and includes hydropower, solar, wind, tide, geothermal, solid biomass and animal products, biomass gas and liquids, industrial and municipal wastes. Sectoral shares of energy consumption and carbon emissions are also based on IEA data.

Energy Industries

Major Energy Companies: Oil: Deutsche Shell, Esso, Ruhr Oel; *Natural Gas:* Ruhrgas, Wintershall/ Wingas; *Coal:* DSK, RAG; *Electricity:* RWE, E.On, EnBW, Vattenfall Europe

Major Refineries (crude capacity, bbl/d): Karlsruhe (285,800), Bayernoil (262,300), Schwedt (210,000), Gelsenkirchen (246,000), Leuna (222,800), Wilhelmshaven (220,000), Godorf (162,000), Wesseling (140,000), Ingolstadt (262,300)